Trends in Mathematics

Trends in Mathematics is a book series devoted to focused collections of articles arising from conferences, workshops or series of lectures.

Topics in a volume may concentrate on a particular area of mathematics, or may encompass a broad range of related subject matter. The purpose of this series is both progressive and archival, a context in which to make current developments available rapidly to the community as well as to embed them in a recognizable and accessible way.

Volumes of TIMS must be of high scientific quality. Articles without proofs, or which do not contain significantly new results, are not appropriate. High quality survey papers, however, are welcome. Contributions must be submitted to peer review in a process that emulates the best journal procedures, and must be edited for correct use of language. As a rule, the language will be English, but selective exceptions may be made. Articles should conform to the highest standards of bibliographic reference and attribution.

The organizers or editors of each volume are expected to deliver manuscripts in a form that is essentially "ready for reproduction." It is preferable that papers be submitted in one of the various forms of TEX in order to achieve a uniform and readable appearance. Ideally, volumes should not exceed 350-400 pages in length.

Proposals to the Publisher are welcomed at either:
Birkhäuser Boston, 675 Massachusetts Avenue, Cambridge, MA 02139, U.S.A.
math@birkhauser.com
or
Birkhäuser Verlag AG, PO Box 133, CH-4010 Basel, Switzerland
math@birkhauser.ch

Recent Progress in Intersection Theory

Geir Ellingsrud
William Fulton
Angelo Vistoli
Editors

Birkhäuser
Boston • Basel • Berlin

Geir Ellingsrud
Department of Mathematics
University of Oslo
N-0316 Oslo
Sweden

William Fulton
Department of Mathematics
University of Michigan
Ann Arbor, MI 48109
USA

Angelo Vistoli
Departimento di Matematica
Università di Bologna
40127 Bologna
Italy

Library of Congress Cataloging-in-Publication Data

Recent progress in intersection theory / Geir Ellingsrud, William Fulton, Angelo Vistoli, editors.

p. cm. – (Trends in Mathematics)

Includes bibliographical references.

ISBN 0-8176-4122-X (alk. paper) – ISBN 3-7643-4122-X

1. Intersection theory. I. Ellingsrud, Geir. II. Fulton, William, 1939- II. Vistoli, Angelo, 1958- IV. Series.

QA564.R44 2000

516.3'5–dc21 99-053410

CIP

AMS Subject Classifications: 14B12, 14Cxx, 14C17, 14C35, 14C25, 14Kxx

Printed on acid-free paper.

Birkhäuser

ISBN 0-8176-4122-X SPIN 10716467
ISBN 3-7643-4122-X

Reformatted from electronic files in LaTeX by TeXniques, Inc., Cambridge, MA.
Printed and bound by Hamilton Printing, Rensselaer, NY.
Printed in the United States of America.

9 8 7 6 5 4 3 2 1

Contents

Preface

The articles in this volume are an outgrowth of an International Conference in Intersection Theory that took place in Bologna, Italy (December 1997). In a somewhat unorthodox format aimed at both the mathematical community as well as summer school students, talks were research-oriented as well as partly expository. There were four series of expository talks by the following people:

M. Brion, University of Grenoble, on *Equivariant Chow groups and applications*;

H. Flenner, University of Bochum, on *Joins and intersections*;

E. M. Friedlander, Northwestern University, on *Intersection products for spaces of algebraic cycles*;

R. Laterveer, University of Strasbourg, on *Bigraded Chow (co)homology.*

Four introductory papers cover the following topics and bring the reader to the forefront of research:

1) the excess intersection algorithm of Stückrad and Vogel, combined with the deformation to the normal cone, together with many of its geometric applications;

2) new and very important homotopy theory techniques that are now used in intersection theory;

3) the Bloch-Beilinson filtration and the theory of motives;

4) algebraic stacks, the modern language of moduli theory.

Other research articles concern such active fields as stable maps and Gromov-Witten invariants, deformation theory of complex varieties, and others.

Organizers of the conference were Rüdiger Achilles, Mirella Manaresi, and Angelo Vistoli, all from the University of Bologna; the scientific committee consisted of Geir Ellingsrud, University of Oslo, William Fulton, University of Michigan at Ann Arbor, and Angelo Vistoli.

The conference was financed by the European Union (contract no. ERBFMMACT970198), the Italian Ministry MURST, the Italian National Council of Research CNR, and the University of Bologna. We are very grateful to all organizations for their support.

We also heartily thank Silvana Serra for her help with the organization, and Davide Aliffi, who did part of the processing of the TeX files for this volume.

Geir Ellingsrud
William Fulton
Angelo Vistoli
Editors

Complete Moduli for Fibered Surfaces

Dan Abramovich*

Angelo Vistoli**

1 Introduction

All the schemes with which we work will be schemes over the field **Q** of rational numbers.[1]

Most of the results contained here are particular cases of our general theory of stable maps into algebraic stacks, announced in [ℵ-V1]; the details will appear in [ℵ-V2]. Here we include a rather thorough discussion of our particular case, even when it partially overlaps with the above-mentioned papers; however, the proof of the main theorem, which is very long and technical, is omitted.

A previous version of this paper appeared as mathAG/9804097 on the xxx.lanl.gov Mathematics e-Print archive.

1.1 The problem

Fix a base field k of characteristic 0, and let C be a smooth, projective, geometrically integral curve of genus g. By a *fibered surface over* C we mean a morphism $X \to C$, with sections $\sigma_1, \dots, \sigma_\nu : C \to X$ forming a family of stable ν-pointed curves of some genus γ. (This notion will be generalized below for singular C.)

A fibered surface naturally corresponds to a morphism $C \to \overline{\mathcal{M}}_{\gamma,\nu}$ into the moduli stack of stable ν-pointed curves of genus γ. In [Vistoli2], the second author showed that $Hom(C, \overline{\mathcal{M}}_{\gamma,\nu})$ has the structure of a Deligne–Mumford stack. This stack is, in general, clearly not complete. A natural question to ask is, can one complete it in a meaningful way? Moreover, can

*Partially supported by NSF grant DMS-9503276 and an Alfred P. Sloan Research Fellowship.

**Partially supported by the University of Bologna, funds for selected research topics.

[1]The theory below could be extended to the case that the characteristic is large with respect to the degrees and genera of the curves involved, but we will not pursue this here.

one do this as the source curve C itself moves in a family?

1.2 Our approach

One construction which may spring to mind is that of *stable maps*. Recall that if $Y \subset \mathbf{P}^r$ is a projective variety, there is a complete Deligne–Mumford stack $\overline{\mathcal{M}}_g(Y, d)$ parametrizing Kontsevich stable maps of degree d from curves of genus g to Y (see [Kont], [B-M], [F-P]). In order to reduce the possibility of later confusion, we will use the notation $\mathcal{K}_g^d(Y)$ rather than $\overline{\mathcal{M}}_g(Y, d)$ for this stack. It admits a projective coarse moduli space $\mathbf{K}_g^d(Y)$.

For instance, if $\gamma = 0$, the moduli stack $\overline{\mathcal{M}}_{0,\nu}$ is actually a projective variety. Fixing a projective embedding, we have natural, complete Deligne–Mumford stacks $\mathcal{K}_g^d(\overline{\mathcal{M}}_{0,\nu})$ parametrizing families of stable ν-pointed curves of genus 0 over nodal curves of genus g, with a suitable stability condition. We can think of d, the degree of the image of C via the fixed projective embedding, as an additional measure of complexity of the family of curves. One is led to ask, is there a complete Deligne–Mumford stack of stable maps $\mathcal{K}_g^d(\overline{\mathcal{M}}_{\gamma,\nu})$ in general?

We need to define our terms: given a nodal curve C, define a morphism $f : C \to \overline{\mathcal{M}}_{\gamma,\nu}$ to be a stable map of degree d if, once projected into the coarse moduli space, we obtain a stable map $f' : C \to \overline{\mathbf{M}}_{\gamma,\nu}$ of degree d:

$$\begin{array}{ccc} C & \xrightarrow{f} & \overline{\mathcal{M}}_{\gamma,\nu} \\ & \searrow^{f'} & \downarrow \\ & & \overline{\mathbf{M}}_{\gamma,\nu} \end{array}$$

As we will see later, these stable maps are parametrized by a Deligne–Mumford stack. However, this stack fails to perform its first goal: it is not always complete! This can be seen via the following example:

Consider the case $\gamma = 1, \nu = 1$ of elliptic curves. It is well known (and easily follows from the formula of Grothendieck–Riemann–Roch) that given a family of stable elliptic curves $E \to C$ over a curve C, the degree of the corresponding moduli map $j_E : C \to \overline{\mathbf{M}}_{1,1} \simeq \mathbf{P}^1$ is divisible by 12. However, it is very easy to construct a family of moduli maps $j_t : C_t \to \mathbf{P}^1, \quad t \neq 0$, where when $t \to 0$ the curve C_t breaks into two components, and the degrees of the limit map j_0 on these components are not divisible by 12. Thus j_0 is not a moduli map of a family of stable curves.

1.3 Compactifying the space of stable maps

Our main goal in this paper is to correct this deficiency. In order to do so, we will enlarge the category of stable maps into $\overline{\mathcal{M}}_{\gamma,\nu}$. The source curve C

of a new stable map $\mathcal{C} \to \overline{\mathcal{M}}_{\gamma,\nu}$ will acquire an orbispace structure at its nodes. Specifically, we endow it with the structure of a Deligne–Mumford stack.

To see how these come about, consider again the example of a one-parameter family of elliptic curves sketched above. Let $C \to S$ be the one-parameter family of base curves, and let C_{sm} be the smooth locus of this morphism. A fundamental purity lemma (see 2.1 below) will show that, after a suitable base change, we can extend the family E of stable elliptic curves over C_{sm}. On the other hand, if $p \in C$ is a node, then on an étale neighborhood U of p, the curve C looks like

$$uv = t^r,$$

where t is the parameter on the base. By taking r-th roots,

$$u = u_1^r;\ v = v_1^r$$

we have a nonsingular covering $V_0 \to U$ where $u_1v_1 = t$. The fundamental purity lemma applies to V_0, so the pullback of E to V_0 extends over all of V_0. There is a minimal intermediate covering $V_0 \to V \to U$ such that the family extends over V. This V gives the orbispace structure $\mathcal{C}$ over C.

The precise definitions will follow in section 4. We can define stable maps $f : \mathcal{C} \to \overline{\mathcal{M}}_{\gamma,\nu}$ where $\mathcal{C}$ is a Deligne–Mumford stack and f is a representable morphism satisfying certain properties; such objects naturally form a 2-category. Our main theorem states that this 2-category is equivalent to a 1-category $\mathcal{F}_g^d(\gamma,\nu)$, which we call *the category of fibered surfaces*; this category forms a complete Deligne–Mumford stack (over schemes over $\mathbf{Q}$), admitting a projective coarse moduli space. Thus the original moduli problem has a solution of the same nature as that of stable curves.

We provide an explicit description of the category $\mathcal{F}_g^d(\gamma,\nu)$ of fibered surfaces in terms of charts and atlases over schemes, in analogy to Mumford's treatment of $\mathbf{Q}$-varieties in [Mum]. Thus given a stable map $\mathcal{C} \to \overline{\mathcal{M}}_{\gamma,\nu}$ we have a stable pointed curve $\mathcal{X} \to \mathcal{C}$, which can be described via an atlas of charts over the associated coarse moduli scheme $X \to C$.

1.4 Comparison with Alexeev's work

The latter, coarse object $X \to C$, which we call a *coarse fibered surface*, has another interpretation: the associated morphism $X \to \overline{\mathbf{M}}_{\gamma,\nu}$ turns out to be a stable map in the sense of Alexeev (see [Al3]). Alexeev has shown the existence of complete moduli of (smoothable) surface stable maps, using the theory of semi-log-canonical surfaces developed in [K-SB] and [Al2]. In the last section we show that, in the case of fibered surfaces, one can use Alexeev's approach to obtain a space which coincides with the space of (smoothable) stable coarse fibered surfaces. We sketch a proof of the existence of the latter space (even as a stack) which is independent of Alexeev's boundedness result.

1.5 Natural generalizations in forthcoming work

It should be evident from our work that this approach should apply to stable maps $\mathcal{C} \to \overline{\mathcal{M}}$ into any Deligne–Mumford stack which admits a projective coarse moduli scheme; moreover, the source "curve" $\mathcal{C}$ should be allowed to be "pointed" as well. See [ℵ-V1] for a discussion of the general setup and various applications that will be worked out in [ℵ-V2] and in [ℵ-C-J-V].

1.6 Gromov–Witten invariants

Originally, the Kontsevich spaces of stable maps were introduced for the purpose of defining Gromov–Witten invariants. It seems likely that, using our construction, one could extend the formalism of [BF] and [B] of virtual fundamental classes to the case of stable maps into an arbitrary Deligne–Mumford stack $\overline{\mathcal{M}}$ admitting a projective coarse moduli space. This should allow one to define Gromov–Witten invariants and quantum cohomology in this generality, which may have interesting applications for specific choices of stacks $\overline{\mathcal{M}}$.

1.7 Acknowledgements

We would like to thank Johan de Jong and Rahul Pandharipande for crucial discussions of ideas in this paper.

2 The purity lemma

There are several results in the literature which give conditions under which a family of curves can fail to be stable only in pure codimension 1 (see [MB], [dJ-O]). For our purposes, the following case will be most useful:

Purity Lemma 2.1. *Let $\overline{\mathcal{M}}$ be a separated Deligne–Mumford stack, $\overline{\mathcal{M}} \to \overline{\mathbf{M}}$ the coarse moduli space. Let X be a separated scheme of dimension 2 satisfying Serre's condition S_2. Let $P \subset X$ be a finite subset consisting of closed points, $U = X \smallsetminus P$. Assume that the local fundamental groups of U around the points of P are trivial.*

Let $f : X \to \overline{\mathbf{M}}$ be a morphism. Suppose there is a lifting $\tilde{f}_U : U \to \overline{\mathcal{M}}$:

$$\begin{array}{ccccc} & & & & \overline{\mathcal{M}} \\ & & \nearrow^{\tilde{f}_U} & & \downarrow \\ U & \longrightarrow & X & \xrightarrow{f} & \overline{\mathbf{M}} \end{array} \tag{1}$$

Then the lifting extends uniquely to X:

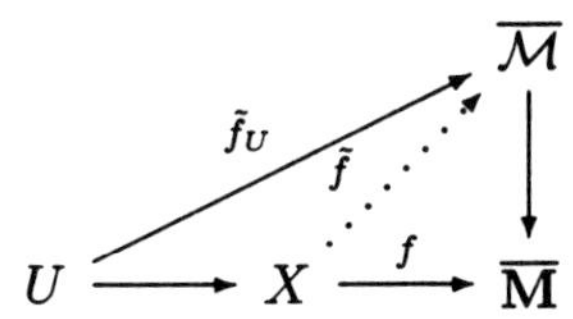

(2)

Remark 2.2. A closely related statement has been obtained by Mochizuki in [Mo1], [Mo2].

Proof. By descent theory the problem is local in the étale topology, so we may replace X and $\overline{\mathbf{M}}$ with the spectra of their strict henselizations at a geometric point; then we can also assume that we have a universal deformation space $V \to \overline{\mathcal{M}}$ which is *finite.* Now U is the complement of the closed point, U maps to $\overline{\mathcal{M}}$, and the pullback of V to U is finite and étale, so it has a section, because U is simply connected; consider the corresponding map $U \to V$. Let Y be the scheme-theoretic closure of the graph of this map in $X \times_{\overline{\mathbf{M}}} V$. Then $Y \to X$ is finite and is an isomorphism on U. Since X satisfies S_2, the morphism $Y \to X$ is an isomorphism. □

Corollary 2.3. *Let X be a smooth surface, $p \in X$ a closed point with complement U. Let $X \to \overline{\mathbf{M}}$ and $U \to \overline{\mathcal{M}}$ be as in the purity lemma. Then there is a unique lifting $X \to \overline{\mathcal{M}}$.*

Corollary 2.4. *Let X be a normal crossings surface, namely a surface which is étale locally isomorphic to* Spec $k[u,v,t]/(uv)$. *Let $p \in X$ be a closed point with complement U. Let $X \to \overline{\mathbf{M}}$ and $U \to \overline{\mathcal{M}}$ be as in the purity lemma. Then there is a unique lifting $X \to \overline{\mathcal{M}}$.*

Proof. In both cases X satisfies condition S_2 and the local fundamental group around p is trivial, hence the purity lemma applies. □

3 Group actions on nodal curves

Fix two nonnegative integers γ and ν. An important ingredient in the theory will be formed of data as follows:

1. a diagram

$$\begin{array}{c} Y \\ \downarrow \circlearrowright \\ V \\ \downarrow \\ S \end{array}$$

(we think of Y and V as S-schemes);

2. the morphism $\rho\colon Y \to V$ comes with sections $\tau_1\colon V \to Y, \dots, \tau_\nu\colon V \to Y$, forming a family of stable ν-pointed curves of genus γ on V.

Given a finite group Γ, by *an action of* Γ *on* ρ we mean a pair of actions of Γ on V and Y as S-schemes such that the morphisms $\rho, \tau_1, \dots, \tau_\nu$ are Γ-equivariant. Such an action induces a morphism $Y/\Gamma \to V/\Gamma$ together with sections $V/\Gamma \to Y/\Gamma$.

We adopt the convention that all families of curves are assumed to be of *finite presentation* over the base.

Definition 3.1. Let $\rho\colon Y \to V$ be a family of ν-pointed curves of genus γ, Γ a finite group. An action of Γ on ρ is called *essential* if no nontrivial element of Γ leaves a geometric fiber of ρ fixed.

Another way to state this condition is to require that if γ is an element of Γ leaving a geometric point v_0 of V fixed, then γ acts nontrivially on the fiber of ρ on v_0.

The following lemma will help us replace an action of Γ on ρ by an essential action on a related family of curves.

Lemma 3.2. *Let V be the spectrum of a local ring R, $Y \to V$ a flat family of nodal curves, Γ_0 a group acting compatibly on Y and V. Suppose that the action of Γ_0 on the residue field k of R and on the fiber Y_0 of Y over* Spec k *is trivial. Then $Y/\Gamma_0 \to V/\Gamma_0$ is again a flat family of nodal curves.*

Remark 3.3. It is easy to give an example showing that this fails in positive characteristic.

Proof. We need to show that the map $\rho\colon Y/\Gamma_0 \to V/\Gamma_0$ is flat, and the natural map $Y_0 \to \rho^{-1}(\operatorname{Spec} k)$ is an isomorphism. Replacing R with the completion of its strict henselization we may assume that R is complete and k is algebraically closed. Choose a rational point $p \in X(k)$, and let M be the completion of the local ring of Y at p. Write $M \otimes_R k$ as a quotient of a power series algebra $k[[t_1, \dots, t_r]] = k[[\mathbf{t}]]$, and lift the images of the t_i in $M \otimes_R k$ to invariant elements in M. We get a surjective homomorphism $R[[\mathbf{t}]] \to M$ which is equivariant, if we let Γ_0 act on $R[[\mathbf{t}]]$ leaving the t_i fixed. Then the result follows from the lemma below.

Lemma 3.4. *Let M be a finitely generated $R[[\mathbf{t}]]$-module which is flat over R, such that Γ_0 acts trivially on $M \otimes_R k$. Then M^{Γ_0} is flat over R^{Γ_0}, and the natural homomorphism $M^{\Gamma_0} \otimes_{R^{\Gamma_0}} R \to M$ is an isomorphism.*

Proof. We have $R[[\mathbf{t}]]^{\Gamma_0} = R^{\Gamma_0}[[\mathbf{t}]]$, and from this we see that the statement holds when M is free over $R[[\mathbf{t}]]$. In general, take a finite set of generators of the finite $k[[\mathbf{t}]]$-module $M \otimes_R k$ and lift them to a set of invariant generators of M. We obtain an equivariant surjective homomorphism $F \to M$, where F is a free $R[[\mathbf{t}]]$-module. Let K be the kernel:

$$0 \to K \to F \to M \to 0.$$

Note that by the finite presentation assumption on $Y \to V$, K is finitely generated. Since M is flat we get $Tor_i(K, \cdot) = Tor_{i+1}(M, \cdot) = 0$, thus K is also flat. Moreover, from the exact sequence

$$0 \to K \otimes_R k \to F \otimes_R k \to M \otimes_R k \to 0$$

we get that Γ_0 acts trivially on $K \otimes_R k$. We obtain a commutative diagram with exact rows

$$\begin{array}{ccccccccc} & & K^{\Gamma_0} \otimes_{R^{\Gamma_0}} R & \longrightarrow & F^{\Gamma_0} \otimes_{R^{\Gamma_0}} R & \longrightarrow & M^{\Gamma_0} \otimes_{R^{\Gamma_0}} R & \longrightarrow & 0 \\ & & \downarrow & & \downarrow & & \downarrow & & \\ 0 & \longrightarrow & K & \longrightarrow & F & \longrightarrow & M & \longrightarrow & 0 \end{array}$$

where the middle column is an isomorphism. It follows that the right column is surjective; since K is also flat and Γ_0 acts trivially on $K \otimes_R k$ we can apply the same argument to K, and it follows that the left column is also surjective. Therefore the right column is also injective, as desired.

Applying the same argument to the module K, we have that the left column is also an isomorphism, and the first arrow in the top row is injective; the flatness of M^{Γ_0} over R^{Γ_0} then follows from Grothendieck's local criterion of flatness. $\square$

The following lemma will help us identify a situation where an essential action is free:

Lemma 3.5. *Let G be a group of automorphisms of a stable ν-pointed curve X of genus γ over an algebraically closed field, and assume that the quotient X/G is also a stable ν-pointed curve of genus γ. Then G is trivial.*

Proof. If C is a complete curve we denote by $\mathrm{CH}^i(C)$ the Chow group of classes of cycles of codimension i on C; so $\mathrm{CH}^0(C)$ is a direct sum of n copies of $\mathbf{Z}$, where n is the number of irreducible components of C, while $\mathrm{CH}^1(C)$ is the divisor class group of C. If $\mathcal{F}$ is a coherent sheaf on C we denote by $\tau^C(\mathcal{F}) \in \mathrm{CH}(C)$ its Riemann–Roch class, and by $\tau_i^C(\mathcal{F}) \in \mathrm{CH}^i(C)$ its component of codimension i; $\tau_0^C(\mathcal{F})$ is the cycle associated with $\mathcal{F}$.

Set $Y = X/G$, and call $\pi \colon X \to Y$ the projection. Call Δ_X and Δ_Y the divisors corresponding to the distinguished points in X and Y respectively, ω_X and ω_Y the dualizing sheaves.

A local calculation reveals that $\chi\big(\omega_X(\Delta_X)\big) \geq \chi\big(\pi^*\omega_Y(\Delta_Y)\big)$. Indeed, for this calculation we may replace X and Y by their normalization, where we add markings on the normalized curves at the points above the nodes. Then the natural pullback $\pi^*\omega_{Y^{nor}} \to \omega_{X^{nor}}$ extends to a homomorphism of invertible sheaves $\pi^*\omega_{Y^{nor}}(\Delta_{Y^{nor}}) \to \omega_{X^{nor}}(\Delta_{X^{nor}})$ which is injective, and whose cokernel is supported on the points of ramification of π not in $\Delta_{X^{nor}}$.

Thus we get an inequality of Euler characteristics of sheaves:

$$\begin{aligned}\chi\big(\omega_X(\Delta_X)\big) &\geq \chi\big(\pi^*\omega_Y(\Delta_Y)\big)\\ &= \chi\big(\pi_*\pi^*\omega_Y(\Delta_Y)\big)\\ &= \chi\big(\omega_Y(\Delta_Y)\otimes\pi_*\mathcal{O}_X\big)\\ &= \chi\big(\omega_Y(\Delta_Y)\big)+\chi\big(\omega_Y(\Delta_Y)\otimes(\pi_*\mathcal{O}_X/\mathcal{O}_Y)\big).\end{aligned}$$

By hypothesis

$$\chi\big(\omega_X(\Delta_X)\big)=\gamma+\nu-1=\chi\big(\omega_Y(\Delta_Y)\big).$$

From this, together with the theorem of Grothendieck–Riemann–Roch, we obtain

$$\begin{aligned}0 &\geq \chi\big(\omega_Y(\Delta_Y)\otimes(\pi_*\mathcal{O}_X/\mathcal{O}_Y)\big)\\ &= \int_Y \mathrm{ch}\big(\omega_Y(\Delta_Y)\big)\tau^Y(\pi_*\mathcal{O}_X/\mathcal{O}_Y)\\ &= \int_Y \mathrm{c}_1\big(\omega_Y(\Delta_Y)\big)\cap\tau_0^Y(\pi_*\mathcal{O}_X/\mathcal{O}_Y)+\int_Y \tau_1^Y(\pi_*\mathcal{O}_X/\mathcal{O}_Y)\\ &= \int_Y \mathrm{c}_1\big(\omega_Y(\Delta_Y)\big)\cap(\pi_*[X]-[Y])+\int_X\tau_1^X(\mathcal{O}_X)-\int_Y\tau^Y(\mathcal{O}_Y)\\ &= \int_Y \mathrm{c}_1\big(\omega_Y(\Delta_Y)\big)\cap(\pi_*[X]-[Y])+\chi(\mathcal{O}_X)-\chi(\mathcal{O}_Y)\\ &= \int_Y \mathrm{c}_1\big(\omega_Y(\Delta_Y)\big)\cap(\pi_*[X]-[Y]).\end{aligned}$$

But $\omega_Y(\Delta_Y)$ is an ample line bundle on Y and $\pi_*[X]-[Y]$ is an effective cycle; the only possibility is that $\pi_*[X]-[Y]=0$. This implies that π is birational, so G is trivial. □

4 Fibered surfaces: definitions

If $C \to S$ is a flat family of nodal curves, we denote by C_{sm} the open subscheme of C consisting of points where the morphism $C \to S$ is smooth.

Definition 4.1. Let $C \to S$ be a flat (not necessarily proper) family of nodal curves, $X \to C$ a proper morphism with one-dimensional fibers, and $\sigma_1, \dots, \sigma_\nu \colon C \to X$ sections of ρ. We will say that

$$\begin{array}{c}X\\ \downarrow\\ C\\ \downarrow\\ S\end{array}$$

is a *family of generically fibered surfaces* if X is flat over S, and the restriction of ρ to C_{sm} is a flat family of stable pointed curves. If S is the spectrum of a field we will refer to $X \to C$ as a generically fibered surface.

Remark 4.2. Notice that we do not require the morphism $X \to C$ to be flat.

Definition 4.3. A *chart* $(U, Y \to V, \Gamma)$ for a family of generically fibered surfaces $X \to C \to S$ consists of a diagram

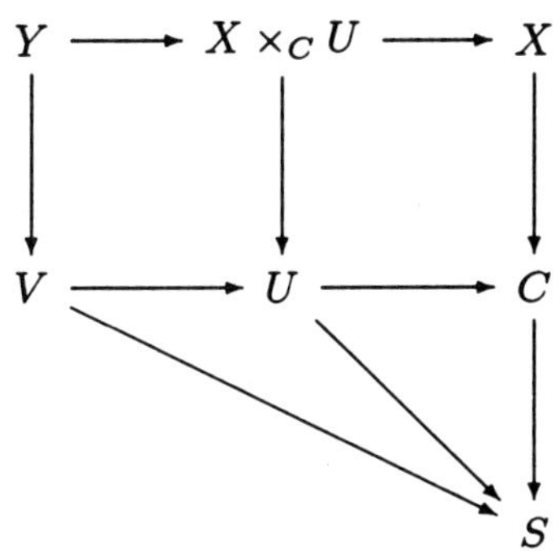

together with a group action $\Gamma \subset \mathrm{Aut}_S(Y \to V)$ satisfying:

1. The morphism $U \to C$ is étale;
2. $V \to S$ is a flat (but not necessarily proper) family of nodal curves;
3. $\rho \colon Y \to V$ is a flat family of stable ν-pointed curves of genus γ,
4. the action of Γ on ρ is essential;
5. we have isomorphisms of S-schemes $V/\Gamma \simeq U$ and $Y/\Gamma \simeq U \times_C X$ compatible with the projections $Y/\Gamma \to V/\Gamma$ and $U \times_C X \to U$, such that the sections $U \to U \times_C X$ induced by the σ_i correspond to the sections $V/\Gamma \to Y/\Gamma$.

Lemma 4.4. *Let $(U, Y \to V, \Gamma)$ be a chart for a family of generically fibered surfaces $X \to C \to S$. If $S' \to S$ is an arbitrary morphism, then $(S' \times_S U, S' \times_S Y \to S' \times_S V, \Gamma)$ (with the obvious definitions of the various maps and of the action of Γ on $S' \times_S Y$ and $S' \times_S V$) is a chart for the family of generically fibered surfaces $S' \times_S X \to S' \times_S C \to S'$.*

Proof. Conditions (1) through (4) in the definition are immediately verified for the pullback diagram. The only point that requires a little care is to check that $(S' \times_S V)/\Gamma \simeq S' \times_S U$ and $(S' \times_S Y)/\Gamma \simeq (S' \times_S U) \times_C X$, which requires the hypothesis that S be a scheme over $\mathbf{Q}$. □

Proposition 4.5. 1. *Let* $(U, Y \to V, \Gamma)$ *be a chart for a family of generically fibered surfaces* $X \to C \to S$. *Let* $V' \subset V$ *be the inverse image of* C_{sm}. *Then*

(a) *the action of* Γ *on* V' *is free; and*

(b) *the natural morphism* $Y|_{V'} \to V' \times_S X$ *is an isomorphism.*

2. *Furthermore, if* t_0 *is a geometric point of* S *and* v_0 *a nodal point of the fiber* V_{t_0} *of* V *over* t_0, *then*

(a) *the stabilizer* Γ' *of* v_0 *is a cyclic group which sends each of the branches of* V_{t_0} *to itself.*

(b) *If* n *is the order of* Γ', *then a generator of* Γ' *acts on the tangent space of each branch by multiplication with a primitive* n*-th root of 1.*

In particular, $V' = V_{\mathrm{sm}}$.

Proof. The claim (1b), that the natural morphism $Y|_{V'} \to V' \times_S X$ is an isomorphism, is a consequence of the statement (1a).

Let v_0 be a geometric point of the inverse image of U_{sm} in V, Γ' its stabilizer. By definition Γ' acts faithfully on the fiber Y_0 of Y on v_0. The fiber X_0 of X over the image of v_0 in C is the quotient of Y_0 by Γ'; from Lemma 3.5 above we get that Γ' is trivial, as claimed.

For part (2) of the proposition we observe that if the stabilizer Γ' of v_0 did not preserve the branches of V_{t_0} then the quotient V_{t_0}/Γ', which is étale at the point v_0 over the fiber U_{t_0}, would be smooth over S at v_0, so v_0 would be in the inverse image of U_{sm}. From part (1) of the proposition it would follow that Γ' is trivial, a contradiction.

So Γ' acts on each of the two branches individually. The action on each branch must be faithful because it is free on the complement of the set of nodes; this means that the representation of Γ' in each of the tangent spaces to the branches is faithful, and this implies the final statement. □

Definition 4.6. A chart is called *balanced* if for any nodal point of any geometric fiber of V, the two roots of 1 describing the action of a generator of the stabilizer on the tangent spaces to each branch of V are inverse to each other.

It is easy to see that a chart is balanced if and only if it admits a deformation to a smooth curve.

4.7 The transition scheme of two charts

Let $X \to C \to S$ be a family of generically fibered surfaces, $(U_1, Y_1 \to V_1, \Gamma_1)$, $(U_2, Y_2 \to V_2, \Gamma_2)$ two charts; call $\mathrm{pr}_i \colon V_1 \times_C V_2 \to V_i$ the ith

projection. Consider the scheme

$$I = \operatorname*{Isom}_{V_1 \times_C V_2} (\mathrm{pr}_1^* Y_1, \mathrm{pr}_2^* Y_2)$$

over $V_1 \times_C V_2$ representing the functor of isomorphisms of the two families $\mathrm{pr}_1^* Y_1$ and $\mathrm{pr}_2^* Y_2$. There is a section of I over the inverse image $\widetilde{V}$ of C_{sm} in $V_1 \times_C V_2$ which corresponds to the isomorphism $\mathrm{pr}_1^* Y_1|_{\widetilde{V}} \simeq \mathrm{pr}_2^* Y_2|_{\widetilde{V}}$ coming from the fact that both $\mathrm{pr}_1^* Y_1$ and $\mathrm{pr}_2^* Y_2$ are pullbacks to $\widetilde{V}$ of the restriction of X to C_{sm}. We will call the scheme-theoretic closure R of this section in I the *transition scheme* from $(U_1, Y_1 \to V_1, \Gamma_1)$ to $(U_2, Y_2 \to V_2, \Gamma_2)$; it comes equipped with two projections $R \to V_1$ and $R \to V_2$. There is also an action of $\Gamma_1 \times \Gamma_2$ on I, defined as follows. Let $(\gamma_1, \gamma_2) \in \Gamma_1 \times \Gamma_2$, and $\phi\colon \mathrm{pr}_1^* Y_1 \simeq \mathrm{pr}_2^* Y_2$ an isomorphism over $V_1 \times_C V_2$; then define $(\gamma_1, \gamma_2) \cdot \phi = \gamma_2 \circ \phi \circ \gamma_1^{-1}$. This action of $\Gamma_1 \times \Gamma_2$ on I is compatible with the action of $\Gamma_1 \times \Gamma_2$ on $V_1 \times_C V_2$, and leaves R invariant. It follows from the definition of an essential action that the action of $\Gamma_1 = \Gamma_1 \times \{1\}$ and $\Gamma_2 = \{1\} \times \Gamma_2$ on I is free.

Definition 4.8. Two charts $(U_1, Y_1 \to V_1, \Gamma_1)$ and $(U_2, Y_2 \to V_2, \Gamma_2)$ are compatible if their transition scheme R is étale over V_1 and V_2.

Let us analyze this definition. Start from two charts $(U_1, Y_1 \to V_1, \Gamma_1)$ and $(U_2, Y_2 \to V_2, \Gamma_2)$. Fix two geometric points $v_1\colon \operatorname{Spec} \Omega \to V_1$ and $v_2\colon \operatorname{Spec} \Omega \to V_2$ mapping to the same geometric point $v_0\colon \operatorname{Spec} \Omega \to C$, and call $\Gamma_i' \subseteq \Gamma_i$ the stabilizer of v_i. Also call V_1^{sh}, V_2^{sh} and C^{sh} the spectra of the strict henselizations of V_1, V_2 and C at the points v_1, v_2 and v_0 respectively. The action of Γ_i on V_i induces an action of Γ_i' on V_i^{sh}. Also, call Y_i^{sh} the pullback of Y_i to V_i^{sh}; there is an action of Γ_i' on Y_i^{sh} compatible with the action of Γ_i' on V_i.

Proposition 4.9. *The two charts are compatible if and only if for any pair of geometric points v_1 and v_2 as above there exist an isomorphism of groups $\eta\colon \Gamma_1' \simeq \Gamma_2'$ and two compatible η-equivariant isomorphisms $\phi\colon V_1^{\mathrm{sh}} \simeq V_2^{\mathrm{sh}}$ and $\psi\colon Y_1^{\mathrm{sh}} \to Y_2^{\mathrm{sh}}$ of schemes over C^{sh}.*

Proof. Consider the spectrum $(V_1 \times_C V_2)^{\mathrm{sh}}$ of the strict henselization of $V_1 \times_C V_2$ at the point $(v_1, v_2)\colon \operatorname{Spec} \Omega \to V_1 \times_C V_2$, and call R^{sh} the pullback of R to $(V_1 \times_C V_2)^{\mathrm{sh}}$. Assume that the two charts are compatible. The action of $\Gamma_1 \times \Gamma_2$ on I described above induces an action of $\Gamma_1' \times \Gamma_2'$ on R^{sh}, compatible with the action of $\Gamma_1' \times \Gamma_2'$ on $(V_1 \times_C V_2)^{\mathrm{sh}}$. The action of $\Gamma_1' = \Gamma_1' \times \{1\}$ on the inverse image of C_{sm} in R^{sh} is free, and its quotient is the inverse image of C_{sm} in V_2^{sh}; but R^{sh} is finite and étale over V_2^{sh}, so the action of Γ_1' on all of R^{sh} is free, and $R^{\mathrm{sh}}/\Gamma_1' = V_2$. Analogously the action of Γ_2' on R^{sh} is free, and $R^{\mathrm{sh}}/\Gamma_2' = V_1$.

Now, each of the connected components of R^{sh} maps isomorphically onto both V_1 and V_2, because V_i is the spectrum of a strictly henselian ring and the projection $R^{\mathrm{sh}} \to V_i$ is étale; this implies in particular that the order of Γ_1 is the same as the number n of connected components, and likewise for Γ_2. Fix one of these components, call it R_0^{sh}; then we get isomorphisms $R_0^{\mathrm{sh}} \simeq V_i$, which yield an isomorphism $\phi\colon V_1 \simeq V_2$.

Call Γ' the stabilizer of the component R_0^{sh} inside $\Gamma_1' \times \Gamma_2'$; the order of Γ' is at least $|\Gamma_1' \times \Gamma_2'|/n = n^2/n = n$. But the action of Γ_2' on R^{sh} is free, and so $\Gamma' \cap \Gamma_2 = \{1\}$; this implies that the order of Γ' is n, and the projection $\Gamma' \to \Gamma_1$ is an isomorphism. Likewise the projection $\Gamma' \to \Gamma_2$ is an isomorphism, so from these we get an isomorphism $\eta\colon \Gamma_1 \to \Gamma_2$, and it is easy to check that the isomorphism of schemes $\phi\colon V_1 \simeq V_2$ is η-equivariant.

There is also an isomorphism of the pullbacks of Y_1^{sh} and Y_2^{sh} to R_0^{sh}, coming from the natural morphism $R_0^{\mathrm{sh}} \to I$, which induces an isomorphism $\psi\colon Y_1^{\mathrm{sh}} \to Y_2^{\mathrm{sh}}$. This isomorphism is compatible with ϕ, and is it also η-equivariant.

Let us prove the converse. Suppose that there exist η, ϕ and ψ as above. Then there is a morphism $\sigma\colon V_1^{\mathrm{sh}} \times \Gamma_1' \to I$ which sends a point (v_1, γ_1) of $V_1^{\mathrm{sh}} \times \Gamma_1'$ into the point of I lying over the point $(v_1, \phi\gamma_1 v_1) = (v_1, \eta(\gamma_1)\phi v_1)$ corresponding to the isomorphism $\gamma_1\psi$ of the fiber of Y_1 over v_1 with the fiber of Y_2 over $\phi\gamma_1 v_1$. The morphism σ is an isomorphism of $V_1^{\mathrm{sh}} \times \Gamma_1'$ with R^{sh} in the inverse image of C_{sm}; it also follows from the fact that the action of Γ' on $Y_1 \to V_1$ is essential that σ is injective. Since the inverse image of C_{sm} is scheme-theoretically dense in R^{sh} and $V_1^{\mathrm{sh}} \times \Gamma_1$ is unramified over V_1 we see that σ is an isomorphism; it follows that R^{sh} is étale over V_1^{sh}; analogously it is étale over V_2^{sh}. So R is étale over V_1 and V_2 at the points v_1 and v_2; since this holds for all v_1 and v_2 mapping to the same point of C, the conclusion follows. □

In 6.14 and 6.15 below, we give two examples of incompatible charts on the same coarse fibered surface.

4.10 The product chart

Given two compatible charts $(U_1, Y_1 \to V_1, \Gamma_1)$ and $(U_2, Y_2 \to V_2, \Gamma_2)$, the graph $Y \subseteq (Y_1 \times_{V_1} R) \times_R (R \times_{V_2} Y_2)$ over R of the canonical isomorphism of the two families $Y_1 \times_{V_1} R$ and $R \times_{V_2} Y_2$ is invariant under the action of $\Gamma_1 \times \Gamma_2$, and it is a family of pointed curves on R. Then

$$(U_1 \times_C U_2, Y \to R, \Gamma_1 \times \Gamma_2)$$

is a chart, called the *product chart.* It is compatible with both the original charts.

Compatibility of charts is stable under base change:

Proposition 4.11. *Let $(U_1, Y_1 \to V_1, \Gamma_1)$ and $(U_2, Y_2 \to V_2, \Gamma_2)$ be two compatible charts for a family of generically fibered surfaces $X \to C \to S$. If $S' \to S$ is an arbitrary morphism, then*

$$(S' \times_S U_1, S' \times_S Y_1 \to S' \times_S V_1, \Gamma_1)$$

and

$$(S' \times_S U_2, S' \times_S Y_2 \to S' \times_S V_1, \Gamma_2)$$

are compatible charts for the the family $S' \times_S X \to S' \times_S C \to S'$.

This is easy.

We now come to the definition of our basic object:

Definition 4.12. A *family of fibered surfaces*

$$\begin{array}{c} \mathcal{X} \\ \downarrow \\ \mathcal{C} \\ \downarrow \\ S \end{array}$$

is a family of generically fibered surfaces $X \to C \to S$ such that $C \to S$ is proper, together with a collection $\{(U_\alpha, Y_\alpha \to V_\alpha, \Gamma_\alpha)\}$ of mutually compatible charts, such that the images of the U_α cover C.

Such a collection of charts is called an *atlas.*

A family of fibered surfaces is called *balanced* if each chart in its atlas is balanced.

The family of generically fibered surfaces $X \to C \to S$ supporting the family of fibered surfaces $\mathcal{X} \to \mathcal{C} \to S$ will be called a family of *coarse fibered surfaces.*

Lemma 4.13. *If two charts for a family of fibered surfaces are compatible with all the charts in an atlas, they are mutually compatible.*

Furthermore, if the family is balanced, then any chart which is compatible with every chart of the atlas is balanced.

The proof is straightforward.

Remark 4.14. The lemma above allows us to define a family of fibered surfaces using a maximal atlas, if we want.

Definition 4.15. A *morphism of fibered surfaces* $\mathcal{X} \to \mathcal{C} \to S$ to $\mathcal{X}' \to \mathcal{C}' \to S'$ is a *cartesian* diagram of coarse fibered surfaces

$$\begin{array}{ccc} X & \longrightarrow & X' \\ \downarrow & & \downarrow \\ C & \longrightarrow & C' \\ \downarrow & & \downarrow \\ S & \longrightarrow & S', \end{array}$$

such that the pullback of the charts of an atlas of $\mathcal{X}' \to \mathcal{C}' \to S$ are all compatible with the atlas of $\mathcal{X} \to \mathcal{C} \to S$.

The composition of morphisms is the obvious one.

We will soon reinterpret this definition of a morphism.

4.16 Fibered surfaces as stacks

Consider a family of fibered surfaces $\mathcal{X} \to \mathcal{C} \to S$ with an atlas $\{(U_\alpha, Y_\alpha \to V_\alpha, \Gamma_\alpha)\}$. For each pair of indices (α, β) let $R_{\alpha\beta}$ be the transition scheme from $(U_\alpha, Y_\alpha \to V_\alpha, \Gamma_\alpha)$ to $(U_\beta, Y_\beta \to V_\beta, \Gamma_\beta)$. Let V be the disjoint union of the V_α, and let R be the disjoint union of the $R_{\alpha,\beta}$. These have the following structure:

- there are two projections $R \to V$, which are étale;
- there is a natural diagonal morphism $V \to R$ which sends each V_α to $R_{\alpha\alpha}$; and
- there is a product $R \times_V R \to R$, sending each $R_{\alpha\beta} \times_{V_\beta} R_{\beta\gamma}$ to $R_{\alpha\gamma}$ via composition of isomorphisms.

These various maps give $R \rightrightarrows V$ the structure of a groupoid. Since the diagonal map $R \to V \times_C V$ is unramified the groupoid defines a quotient Deligne–Mumford stack. In a slight abuse of notation, we call this stack $\mathcal{C}$ as well.

A similar groupoid structure can be formed using Y_α and their pullback to $R_{\alpha\beta}$, endowing $\mathcal{X}$ with the structure of a Deligne–Mumford stack as well. Note that $\mathcal{X} \to \mathcal{C}$ is representable, and the stack $\mathcal{X}$ is a family of stable ν-pointed curves of genus γ over the stack $\mathcal{C}$.

Let us list some properties of $\mathcal{X}$ and $\mathcal{C}$:

- There is a moduli morphism $\mathcal{C} \to \overline{\mathcal{M}}_{\gamma,\nu}$ associated to the family $\mathcal{X} \to \mathcal{C}$.
- The stack $\mathcal{C}$ is a proper nodal stack over S, namely, it has an étale cover, given by the schemes V underlying the charts, which are nodal over S.
- Over the inverse image of C_{sm} the scheme R is isomorphic to the fibered product $V \times_C V$, so the quotient stack $\mathcal{C}_{\text{sm}}$ coincides with C_{sm}. A similar statement holds for $\mathcal{X}$.
- Since U_α is the schematic quotient of V_α by the action of Γ_α, and $U \to C$ is étale, it is immediate that C is the schematic quotient of the groupoid $R \rightrightarrows V$, in other words, the stack morphism $\mathcal{C} \to C$ exhibits C as the coarse moduli scheme of $\mathcal{C}$. Similarly, X is the coarse moduli scheme of $\mathcal{X}$.

As mentioned above, a fibered surface being *balanced* is tantamount to the existence of local smoothing of the fibered surface. It is interesting to interpret the requirement that the action of Γ is *essential* within the language of stacks.

We claim that the action being essential is equivalent to the condition that the moduli morphism $\mathcal{C} \to \overline{\mathcal{M}}_{\gamma,\nu}$ be *representable.* This follows from the definition of essential action, using the following well known lemma:

Lemma 4.17. *Let $g : \mathcal{G} \longrightarrow \mathcal{F}$ be a morphism of Deligne–Mumford stacks. The following two conditions are equivalent:*

1. *The morphism $g : \mathcal{G} \longrightarrow \mathcal{F}$ is representable.*

2. *For any algebraically closed field k and any $\xi \in \mathcal{G}(k)$, the natural group homomorphism $\mathrm{Aut}(\xi) \longrightarrow \mathrm{Aut}(g(\xi))$ is a monomorphism.*

4.18 A stack-theoretic formulation of the category of fibered surfaces

We will now formulate stack-theoretic data similar to that obtained above, and then compare them to the data of a fibered surface.

Let S be a scheme over $\mathbf{Q}$. Consider a proper Deligne–Mumford stack $\mathcal{C} \longrightarrow S$, such that its fibers are purely one-dimensional and geometrically connected, with nodal singularities. Call C the moduli space of $\mathcal{C}$; this automatically exists as an algebraic space.

Proposition 4.19. *The morphism $C \longrightarrow S$ is a flat family of nodal curves.*

Proof. First of all let us show that C is flat over S. We may assume that S is affine; call R its ring. Fix a geometric point $c_0 \longrightarrow C$, and call C^{sh} the strict henselization of C at c_0. Let U be an étale cover of $\mathcal{C}$, let u_0 be a geometric point of U lying over c_0, and call U^{sh} the strict henselization of U at u_0. If Γ is the automorphism group of the object of $\mathcal{C}$ corresponding to u_0, then Γ acts on U^{sh}, and C^{sh} is the quotient U^{sh}/Γ. Because the schemes are defined over $\mathbf{Q}$, the ring of C^{sh} is a direct summand of the ring of Y^{sh}, as an R-module, so it is flat over R.

The fact that the fibers are nodal follows from the fact that, over an algebraically closed field, the quotient of a nodal curve by a group action is again a nodal curve. □

Definition 4.20. A stack-like family of curves of genus g over a scheme S consists of a proper stack $\mathcal{C} \longrightarrow S$, whose fibers are purely one-dimensional and geometrically connected, with nodal singularities, such that:

1. the fibers of the morphism $C \longrightarrow S$, where C is the moduli space of $\mathcal{C}$, have genus g;

2. over the inverse image of the smooth locus C_{sm} of the map $C \longrightarrow X$, the projection $\mathcal{C} \longrightarrow C$ is an isomorphism.

If $\mathcal{C} \longrightarrow S$ and $\mathcal{C}' \longrightarrow S'$ are stack-like families of curves of genus g, a morphism $F\colon \mathcal{C} \longrightarrow \mathcal{C}'$ consists of a cartesian diagram

$$\begin{array}{ccc} \mathcal{C} & \xrightarrow{F} & \mathcal{C}' \\ \downarrow & & \downarrow \\ S & \xrightarrow{f} & S' \end{array} .$$

The composition of morphisms of stack-like families of curves is defined in the obvious way. In this way stack-like families of curves of genus g form a 2-category, 2-arrows being defined in the usual way.

Proposition 4.21. *The 2-category of stack-like families of curves of genus g is equivalent (in the lax sense) to a 1-category.*

Proof. This is the same as saying that a 1-arrow in the category cannot have nontrivial automorphisms. The point here is that the stack $\mathcal{C}$ has an open subscheme which is stack-theoritically dense in $\mathcal{C}$, which is sufficient by the following lemma.

Lemma 4.22. *Let $F\colon \mathcal{X} \to \mathcal{Y}$ be a representable morphism of Deligne–Mumford stacks over a scheme S. Assume that there exists an open representable substack (i.e. an algebraic space) $U \subseteq \mathcal{X}$ and a dense open representable substack $V \subseteq \mathcal{Y}$ such that F maps U into V. Further assume that the diagonal $\mathcal{Y} \to \mathcal{Y} \times_S \mathcal{Y}$ is separated. Then any automorphism of F is trivial.*

Definition 4.23. A stack-like fibered surface $f\colon \mathcal{C} \to \overline{\mathcal{M}}_{\gamma,\nu}$ over a scheme S is a stack-like family $\mathcal{C} \to S$ of curves of genus g over S, with projective coarse moduli space $C \to S$, together with a representable morphism $\mathcal{C} \to \overline{\mathcal{M}}_{\gamma,\nu}$.

As pointed out above, stack-like fibered surfaces form a 2-category. A 1-arrow $\Phi = (\phi, \alpha_\Phi)$ from $f\colon \mathcal{C} \to \overline{\mathcal{M}}_{\gamma,\nu}$ to $f'\colon \mathcal{C}' \to \overline{\mathcal{M}}_{\gamma,\nu}$ consists of a morphism of stack-like families of curves $\phi\colon \mathcal{C} \to \mathcal{C}'$, together with an isomorphism α_Φ of the composition of $f' \circ \phi$ with f; the compostion of 1-arrows is defined in the obvious way. A 2-arrow from a 1-arrow Φ from $f\colon \mathcal{C} \to \overline{\mathcal{M}}_{\gamma,\nu}$ to $f'\colon \mathcal{C}' \to \overline{\mathcal{M}}_{\gamma,\nu}$ to another arrow Ψ with the same domain and codomain consists of an isomorphism β of the two functors ϕ and ψ, with the usual compatibility condition of the isomorphism β with α_Φ and α_Ψ.

Because of the proposition, this category is equivalent to a nice unproblematic 1-category, which we can think of as the category of stack-like fibered surfaces, the arrows being given by isomorphism classes of morphisms.

We can now compare the two categories we have defined.

Theorem 4.24. *The category of stack-like fibered surfaces over schemes over $\mathbf{Q}$ is equivalent (in the lax sense) to the category of fibered surfaces.*

Sketch of proof. The construction in 4.16 shows that a fibered surface gives rise to a stack-like fibered surface in a functorial way.

To go from stack-like fibered surfaces to honest fibered surfaces, let us take a stack-like fibered surface $\mathcal{C} \to \overline{\mathcal{M}}_{\gamma,\nu}$, and then take as a family of pointed nodal curves its coarse moduli space $C \to S$. The charts are given by étale morphism from schemes to $\mathcal{C}$. We leave the treatment of morphisms to the reader. $\square$

From now on we will use *fibered surfaces* and *stack-like fibered surfaces* interchangeably.

4.25 Stable fibered surfaces

We first define a moduli morphism on a coarse fibered surface:

Lemma 4.26. *Let $\mathcal{X} \to \mathcal{C} \to S$ be a family of fibered surfaces. Then the morphism $C_{\mathrm{sm}} \to \overline{\mathbf{M}}_{\gamma,\nu}$ induced by the restriction of $X \to C$ to C_{sm} extends uniquely to a morphism $C \to \overline{\mathbf{M}}_{\gamma,\nu}$.*

Proof. The unicity is clear from the fact that $\overline{\mathbf{M}}_{\gamma,\nu}$ is separated and C_{sm} is scheme-theoretically dense in C. To prove the existence of an extension is a local question in the étale topology; but if $\{(U_\alpha, Y_\alpha \to V_\alpha, \Gamma_\alpha)\}$ is an atlas then the families $Y_\alpha \to V_\alpha$ induce morphisms $V_\alpha \to \overline{\mathbf{M}}_{\gamma,\nu}$, which are Γ_α-equivariant, yielding morphisms $U_\alpha \to \overline{\mathbf{M}}_{\gamma,\nu}$. These morphisms are extensions of the restriction to the U_α of the morphism $C_{\mathrm{sm}} \to \overline{\mathbf{M}}_{\gamma,\nu}$. Therefore they descend to C. $\square$

We can use this lemma to define stable fibered surfaces:

Definition 4.27. A family of fibered surfaces $\mathcal{X} \to \mathcal{C} \to S$ is *stable* if the associated morphism $C \to \overline{\mathbf{M}}_{\gamma,\nu}$ is Kontsevich stable.

4.28 Natural line bundles on a coarse fibered surface

Proposition 4.29. *On each family of fibered surfaces $\mathcal{X} \to \mathcal{C} \to S$ there exists a canonically defined line bundle $\mathcal{L}_X$ on the coarse fibered surface X, which is relatively ample along the map $X \to C$. This line bundle satisfies:*

1. *For any morphism of fibered surfaces*

$$\begin{array}{ccc} \mathcal{X}' & \xrightarrow{\phi} & \mathcal{X} \\ \downarrow & & \downarrow \\ \mathcal{C}' & \longrightarrow & \mathcal{C} \\ \downarrow & & \downarrow \\ S' & \longrightarrow & S \end{array}$$

   there is an isomorphism of line bundles $\sigma_\phi \colon \mathcal{L}_{X'} \simeq \phi^* \mathcal{L}_X$.

2. *These isomorphisms satisfy the cocycle condition, in the sense that $\sigma_{\mathrm{id}_X} = \mathrm{id}_{\mathcal{L}_X}$ for all families of fibered surfaces $X \to S$, and if we*

have two morphisms

$$\begin{array}{ccccc} \mathcal{X}'' & \xrightarrow{\psi} & \mathcal{X}' & \xrightarrow{\phi} & \mathcal{X} \\ \downarrow & & \downarrow & & \downarrow \\ \mathcal{C}'' & \longrightarrow & \mathcal{C}' & \longrightarrow & \mathcal{C} \\ \downarrow & & \downarrow & & \downarrow \\ S'' & \longrightarrow & S' & \longrightarrow & S \end{array}$$

then

$$\sigma_{\phi\psi} = \psi^*\sigma_\phi \circ \sigma_\psi \colon \mathcal{L}_{X''} \longrightarrow (\phi\psi)^*\mathcal{L}_X = \psi^*\phi^*\mathcal{L}_X \,.$$

Proof. Fix an integer N which is divisible by the order of the automorphism group of any ν-pointed stable curve of genus γ. The construction below will depend on the choice of the integer N. For each index α consider the line bundle $\omega_\alpha = \omega_{Y_\alpha/V_\alpha}(\Delta_\alpha)$, the relative dualizing sheaf of Y_α over V_α twisted by the divisor Δ_α of marked points. This line bundle has a natural action of Γ_α, and the stabilizers of all the geometric points act trivially on the fiber of $\omega_\alpha^{\otimes N}$; therefore $\omega_\alpha^{\otimes N}$ descends to a line bundle $\mathcal{L}_\alpha$ on $X \times_C U_\alpha$.

Given two indices α and β, let $R_{\alpha\beta}$ be the transition scheme of two corresponding charts and $Y_{\alpha\beta} \to R_{\alpha\beta}$ the family of pointed stable curves constructed above; there is a canonical isomorphism between the pullback of ω_α to $Y_{\alpha\beta}$ and the analogous bundle $\omega_{\alpha\beta}$ on the family $Y_{\alpha\beta}$, so we can compose these to get isomorphisms between the pullbacks of ω_α and ω_β. These isomorphisms satisfy the cocycle condition and give the collection of the ω_α the structure of a line bundle on the groupoid $R \rightrightarrows V$.

Passing to the N-th power, the isomorphisms above also descend to isomorphisms between the pullbacks of $\mathcal{L}_\alpha$ and $\mathcal{L}_\beta$ on the fiber product $X \times_C (U_\alpha \times_C U_\beta)$ satisfying the cocycle condition, so the $\mathcal{L}_\alpha$ are the pullback of a well defined line bundle $\mathcal{L}_X$, which is the one we want. □

Proposition 4.30. *On each family of* stable *fibered surfaces $\mathcal{X} \to \mathcal{C} \to S$ there exists a canonically defined line bundle $\mathcal{A}_X$ on the coarse fibered surface X, which is relatively ample along the map $X \to S$. This line bundle satisfies the properties listed in Proposition 4.29.*

Proof. Fix N as in the proof of Proposition 4.29. Fix an ample line bundle H_0 on $\overline{\mathbf{M}}_{\gamma,\nu}$, and let $H = H_0^{\otimes 3}$. Let $\mathcal{X} \to \mathcal{C} \to S$ be a stable fibered surface, and let $m_C : C \to \overline{\mathbf{M}}_{\gamma,\nu}$ be the associated moduli morphism. Consider the line bundle $A_C := (\omega_{C/S} \otimes m_C^* H)^{\otimes N}$ on C. By the theory of stable maps, the line bundle A_C is ample relative to the morphism $C \to S$. It clearly satisfies the invariance condition in the proposition. Also, by Kollár's semipositivity lemma (see [Kollár1]) the line bundle $\mathcal{L}_X$ constructed above is nef. Set $\mathcal{A}_X = \mathcal{L}_X \otimes f^* A_C$. By Kleiman's criterion for ampleness we clearly have that $\mathcal{A}_X$ is ample. The invariance conditions follow by construction.

Remark 4.31. It is important to note that both $\mathcal{L}_X$ and $\mathcal{A}_X$ are invariants of the *coarse* fibered surfaces, independent of the given atlas. This is because, on the open subscheme $U \subset X$ where $X \to S$ is Gorenstein, these coincide with $\omega_{U/C}^N$ and $\omega_{U/S}^N$. Since U has a complement of codimension ≥ 2, the extensions to X are unique.

5 The stack of fibered surfaces

Consider the stack $\overline{\mathcal{M}}_{\gamma,\nu}$ of stable ν-pointed curves of genus γ over $\mathbf{Q}$, and the associated moduli space $\overline{\mathbf{M}}_{\gamma,\nu}$; choose an ample line bundle $\mathcal{H}$ on $\overline{\mathbf{M}}_{\gamma,\nu}$. Fix two nonnegative integers g and d.

We define a category $\mathcal{F}_g^d(\gamma,\nu)$, fibered over the category Sch $/\mathbf{Q}$ of schemes over $\mathbf{Q}$, as follows. The objects are stable families $\mathcal{X} \to \mathcal{C} \to S$ such that for the associated morphism $f\colon C \to \overline{\mathbf{M}}_{\gamma,\nu}$, the degree of the line bundle $f^*\mathcal{H}$ on each fiber of C over S is d. The arrows are morphisms of fibered surfaces. We also have a subcategory $\mathcal{F}_g^d(\gamma,\nu)^{balanced}$ of stable *balanced* families.

There is an obvious morphism from $\mathcal{F}_g^d(\gamma,\nu)$ to the stack $\mathcal{K}_g^d(\overline{\mathbf{M}}_{\gamma,\nu})$ of Kontsevich-stable maps of genus g and degree d into $\overline{\mathbf{M}}_{\gamma,\nu}$ which sends each stable family of fibered surfaces $\mathcal{X} \to \mathcal{C} \to S$ to the associated morphism $C \to \overline{\mathbf{M}}_{\gamma,\nu}$.

Our main result is:

Theorem 5.1. *The category $\mathcal{F}_g^d(\gamma,\nu)$ is a complete Deligne–Mumford stack, admitting a projective coarse moduli space $\mathbf{F}_g^d(\gamma,\nu)$. The subcategory $\mathcal{F}_g^d(\gamma,\nu)^{balanced}$ forms an open and closed substack.*

This is a very particular case of our general theorem of existence for the Deligne–Mumford stack of twisted stable maps from a curve into a general Deligne–Mumford stack with projective moduli space (see [ℵ-V1], [ℵ-V2]). A proof (a little sketchy at some technical points) can be found in the previous version of this paper. This is substantially simpler than in the general case, because it takes advantage of many simplifications afforded by dealing with families of algebraic curves, rather than by abstract objects in a Deligne–Mumford stack.

6 Fibered surfaces and Alexeev stable maps

6.1 Semi-log-canonical surfaces

Semi-log-canonical surfaces are introduced in [K-SB] and further studied in [Al2] and [Al3]. We now review their definition. First, let us define log-canonical pairs.

Definition 6.2. Let (X, D) be a pair consisting of a normal variety X and an effective, reduced Weil divisor D. Denote by $X_{\rm sm}$ the nonsingular locus of X. We say that (X, D) is a *log canonical pair* if the following conditions hold:

1. (X, D) is log-**Q**-Gorenstein, namely: for some positive integer m, we assume that the invertible sheaf $(\omega_{X_{\rm sm}}(D))^m$ extends to an invertible sheaf $(\omega_{X_{\rm sm}}(D))^{[m]}$ on X.

2. Let $r : Y \to X$ be a desingularization, such that the proper transform of D together with the exceptional locus of $Y \to X$ form a normal crossings divisor. Call this divisor, taken with reduced structure, D'. Then, we assume that the natural pullback of rational differentials gives a morphism of sheaves: $r^*(\omega_{X_{\rm sm}}(D))^{[m]} \to (\omega_Y(D'))^m$. That is, a logarithmic differential on X pulls back to a logarithmic differential on Y.

A complete description of log-canonical singularities of surface pairs is given in [Al1].

Semi-log-canonical surfaces are a natural generalization of the above to the case of non-normal surfaces.

Definition 6.3. Let X be a surface and D a reduced, pure codimension 1 subscheme such that no component of D lies in $Sing(X)$. We say that (X, D) is a semi-log-canonical if the following conditions hold.

1. X is Cohen Macaulay and normal crossings in codimension 1. Let X_{nc} be the locus where X is either nonsingular or normal crossings.

2. (X, D) is log-**Q**-Gorenstein, namely: for some positive integer m, we assume that the invertible sheaf $(\omega_{X_{nc}}(D))^m$ extends to an invertible sheaf on X.

3. Let $X' \to X$ be the normalization, D' the Weil divisor corresponding of D on X', and $C \subset X'$ the conductor divisor, namely the divisor where $X' \to X$ is not one-to-one, taken with reduced structure. Then the pair $(X', D' + C)$ is log-canonical.

The unique locally free extension of $(\omega_{X_{nc}}(D))^m$ is denoted by $(\omega_X(D))^{[m]}$.

In case D is empty, we just say that X is semi-log-canonical (or "has semi-log-canonical singularities").

We will use the following lemma about quotients of semi-log-canonical surfaces:

Lemma 6.4. *Let (Y, D) be a semi-log-canonical surface pair. Let $\Gamma \subset \mathrm{Aut}(Y, D)$ be a finite subgroup, and let $X = Y/\Gamma$, $D_X = D/\Gamma$. Then the following conditions are equivalent:*

1. *The quotient (X, D_X) is semi-log-canonical;*

2. *The pair* (X, D_X) *is log-*$\mathbf{Q}$*-Gorenstein.*

The proof is straightforward, see [Kollár2].

The reader is referred to [K-SB] for the refined notions of semismooth, semi-canonical and semi-log-terminal singularities.

6.5 *Nodal families and semi-log-canonical singularities.*

A family of nodal curves over a nodal curve is always semi-log-canonical:

Lemma 6.6. *Let* V *be a nodal curve and* $f : Y \to V$ *a nodal family. Then* Y *has Gorenstein semi-log-canonical singularities. If, furthermore,* $D \subset Y$ *is a section which does not meet* $\operatorname{Sing}(f)$, *then* (Y, D) *is a semi-log-canonical pair.*

Proof. Since any family of nodal curves is a Gorenstein morphism, we have that both $V \to \operatorname{Spec} k$ and $Y \to V$ are Gorenstein morphisms, therefore Y is Gorenstein.

Let $y \in Y$ and $p = f(x) \in V$. It is convenient to replace Y and V by their respective formal completions at y and p. The situation falls into one of the following cases:

1. Both $p \notin \operatorname{Sing}(V)$ and $y \notin Sing(f)$. Then $y \in Y_{ns}$, and there is nothing to prove.

2. $p \notin \operatorname{Sing}(V)$ and $y \in Sing(f)$. Choose a regular parameter t at p. By the deformation theory of a node, we have the description

$$Y \simeq \operatorname{Spf} k[[u, v, t]]/(uv - h(t)),$$

with $h(0) = 0$. If $h(t) \not\equiv 0$ then we may write $uv = \mu t^k$ for some unit $\mu \in \mathcal{O}_{V,p}$, in which case (Y, y) is a canonical singularity. If $h(t) \equiv 0$ then (Y, y) is a normal crossing point, which is semismooth.

3. $p \in \operatorname{Sing}(V)$ and $y \notin Sing(f)$. Then (Y, y) is a normal crossing point again.

4. $p \in \operatorname{Sing}(V)$ and $y \in Sing(f)$. We have

$$V \simeq \operatorname{Spf} k[[t_1, t_2]]/(t_1 t_2) \quad \text{and} \quad Y \simeq \operatorname{Spf} O_V[[u, v]]/(uv - h(t_1, t_2)).$$

We can write $h(t_1, t_2) = h_1(t_1) + h_2(t_2)$, with $h_i(0) = 0$.

(a) If neither h_i is 0, write $h_i(t_i) = \mu_i t_i^{k_i}$, with μ_i units, then (Y, y) is a degenerate cusp with exceptional locus a cycle of rational curves with exactly $\max(k_1 - 1, 1) + \max(k_2 - 1, 1)$ components.

(b) If, say, only $h_2 \equiv 0$, then we have a degenerate cusp with $\max(k_1 - 1, 1) + 2$ exceptional components.

(c) If $h_1 \equiv 0 \equiv h_2$, then we have a degenerate cusp with 4 components.

(See [K-SB] §4, for (4a)–(4c).)

The statement about the pair (Y, D) follows easily, since D meets $\mathrm{Sing}(Y)$ transversally at normal crossing points: locally in the étale topology it is isomorphic to the pair

$$(\mathrm{Spec}\ k[t_1, t_2, u]/(t_1t_2), \mathrm{Spec}\ k[t_1, t_2, u]/(t_1t_2, u)),$$

which is semi-log-canonical. □

For the following lemma assume that the base field is algebraically closed. Here we are interested in balanced quotients of families of nodal curves over a nodal base. Studying the resulting singularities becomes easy when we show that the situation can be deformed, which is shown in this lemma.

Lemma 6.7. *Let V be the formal completion of a nodal curve and $f : Y \to V$ the formal completion of a nodal family at a closed point. Let $y \in Y$ be the closed point and assume $f(y) = p$ is a node. Suppose a finite cyclic group $\Gamma_y \subset \mathrm{Aut}(Y \to V)$ of order r fixes y and acts faithfully on V stabilizing the two branches of V at p with complementary eigenvalues (in other words, the action is balanced). Then there exists a smoothing*

$$\begin{array}{ccc} Y & \subset & Y' \\ \downarrow & & \downarrow \\ V & \subset & V' \\ \downarrow & & \downarrow \\ \mathrm{Spec}\ k & \subset & \mathrm{Spf}\ k[[s]] \end{array}$$

and a lifting of the action of Γ_y to $Y' \to V'$.

Proof.

1. Suppose $y \notin Sing(f)$. We may choose a parameter u along the fiber so that $Y_y = \mathrm{Spf}\ k[[t_1, t_2, u]]/(t_1t_2)$. It is not hard to choose u as an eigenvector for Γ_y. The action is

$$(t_1, t_2, u) \mapsto (\zeta t_1, \zeta^{-1} t_2, \zeta^a u).$$

This clearly lifts to the family given by $t_1t_2 = s$.

2. Suppose now $y \in Sing(f)$. We have the local equation $uv = h_1(t_1) + h_2(t_2)$. It is easy to choose u, v so that uv is an eigenvector.

   (a) Suppose neither h_i is 0. After a change of coordinates we may assume our local equation is $uv = t_1^{k_1} + t_2^{k_2}, t_1t_2 = 0$ (so r divides $k_1 + k_2$). We can analyse the action of Γ_y via its action on the fiber $t_1 = t_2 = 0$. Depending upon whether Γ_y stabilizes the branches of u and v or switches them, the action is either $(u,v) \mapsto (\zeta^a u, \zeta^{k_1-a} v)$ or $(u,v) \mapsto (v, \zeta^{k_1} u)$. In either case this action lifts to the deformation $uv = t_1^{k_1} + t_2^{k_2}, t_1 t_2 = s$.

(b) Suppose only $h_2 \equiv 0$. We have $uv = t_1^{k_1}, t_1 t_2 = 0$. Again, the possible actions are either $(u,v) \mapsto (\zeta^a u, \zeta^{k_1-a} v)$ or $(u,v) \mapsto (v, \zeta^{k_1} u)$. As before, the action lifts to the deformation.

(c) If we have $uv = t_1 t_2 = 0$, the possible actions are either $(u,v) \mapsto (\zeta^a u, \zeta^{k-a} v)$ or $(u,v) \mapsto (v, \zeta^k u)$ for some integers k and a. For any choice of positive k_1, k_2 such that $k_1 - k_2 \equiv k \mod r$ we have that this action lifts to the family $uv = t_1^{k_1} t_2^{k_2}; t_1 t_2 = s$.

Now we look at quotients. First the case of a fixed point which lies in the smooth locus of a fiber:

Lemma 6.8. *Let V be a nodal curve and $f : Y \to V$ a nodal family, and let $y \in Y, y \notin \mathrm{Sing}(f)$. Let $\Gamma \subset \mathrm{Aut}Y$ be a balanced finite subgroup, fixing y. Let $q : Y/\Gamma \to X$ be the quotient map and let $x = q(y)$. Then X has a semi-log-terminal singularity at x. If, furthermore, $D \subset Y$ is a section through y which is Γ-stable, then $(X, D/\Gamma)$ is a semi-log-canonical pair.*

The fact that X has a semi-log-terminal singularity is in [K-SB] 4.23(iii). The statement with D follows as in Lemma 6.4.

We are left to deal with a fixed point which is a node:

Lemma 6.9. *Let V be a nodal curve and $f : Y \to V$ a nodal family, and let $y \in Y, y \in \mathrm{Sing}(f)$. Let $\Gamma \subset \mathrm{Aut}Y$ be a balanced finite subgroup, fixing y. Let $q : Y/\Gamma \to X$ be the quotient map and let $x = q(y)$. Then X has a Gorenstein semi-log-canonical singularity at x.*

By Lemma 6.4 it is enough to show that X is Gorenstein, and for this it suffices to show that the quotient of the smoothing Y' in Lemma 6.7 is Gorenstein. The quotient variety Y'/Γ is clearly Cohen–Macaulay; therefore it suffices to show that its canonical divisor class is Cartier. We are in case (2) of Lemma 6.7. In the cases (2a) and (2b) (respectively (2c)), the sheaf $\omega_{Y'}$ is generated at y by $\frac{du \wedge dv \wedge dt_2}{t_1^{k_1-1}}$ (respectively, $\frac{du \wedge dv \wedge dt_2}{t_1^{k_1-1} t_2^{k_2}}$). The generator is easily seen to be Γ_p-invariant. □

We have thus obtained:

Proposition 6.10. *Let $X \to C$ with sections $s_i : C \to X$ be a coarse fibered surface, $S_i = Im(s_i)$ and $D = \sum S_i$. Then (X, D) is a semi-log-canonical pair.*

6.11 Alexeev stable maps

In [Al3], V. Alexeev defined *surface stable maps*, for which he constructed complete moduli spaces. Our goal here is to compare our moduli of balanced fibered surfaces and coarse balanced fibered surfaces with Alexeev's moduli spaces. In particular, we would like to construct a stack of stable balanced coarse fibered surfaces.

Let X be a reduced, connected projective surface and $D \subset X$ a reduced subscheme of codimension 1. Let $M \subset \mathbf{P}^r$ be a projective scheme.

A morphism $f : X \to M$ is called a *stable map* of the pair (X, D) to M, if

1. the pair (X, D) has only semi-log-canonical singularities; in particular, for some integer $m > 0$ the sheaf $(\omega_X(D))^{[m]}$ is invertible.
2. For a sufficiently large integer n, the sheaf $(\omega_X(D))^{[m]} \otimes f^*\mathcal{O}_M(mn)$ is ample.

It is easy to see that the property of a morphism $f : X \to M$ being stable is independent of the choice of the projective embedding of M.

Note that, given a stable map $f : X \to M$, one has a well-defined triple of rational numbers:

$$A = c_1(\omega_X(D))^2; \quad B = c_1(\omega_X(D)) \cdot c_1(f^*\mathcal{O}_M(1)); \quad C = c_1(f^*\mathcal{O}_M(1))^2.$$

One can define a functor of families of stable maps with fixed invariants A, B and C. This is somewhat subtle, since the sheaf $\omega_X(D)$ is not invertible, and saturation does not commute with base change. This can be resolved either by restricting to "allowable" deformations on which the saturation $(\omega_X(D))^{[m]}$ does commute with base change (this is discussed in an unpublished work by Kollár), or by endowing the surface X with the structure of a Deligne–Mumford stack using the log-Gorenstein covers (this has not been carried out in the literature). Once the functor is defined, one can look for a moduli space. The following result of Alexeev gives the answer:

Theorem 6.12. (Alexeev) *Given rationals A, B and C, there is a Deligne–Mumford stack $\mathcal{A}l_{A,B,C}(M)$ admitting a projective coarse moduli space $\mathbf{Al}_{A,B,C}(M)$ for surface stable maps $f : X \to M$ with invariants A, B, C.*

Now let $\mathcal{X} \to \mathcal{C} \to S$ be a balanced fibered surface and $X \to C \to S$ the associated coarse fibered surface. We have a morphism $C \to \overline{\mathbf{M}}_{\gamma,\nu}$, which we can compose with $X \to C$ and obtain a morphism $f : X \to \overline{\mathbf{M}}_{\gamma,\nu}$. In addition to that, we have ν sections $C \to X$. The union of the images of these sections gives rise to a divisor $D \subset X$. We have already seen that (X, D) is a semi-log-canonical pair. We now claim:

Proposition 6.13. *The morphism $f : X \to \overline{\mathbf{M}}_{\gamma,\nu}$ is a stable map of the pair (X, D) to $\overline{\mathbf{M}}_{\gamma,\nu}$.*

Proof. Fix a projective embedding $\overline{\mathbf{M}}_{\gamma,\nu} \in \mathbf{P}^r$. Since $C \to \overline{\mathbf{M}}_{\gamma,\nu}$ is a stable map, we have that $\omega_{C/S} \otimes f^*\mathcal{O}_{\overline{\mathbf{M}}_{\gamma,\nu}}(n)$ is ample for $n \geq 3$. Moreover, the line bundle $\mathcal{L} = \omega_{X/C}^{[m]}$ defined in Lemma 4.30 is nef and relatively ample for $X \to C$. Therefore $\omega_{X/S}^{[m]} \otimes f^*\mathcal{O}_{\overline{\mathbf{M}}_{\gamma,\nu}}(mn)$ is an invertible sheaf, which is relatively ample for $X \to S$. □

It is now easy to see that we have a finite morphism

$$\mathcal{F}_g^d(\gamma,\nu)^{balanced} \to \mathcal{A}l_{A,B,C}(\overline{\mathbf{M}}_{\gamma,\nu})$$

for a suitable A, B, C. The image can be viewed as the stack parametrizing the underlying surfaces X of the coarse balanced fibered surfaces, along with the map to $\overline{\mathbf{M}}_{\gamma,\nu}$. In many cases one can recover the structure map $X \to C$, but not always: consider two non-isomorphic smooth stable curves C_1, C_2 over a field K which become isomorphic over $\bar{K}$, and set $X = C_1 \times C_2$. The surface X has two structures of a stable coarse fibered surface, defined by $X \to C_1$ and $X \to C_2$, which both give rise to the same constant map to $\overline{\mathbf{M}}_{\gamma,\nu}$.

One can avoid this phenomenon in the following manner: the stable map $C \to \overline{\mathbf{M}}_{\gamma,\nu}$ naturally induces a map $C \to \mathbf{K}_{g,1}^d(\overline{\mathbf{M}}_{\gamma,\nu})$ (onto the fiber over the point in $\mathbf{K}_g^d(\overline{\mathbf{M}}_{\gamma,\nu})$ corresponding to $C \to \overline{\mathbf{M}}_{\gamma,\nu}$). The family $\mathcal{X} \to \mathcal{C}$ induces a map to the "universal curve" $X \to \overline{\mathbf{M}}_{\gamma,\nu+1}$. One can replace the morphism $X \to \overline{\mathbf{M}}_{\gamma,\nu}$ in the previous consideration by the morphism $X \to \mathbf{K}_{g,1}^d(\overline{\mathbf{M}}_{\gamma,\nu}) \times \overline{\mathbf{M}}_{\gamma,\nu+1}$. One still needs to show that the morphism $X \to C$ is determined by the map, which is not entirely trivial.

Perhaps the most natural approach is to introduce the morphism $X \to C$ into the definition. Consider a pair of objects $((X \to S, X \to \overline{\mathbf{M}}_{\gamma,\nu}), (C \to S, C \to \overline{\mathbf{M}}_{\gamma,\nu}))$ in $\mathcal{A}l_{A,B,C}(\overline{\mathbf{M}}_{\gamma,\nu}) \times \mathcal{K}_g^d(\overline{\mathbf{M}}_{\gamma,\nu})(S)$. By the theory of Hilbert schemes, the functor

$$\begin{array}{rcl} \mathrm{Sch}/S & \to & \mathrm{Sets} \\ T & \mapsto & Hom_T(X \times_S T, C \times_S T) \end{array}$$

is representable by a scheme with quasi-projective connected components, whose formation commutes with base changes. It follows by étale descent that there is a separated Deligne-Mumford stack $\mathcal{T}$ of triples $\big((X \to S, X \to \overline{\mathbf{M}}_{\gamma,\nu}), (C \to S, C \to \overline{\mathbf{M}}_{\gamma,\nu}), (X \to C)\big)$. There is a morphism $\mathcal{F}_g^d(\gamma,\nu)^{balanced} \to \mathcal{T}$, and we call the image $\mathcal{C}F_g^d(\gamma,\nu)^{balanced}$ the *stack of stable balanced coarse fibered surfaces.*

Since the proof of Alexeev's theorem is quite involved, especially showing boundedness of the stable maps, and since a good resolution of the issue of the "right" deformation space is not available in the literature, it is worthwhile to see that in this particular case we can deduce the existence of the space of stable coarse fibered surfaces from our work so far.

Indeed, the existence of $\mathcal{F}_g^d(\gamma,\nu)^{balanced}$ implies that there is a scheme Z and a finite surjective morphism $Z \to \mathcal{F}_g^d(\gamma,\nu)^{balanced}$. Over Z we have a family of fibered surfaces, and in particular a family of coarse fibered surfaces $X \to C \to Z$. We have constructed a relatively very ample line bundle L on X; we may replace L by a suitable power and thus assume that L is relatively ample. After passing to the frame bundle Z' of sections of L, we may assume that the pushforward of L to Z is a free sheaf V. Since $X \to Z$ is flat and embedded in $\mathbf{P}(V)$, we have a morphism $Z \to Hilb$ to a suitable Hilbert scheme. The projective linear group acts with finite stabilizers, therefore the quotient is a Deligne–Mumford stack, parametrizing the surfaces X underlying our coarse fibered surfaces. We can now apply the construction above to define $\mathcal{C}F_g^d(\gamma,\nu)^{balanced}$.

It is interesting to look for properties of the morphism

$$\mathcal{F}_g^d(\gamma,\nu)^{balanced} \to \mathcal{C}F_g^d(\gamma,\nu)^{balanced}.$$

For instance, it is clearly birational on the closure of the locus of normal fibered surfaces. As it turns out, there is little more that one can say: this morphism fails in general to be one to one, and moreover, it may be ramified, even on the closure of the locus of normal fibered surfaces. To show this, one simply needs to produce examples of (deformable) coarse fibered surfaces admitting incompatible atlases. The following examples are local, but can be easily globalized.

Example 6.14. Here we give an example of two incompatible balanced charts for a coarse fibered surface. Fix an algebraically closed field k of characteristic 0, and set $S = \operatorname{Spec} k$. Take a smooth projective curve W over k with an automorphism s of order 2 with a fixed point $p \in W(k)$, and let Y_0 be the curve obtained by attaching two copies of W at p. Consider the action of a cyclic group Γ of order 2 on X_0 where a generator acts like s on each copy of W; also let f be the equivariant automorphism of Y_0 which acts like s on one copy of W and as the identity on the other. The point is that f commutes with s, and the automorphism of Y_0/Γ induced by f is the identity.

Let L be a universal deformation space of X_0; Γ acts on L. Let V' be a small étale neighborhood of 0 in $\mathbf{A}^1$, and let Γ act on V', such that a generator sends t to $-t$. Choose a non-constant Γ-equivariant map $V' \to L$; this yields a family Y' of stable curves on V' whose fiber over 0 is exactly X_0; Γ also acts on Y. Let V be the union of two copies of V' glued at 0; there are two families Y_1 and Y_2 on V obtained by attaching two copies of Y' at X_0, one using the identity, the other using f. These will not be isomorphic in general, not after an automorphism of V, and not even after going to an étale neighborhood of 0 in V.

Set $U = V/\Gamma$ and call X the union of two copies of Y'/Γ along X_0/Γ; we claim that the two generically fibered surfaces $X_1 = Y_1/\Gamma \to V/\Gamma = U$ and $X_2 = Y_2/\Gamma \to V/\Gamma = U$ are canonically isomorphic to $X \to U$. In fact the structure sheaf $O_{X_i'}$ fits into an exact sequence

$$0 \longrightarrow O_{X_i'} \longrightarrow O_{Y'} \times O_{Y'} \longrightarrow O_{X_0} \longrightarrow 0,$$

where the map $O_{Y'} \times O_{Y'} \longrightarrow O_{X_0}$ is the difference of the two projections for $i = 1$, and the difference of one projection with the other projection twisted by f for $i = 2$. Now take invariants; we get an exact sequence

$$0 \longrightarrow O_{X'_i/\Gamma} \longrightarrow O_{Y'/\Gamma} \times O_{Y'/\Gamma} \longrightarrow O_{X_0/\Gamma} \longrightarrow 0;$$

where the two maps $O_{Y_1/\Gamma} \times O_{Y_2/\Gamma} \to O_{X_0/\Gamma}$ are equal, because f induces the identity on $O_{X_0/\Gamma}$.

So $(U, Y_1 \to V, \Gamma)$ and $(U, Y_2 \to V, \Gamma)$ are incompatible charts for the surface $X \to U$. If we call $\overline{U}$ the union of two disjoint copies of U and $\overline{V}$ the union of two disjoint copies of V and $\overline{Y} \to V$ the family of stable curves which coincides with Y_1 on one copy and with Y_2 on the other one, we even obtain an example $(\overline{U}, \overline{Y} \to V, \Gamma)$ of a chart that is not compatible with itself. This is not surprising, because we are using the étale topology, so U is not in general embedded in C; this could not happen if we were working over $\mathbf{C}$ with the analytic topology.

Example 6.15. We now give an example of two incompatible charts on a fibered surface over $S = \operatorname{Spec} k[\epsilon]/(\epsilon^2)$, which coincide modulo ϵ. We can globalize this example to a complete curve. This implies, in particular, that the map $\mathcal{F}_g^d(\gamma, \nu) \to \mathcal{CF}_g^d(\gamma, \nu)$ is not always unramified.

Let $F = \operatorname{Spec} k[u, w]/(uw)$. We will use F as a constant fiber in charts $Y \to V$, so that $Y = F \times V$. Specifically, consider the following two curves: $V_0 = \operatorname{Spec} k[z, t, \epsilon]/(zt, \epsilon^2)$ and $V_\epsilon = \operatorname{Spec} k[z, t, \epsilon]/(zt - \epsilon, \epsilon^2)$. Define $Y_0 = F \times V_0$ and $Y_\epsilon = F \times V_\epsilon$. We have obvious morphisms $Y_0 \to V_0 \to S$ and $Y_\epsilon \to V_\epsilon \to S$. The fibers over $\operatorname{Spec} k \subset B$ are clearly isomorphic, but Y_0 and Y_ϵ are clearly non-isomorphic.

We now define an action of C_6, the cyclic group of order 6, on these schemes. Let ζ be a primitive sixth root of 1. We choose a generator of C_6 and call it ζ as well; define its action as follows:

$$(u, w, z, t, \epsilon) \mapsto (\zeta^3 u, \zeta^2 w, \zeta z, \zeta^{-1} t, \epsilon)$$

This clearly defines an action of C_6 on both $Y_0 \to V_0 \to S$ and $Y_\epsilon \to V_\epsilon \to S$.

Let $C_0 = V_0/C_6$. Explicitly,

$$C_0 = \operatorname{Spec} k[z^6, t^6, \epsilon]/(z^6 t^6, \epsilon^2).$$

Similarly let $C_\epsilon = V_\epsilon/C_6$. It is easy to see that we have

$$C_\epsilon = \operatorname{Spec} k[z^6, t^6, \epsilon]/(z^6 t^6, \epsilon^2).$$

Thus we can use the notation

$$C = V_0/C_6 = V_\epsilon/C_6.$$

We denote $X_0 = Y_0/C_6$ and $X_\epsilon = Y_\epsilon/C_6$. Clearly $X_0 \to C \to S$ and $X_\epsilon \to C \to S$ are coarse fibered surfaces over S. Our main claim is that they are isomorphic.

We will produce an isomorphism of coarse fibered surfaces by choosing isomorphisms over suitable open sets, showing that these glue together over a large open set, and arguing that an isomorphism on the large open set must extend. Let us first work out the open sets.

We denote by U_0 the localization of Y_0 at $u, \ldots$, and by T_ϵ the localization of Y_ϵ at t. Explicitly,

$$\begin{aligned}
U_0 &= \operatorname{Spec} k[u, u^{-1}, z, t, \epsilon]/(zt, \epsilon^2) \\
W_0 &= \operatorname{Spec} k[w, w^{-1}, z, t, \epsilon]/(zt, \epsilon^2) \\
Z_0 &= \operatorname{Spec} k[u, w, z, z^{-1}, \epsilon]/(uw, \epsilon^2) \\
T_0 &= \operatorname{Spec} k[u, w, t, t^{-1}, \epsilon]/(uw, \epsilon^2) \\
U_\epsilon &= \operatorname{Spec} k[u, u^{-1}, z, t, \epsilon]/(zt - \epsilon, \epsilon^2) \\
W_\epsilon &= \operatorname{Spec} k[w, w^{-1}, z, t, \epsilon]/(zt - \epsilon, \epsilon^2) \\
Z_\epsilon &= \operatorname{Spec} k[u, w, z, z^{-1}, \epsilon]/(uw, \epsilon^2) \\
T_\epsilon &= \operatorname{Spec} k[u, w, t, t^{-1}, \epsilon]/(uw, \epsilon^2) .
\end{aligned}$$

These open sets are clearly stable under the action of C_6.

The expressions above immediately give isomorphisms $Z_0 \simeq Z_\epsilon$ and $T_0 \simeq T_\epsilon$. These canonically induce isomorphisms $Z_0/C_6 \simeq Z_\epsilon/C_6$ and $Z_0/C_6 \simeq Z_\epsilon/C_6$.

Let us address the open sets U_0 and U_ϵ. These are clearly nonisomorphic, but we claim that their quotients by the subgroup of two elemets are isomorphic. Indeed, note that ζ^3 acts trivially on u, u^{-1}. Therefore

$$\begin{aligned}
U_\epsilon/C_2 &\simeq \operatorname{Spec} k[u, u^{-1}] \times V_\epsilon/C_2 \\
&= \operatorname{Spec} k[u, u^{-1}] \times \operatorname{Spec} k[z^2, t^2, zt, \epsilon]/(zt - \epsilon, \epsilon^2) \\
&= \operatorname{Spec} k[u, u^{-1}] \times \operatorname{Spec} k[z^2, t^2, \epsilon]/(\epsilon^2) \\
&\simeq \operatorname{Spec} k[u, u^{-1}] \times V_0/C_2 \\
&\simeq U_0/C_2.
\end{aligned}$$

Taking the quotient by the residual C_3 action, this isomorphism clearly induces an isomorphism

$$U_\epsilon/C_6 \simeq U_0/C_6.$$

Note that on the intersections $U_\epsilon/C_2 \cap Z_\epsilon/C_2$ and $U_0/C_2 \cap Z_0/C_2$ the isomorphism above coincides with the restriction of the isomorphism $Z_\epsilon/C_2 \simeq Z_0/C_2$. Indeed, both are given by identifiying the variables u, z^2 and ϵ. Therefore these isomorphisms coincide on $U_\epsilon/C_6 \cap Z_\epsilon/C_6$ as well. A similar situation is obtained by replacing Z_ϵ by T_ϵ.

We now address the open sets W_ϵ and W_0 in a similar manner, noting that this time C_3 acts trivially on the variable w. Thus

$$\begin{aligned} W_\epsilon/C_3 &\simeq \operatorname{Spec} k[w,w^{-1}] \times V_\epsilon/C_3 \\ &= \operatorname{Spec} k[w,w^{-1}] \times \operatorname{Spec} k[z^3,t^3,zt,\epsilon]/(zt-\epsilon,\epsilon^2) \\ &= \operatorname{Spec} k[w,w^{-1}] \times \operatorname{Spec} k[z^3,t^3,\epsilon]/(\epsilon^2) \\ &\simeq \operatorname{Spec} k[w,w^{-1}] \times V_0/C_3 \\ &\simeq W_0/C_3. \end{aligned}$$

Again, it is easy to check that this isomorphism agrees on the intersections with the open sets Z_ϵ and T_ϵ. We note that all these isomorphisms are isomorphisms over the curve C, which coincide with the identity modulo ϵ.

Let X'_ϵ be the open subscheme which is the complement of the image of the point $u = w = z = t = 0$ on Y_ϵ; define X'_0 in an analogous way. Thus far we have obtained an isomorphism $\phi' : X'_\epsilon \to X'_0$. The closure $\Sigma \subset X_\epsilon \times_C X_0$ of the graph of ϕ' is a subscheme supported along the graph of the identity $(X_\epsilon)_{\operatorname{Spec} k} = (X_0)_{\operatorname{Spec} k}$. Therefore the projections $p_\epsilon : \Sigma \to X_\epsilon$ and $p_0 : \Sigma \to X_0$ are finite. Note also p_ϵ and p_0 are isomorphic along a dense open set. Since the schemes X_0 and X_ϵ satisfy Serre's condition S_2, these projection admit sections, giving rise to a morphism $\phi : X_\epsilon \to X_0$, which by the same reason is an isomorphism.

References

[ℵ-V1] D. Abramovich and A. Vistoli, *Complete moduli for families over semistable curves*, preprint

[ℵ-V2] D. Abramovich and A. Vistoli, *Compactifying the space of stable maps*, preprint

[ℵ-C-J-V] D. Abramovich, A. Corti, A. J. de Jong and A. Vistoli, *Twisted bundles, admissible covers, and level structures*, in preparation

[Al1] V. Alexeev, Log canonical surface singularities, in: *Flips and Abundance on Algebraic Threefolds*, Astérisque 211, 1992

[Al2] Valery Alexeev, Boundedness and K^2 for log surfaces, *Internat. J. Math.* **5**:6 (1994), 779–810

[Al3] Valery Alexeev, Moduli spaces $M_{g,n}(W)$ for surfaces, in *Higher-dimensional complex varieties (Trento, 1994)*, 1–22, de Gruyter, Berlin

[Artin] M. Artin, Versal deformations and algebraic stacks, *Invent. Math.* **27** (1974), 165–189

[B-M] K. Behrend and Yu. Manin, Stacks of Stable Maps and Gromov-Witten Invariants, *Duke Math. J.* **85**:1 (1996), 1–60

[F-P] W. Fulton and R. Pandharipande, *Notes on stable maps and quantum cohomology*, Algebraic geometry–Santa Cruz 1995, 45–96, Proc. Sympos. Pure Math., 62, Part 2, Amer. Math. Soc., Providence, RI, 1997

[H-M] Joe Harris and David Mumford, On the Kodaira dimension of the moduli space of curves, With an appendix by William Fulton, *Invent. Math.* **67**:1 (1982), 23–88

[Illusie] L. Illusie, *Complexe cotangent et deformations. I, II* Lecture Notes in Mathematics, Vol. 239, p. 283

[dJ-O] A. J. de Jong and F. Oort, On extending families of curves, *J. Alg. Geom.* **6**:3 (1997), 545–562

[Kollár1] J. Kollár Projectivity of complete moduli, *J. Differential Geom.* **32**:1 (1990), 35–268

[Kollár2] J. Kollár, *Log surfaces of general type; some conjectures.* Classification of algebraic varieties (L'Aquila, 1992), 261–275, Contemp. Math., 162, Amer. Math. Soc., Providence, RI, 1994

[K-SB] J. Kollár and N. I. Shepherd-Barron, Threefolds and deformations of surface singularities, *Invent. Math.* **91**:2 (1988), 299–338

[Kont] M. Kontsevich, *Enumeration of rational curves via torus actions.* The moduli space of curves (Texel Island, 1994), 335–368, Progr. Math., 129, Birkhäuser, Boston, 1995

[L] Eduard Looijenga, Smooth Deligne–Mumford compactifications by means of Prym level structures, *J. Algebraic Geom.* **3**:2 (1994), 283–293

[Mo1] S. Mochizuki, *Extending Families of Curves I: Compactification of Hyperbolic Polycurves*, preprint (1998)

[Mo2] S. Mochizuki, Extending families of curves over log regular schemes. *J. Reine Angew. Math.* **511** (1999), 43–71

[MB] L. Moret-Bailly, *Un théorème de pureté pour les familles de courbes lisses.* C. R. Acad. Sci. Paris Sér. I Math. **300**:14 (1985), 489–492

[Mum] D. Mumford, Towards an enumerative geometry of the moduli space of curves, in *Arithmetic and geometry, Vol. II*, 271–328, Progr. Math., 36, Birkhäuser, Boston, 1983

[Vistoli1] A. Vistoli, Intersection theory on algebraic stacks and on their moduli spaces. *Invent. Math.* **97**:3 (1989), 613–670

[Vistoli2] A. Vistoli, *The Hilbert stack and the theory of moduli of families.* Geometry Seminars, 1988–1991 (Bologna, 1988–1991), 175–181, Univ. Stud. Bologna, Bologna, 1991

[Vistoli3] A. Vistoli, *Deformation theory of local complete intersections*, preprint. `alg-geom/9703008`

Department of Mathematics
Boston University
111 Cummington Street
Boston, MA 02215
USA
e-mail: `abrmovic@@math.bu.edu`

Dipartimento di Matematica
Università di Bologna
Piazza di Porta San Donato 5
40127 Bologna, Italy
e-mail: `vistoli@dm.unibo.it`

Normal Differential Operators and Deformation Theory

Paul Burchard
Herb Clemens

ABSTRACT This paper develops the theory of a sheaf of normal differential operators to a submanifold Y of a complex manifold X as a generalization of the normal bundle. We show that the global sections of this sheaf play an analogous role for formal deformations of Y to the role played by the normal bundle with respect to first-order deformations.

1 Introduction and Motivation

Our aim in this paper is to develop a more concrete and geometrically intuitive formulation of the higher-order deformation theory of subvarieties. Our basic strategy is to use local-formal duality to reformulate the standard theory in terms of *local cohomology sheaves.* These sheaves are rich in geometric meaning, as demonstrated by their alternative interpretations as sheaves of *residue operators* or *normal differential operators.* Because these sheaves already carry higher-order information, it is possible to work more concretely (doing more at the level of sheaves, or H^0).

Given a deformation of a complex manifold X_0 over a parameter space T, there is a categorical equivalence between deformations of submanifolds compatible with a given deformation of the ambient manifold and morphisms from the (pullback of the) stalk at 0 of the sheaf of differential operators on T to our new object, the "sheaf of normal differential operators." This is not a statement about the (formal) local ring of T at 0 but rather about its "dual," the differential operators on the local ring. The precise statement of this equivalence and its proof is the content of Theorem 9.3.

The geometric setting in which our local cohomology sheaves become functorial is that of *deformation categories* of subvarieties. In the simplest such category, an object is a smooth subvariety $Y \rightarrowtail X$, and a morphism from $T \rightarrowtail S$ to $Y \rightarrowtail X$ is a deformation diagram

$$\begin{array}{ccccc} Y & \rightarrowtail & W & \longrightarrow & X \\ \downarrow & & \downarrow & & \\ T & \rightarrowtail & S & & \end{array}$$

where the square is a pullback and the maps at S are transversal. Composition of morphisms is pullback.

Thus our main technical result—a special case of which is proved in §9 of this paper—becomes an *equivalence of categories* between a certain deformation category of smooth formal subvarieties, and a corresponding category of local cohomology sheaves. In order to obtain such an equivalence, however, the natural homotopy product structure which these special local cohomology sheaves carry must be respected.

Let us describe the local cohomology sheaves a bit more explicitly, if only to indicate why our global-to-formal-to-local strategy doesn't destroy the cohomological information required for deformation theory. This may not be obvious because, for a smooth subvariety $Y \rightarrowtail X$ of dimension d and codimension c, the interesting cohomology group $H^{2d+1}(\mathbb{C}_X)$ is immediately killed upon restriction to the formal neighborhood $\widehat{X}_Y$. Fortunately, the Abel–Jacobi map involves only a specific Hodge level of this cohomology, and once we restrict our attention to this level, the relevant classes manage to sneak safely across into local cohomology:

$$H^{2d+1}(\mathbb{C}_X)^* \twoheadrightarrow \mathbf{H}^d(\Omega_X^{>d})^* \longleftarrow \mathbf{H}^d(\Omega_{\widehat{X}_Y}^{>d})^* \xrightarrow{\delta^*} \mathbf{H}^{2d}(\Omega_{\widehat{X}_Y}^{\leqslant d})^* = \mathbf{H}_Y^c(\Omega_X^{\geqslant c})$$

Thus, the "local Hodge sheaf" $\mathcal{H}_Y^c(\Omega_X^{\geqslant c})$ is sufficient to express Abel–Jacobi data.

Our ultimate object of study is the local cohomology of the whole ambient de Rham complex

$$\left(\mathcal{H}_Y^c(\Omega_X^{\cdot}),\ \partial\right). \tag{1}$$

To handle relative subvarieties $Y/Y' \rightarrowtail X/X'$ properly, we must further endow this complex with a filtration by base degree of closed forms:

$$F^{p/p'}\mathcal{H}_Y^c(\Omega_X^p) := \mathcal{H}_Y^c\left((\Omega_X^{\cdot} \wedge \Omega_{X'}^{\geqslant p'})^{\geqslant p}\right). \tag{2}$$

For the purposes of this paper, however, we only need a particular piece of this complex—one which can be defined directly in terms of relative forms (§4).

The special local cohomology sheaf (1) carries a natural homotopy product, which can be formalized using the notion of *co-implicial* objects. Recall that co-simplicial objects are representations of the category of finite ordinals and ordered maps; *co-implicial* objects are representations of the richer category which includes partially defined ordered maps. A co-implicial object will therefore carry *co-barycenter* maps in addition to the usual *co-face* and *co-degeneracy* maps. These maps can be thought of as operations on product expressions implementing (respectively) factor deletions, unit insertions, and pairwise factor associations. To avoid extra formalism, however, this paper will take a more ad-hoc approach to the product.

Our main applications will come from studying the pairings between ambient global cohomology and the cohomology of normal differential operators. These pairings essentially come from the local-formal pairings which were used to construct the theory in the first place; for example,

$$H^q(\mathcal{H}^c_Y(\Omega_X^{\geqslant c}))\otimes \mathbf{H}^r(\Omega_X^{\leqslant d}) \longrightarrow \mathbf{H}^{q+c}_Y(\Omega_X^{\geqslant c})\otimes \mathbf{H}^r(\Omega_{\hat{X}_Y}^{\leqslant d}) \xrightarrow{\text{local-formal}} H^{q+r}(\mathbb{C}_Y)$$

By looking at similar pairings in the relative setting, we are able to formulate and prove our second main result, namely to generalize a result of Bloch–Ran–Kawamata on the *semi-regularity map*. Our generalization, given as Theorem 11.4 at the end of §11, produces an *obstruction differential* which is exact if there is no cohomological obstruction to deformation. Although we believe that this result produces a natural, potentially important invariant of the interaction between higher-order deformation theory and Hodge theory, unfortunately we have so far been unable to find the definitive application of our invariant which will justify that belief.

The authors wish to thank Henryk Hecht, János Kollár, and Dragan Milicic of the University of Utah, Morihiko Saito of RIMS, Japan, and Claire Voisin of the CNRS, France, for valuable conversations and extremely useful advice during the preparation of this paper.

2 Relation to the Kuranishi Theory

To explain the results and methods of this paper in more detail, we must first recall, at least in rough form, some basic facts from the Kuranishi theory of deformations of complex manifold as developed by Deligne and Goldman–Millson in [GM90]. Given a smooth family

$$Z/Z' = \{Z_{z'}\}_{z'\in Z'}$$

of connected complex manifolds over a polydisc Z' with coordinate

$$z' = (z'_1, \ldots , z'_n)$$

and distinguished point 0, one can associate a transversely holomorphic trivialization, that is, a C^∞-trivialization

$$\begin{array}{ccc} Z & \xrightarrow{(\sigma, p_{Z'})} & Z_0 \times Z' \\ \Big\downarrow{\scriptstyle p_{Z'}} & & \Big\downarrow \\ Z' & = & Z' \end{array} \qquad (3)$$

for which the fibers of σ are all complex holomorphic polydisks. (See, for example, [Cle77].) Following [GM90], Kuranishi theory then associates a formal deformation as follows (see also [Cle98]).

Let $\mathcal{A}^{0,1}(T_{Z_0})$ denote global $(0,1)$-forms on Z_0 with coefficients in the tangent bundle T_{Z_0}.

Lemma 2.1. *The transversely holomorphic trivialization σ distinguishes a subset of the complex-valued C^∞-functions on Z, namely those functions f which restrict to a holomorphic function on each fiber of σ (i.e., functions with a power-series representation*

$$\sum_I f_I z'^I := \sum_{I\in(\mathbb{Z}^{\geqslant 0})^{n'}} (f_I \circ \sigma)\, z'^I \tag{4}$$

where the f_I are C^∞-functions on Z_0). Furthermore there exist global C^∞-bundle-sections

$$\xi_J \in \mathcal{A}^{0,1}(T_{Z_0}), \quad J \in \left(\left(\mathbb{Z}^{\geqslant 0}\right)^{n'} - \{(0,\dots,0)\}\right)$$

such that a function (4) is holomorphic if and only if

$$\bar{D}_\sigma(f) := \left(\bar{\partial} - \sum_J z'^J \xi_J\right)(f) = \sum_I \bar{\partial} f_I z'^I - \sum_{I,J} \xi_J(f_I)\, z'^{I+J} = 0.$$

If we let τ_i denote the unique lifting of $\frac{\partial}{\partial z'_i}$ to a $(1,0)$ vector field on Z determined by the trivialization (3), we have that as operators on functions on Z_0, for example

$$\xi_K = -\left(|_{Z_0}\right) \circ \left[L_{\tau_i}, \bar{\partial}_Z\right]$$

where $K = (0,\dots,0,1,0,\dots,0)$ denotes the i-th position and L_{τ_i} denotes Lie differentiation with respect to the vector field τ_i. For

$$\xi := \sum_J z'^J \xi_J$$

the integrability condition

$$\bar{\partial}\xi - \frac{[\xi,\xi]}{2} = 0 \tag{5}$$

is satisfied. If two transversely holomorphic trivializations of Z/Z' differ by a holomorphic automorphism defined over Z' and equal to the identity on Z_0, they give the same Kuranishi datum ξ.

Continuing to follow [GM90], we define the differential graded Lie algebra

$$\left(\sum_{i\geqslant 0} \mathcal{L}^i, \bar{\partial}\right)$$

where

$$\mathcal{L}^i = \left\{ \lambda = \sum\nolimits_J z'^J \lambda_J : \lambda_J \in \mathcal{A}^{0,i}_{Z_0}(T_{Z_0}) \right\}.$$

The Lie algebra of vector fields $\mathcal{L}^0$ acts via an exponential map on the subspace $\check{\mathcal{L}}^1$ of $\mathcal{L}^1$ consisting of those elements which satisfy the integrability condition (5). The quotient is the space of formal deformations of Z_0 parametrized by Z'.

Next suppose that we have a commutative diagram

$$\begin{array}{ccc} Y & \xrightarrow{h} & Z \\ \downarrow{\scriptstyle p_{Y'}} & & \downarrow{\scriptstyle p_{Z'}} \\ Y' & \xrightarrow{h'} & Z' \end{array} \tag{6}$$

of complex manifolds where $p_{Y'}$ is also a smooth surjection and h' factors as a smooth surjection followed by a closed immersion

$$Y' \xrightarrow{p_{Y''}} Y'' \xrightarrow{i''} Z'$$

so that we have the fibered product

$$\begin{array}{ccc} Y' & \xrightarrow{i' := (y', h'(y'))} & X' := Y' \times Z' \\ \downarrow{\scriptstyle p_{Y''}} & & \downarrow{\scriptstyle p_{X''} := (p_{Y''}(y'), z')} \\ Y'' & \xrightarrow{i'' := (y'', i''(y''))} & X'' := Y'' \times Z' \end{array}$$

and

$$\begin{array}{ccc} X := X' \times_{Z'} Z & \xrightarrow{j} & Z \\ \downarrow{\scriptstyle p_{X'}} & & \downarrow{\scriptstyle p_{Z'}} \\ X' = Y' \times Z' & \xrightarrow{j'} & Z' \end{array}$$

Suppose that, in the induced diagram

$$\begin{array}{ccccc} Y & \xrightarrow{i} & X & \xrightarrow{j} & Z \\ \downarrow{\scriptstyle p_{Y'}} & & \downarrow{\scriptstyle p_{X'}} & & \downarrow{\scriptstyle p_{Z'}} \\ Y' & \xrightarrow{i'} & X' & \xrightarrow{j'} & Z' \end{array} \tag{7}$$

i is a closed imbedding and $p_{Y'}$ is proper. For many algebro-geometric purposes, it is desirable to characterize a certain class of deformation/extensions of (7), namely to characterize commutative diagrams

$$\begin{array}{ccccc} Y & \longrightarrow & W & \longrightarrow & X \\ \Big\downarrow{\scriptstyle p_{Y'}} & & \Big\downarrow{\scriptstyle q_{X'}} & & \Big\downarrow{\scriptstyle p_{X'}} \\ Y' & \xrightarrow{i'} & X' & = & X' \end{array} \tag{8}$$

with $q_{X'}$ a smooth surjection such that the left-hand square is a fibered product. (If (6) itself is a fibered product and vertical maps are proper, there is a unique such (proper) extension.) From (8) we can construct a smoothly varying family of transversely holomorphic trivializations

$$\{y'\} \times Z \xrightarrow{\sigma_{y'}} Z_{h'(y')}$$

for X/X' as above but with the added property that

$$\sigma_{y'}^{-1}(Y_{y'}) = q_{X'}^{-1}(\{y'\} \times Z').$$

This translates to the condition

$$\xi_{y',J} \in \ker\left(\mathcal{A}^{0,1}_{Z_{h'(y')}}\left(T_{Z_{h'(y')}}\right) \to \mathcal{A}^{0,1}_{Y_{y'}}\left(N_{Y_{y'}\backslash Z_{h'(y')}}\right)\right)$$

on the Kuranishi data in Lemma 2.1, where $N_{Y_{y'}\backslash Z_{h'(y')}}$ denotes the normal bundle.

For each $y' \in Y'$, we define the differential graded Lie algebra

$$\left(\sum\nolimits_{i\geqslant 0} \mathcal{L}^i_{y'}, \bar{\partial}\right)$$

where

$$\begin{aligned} \mathcal{L}^i_{y'} &= \Big\{\lambda = \sum\nolimits_J z'\lambda_J(y') : \lambda_J(y') \\ &\in \ker\left(\mathcal{A}^{0,i}_{Z_{h'(y')}}\left(T_{Z_{h'(y')}}\right) \to \mathcal{A}^{0,i}_{Y_{y'}}\left(N_{Y_{y'}\backslash Z_{h'(y')}}\right)\right)\Big\}. \end{aligned}$$

For a fixed transversely holomorphic trivialization of the deformation Z/Z' given in (3), we have Kuranishi data

$$\xi_{z'} \in \check{\mathcal{L}}^1_{z'} \subseteq \mathcal{L}^1_{z'}$$

for each $z' \in Z'$. As above, the vector fields $\mathcal{L}^0_{z'}$ act on $\mathcal{L}^1_{z'}$ as does the subspace $\mathcal{L}^0_{y'}$ whenever $h'(y') = z'$. If $\xi_{h'(y')} \in \check{\mathcal{L}}^1_{y'}$ and $\alpha \in \mathcal{L}^0_{y'}$, then the infinitesimal action of α on ξ is given by the formula

$$\dot{\xi} = \bar{\partial}\alpha + [\alpha, \xi],$$

so, if $\alpha \in \mathcal{L}^0_{y'}$, then $\dot{\xi} \in \mathcal{L}^1_{y'}$. The orbits of the exponential action of $\mathcal{L}^0_{y'}$ on $\check{\mathcal{L}}^1_{y'}$ form the space of formal deformations of $Y_{y'} \subseteq Z_{h'(y')}$ inside the fixed deformation Z/Z'.

The Kuranishi datum is invariant under holomorphic automorphisms over Z' which restrict to the identity at time zero (see the final assertion of Lemma 2.1). So, if Y in (3) is a small neighborhood of a point, the action of $\mathcal{L}^0_{y'}$ on $\check{\mathcal{L}}^1_{y'}$ is transitive. Thus, in the above formulation, we do not see the local structure of the space of deformations of a small neighborhood of a point in Y. Our purpose is to provide a formalism for that local structure. To the first order, that structure is classical. For example, if $Y' = \{0\}$, the local structure is as follows. The tangent sheaf T_Z of the total space Z has distinguished subsheaves, which we denote as $T_{Z/Z'}$, consisting of those derivations which annihilate $p_{Z'}^{-1}(\mathcal{O}_{Z'})$ where $p_{Z'} : Z \to Z'$ is the natural projection associated with the family, and $\widehat{T}_{Z/Z'}$, consisting of those derivations which are "lifts" of derivations on Z'. The latter forms a $p_{Z'}^{-1}(\mathcal{O}_{Z'})$-module and contains the subsheaf $T_{Z/Z'}$ of the $p_{Z'}^{-1}(\mathcal{O}_{Z'})$-linear derivations. ($T_{Z/Z'}$, on the other hand, is an $\mathcal{O}_Z$-submodule of T_Z.) If Y_0 is a compact, connected complex d-dimensional submanifold of a fiber Z_0 of the family, then all global sections of the normal bundle $N_{Y_0\backslash Z}$ actually lie in a sub-sheaf

$$\widehat{N}_{Y_0\backslash Z} = \frac{\widehat{T}_{Z/Z'}\big|_{Y_0}}{T_{Y_0}}.$$

Morphisms

$$T_{Z',0} \to H^0\left(\widehat{N}_{Y_0\backslash Z}\right) \tag{9}$$

give the first-order deformations of Y_0 in Z with parameter space Z', and $\widehat{N}_{Y_0\backslash Z}$ sits in an exact sequence

$$0 \to N_{Y_0\backslash Z_0} \to \widehat{N}_{Y_0\backslash Z} \to p_{Z'}^{-1}T_{Z'}\big|_{Y_0} \to 0.$$

In this paper, we will develop analogous machinery for higher-order deformations of Y/Y' in Z/Z'. This analogue comes from the theory of $\mathcal{D}$-modules, and is similar to the sheaf of "normal differential operators," the push-forward of the structure sheaf of $Y_{y'}$ in the sense of $\mathcal{D}$-modules. Although the sheaf of Lie algebras $\mathcal{L}_Z$ given by vector fields on Z does not have a natural action on the relative deRham complex $\Omega^*_{Z/Z'}$, the sub-Lie-algebra $\mathcal{L}_{Z/Z'}$ given by sections of $\widehat{T}_{Z/Z'}$ does, via Lie differentiation. Under this action, sections of $\widehat{N}_{Y_0\backslash Z_0}$ do not give well-defined operators at points of $Y_{y'}$, but rather should be thought of as equivalence classes of operators which give well defined morphisms

$$R^j\left(p_{Z'}\right)_*\left(\Omega^*_{Z/Z'}\right) \to H^j\left(\Omega^*_{Y/Y''}\right).$$

These considerations lead us to use $\mathcal{D}$-modules constructed from differential operators on the relative deRham complex of Z/Z' with values in the relative deRham complex of Y/Y''. The fundamental object of study is a certain sheaf $\mathcal{D}_{Y/Y''}^{Z/Z'}$ supported on Y. The fiber of its push-forward $\left((p_{Y'})_* D_{Y/Y''}^{Z/Z'}\right)\Big|_{y'}$ should be thought of as the higher-order analogue of the space

$$(p_{Y'})_* \widehat{N}_{Y\backslash Y'\times Z}$$

where $\widehat{N}_{Y\backslash Y'\times Z}$ sits in the exact sequence

$$0 \to N_{Y\backslash Y'\times_{Z'} Z} \to \widehat{N}_{Y\backslash Y'\times Z} \to p_{Y'}^{-1} h'^{*} T_{Z'} \to 0.$$

The general construction and its basic properties and filtrations occupy §§3–6. The pairing of global sections with ambient cohomology is given in §7.

To illustrate, we return to the case in which $Y' = \{0\}$ is a point. "Almost-multiplicative" morphisms from the differential operators at $0 \in Z'$ with values in $\mathbb{C}$ to global sections of $\mathcal{D}_{Y_0/\{0\}}^{Z/Z'}$ will characterize deformations of Y_0 in Z with parameter space Z' just as morphisms (9) characterize first-order deformations. This equivalence, in its more general, relativized version, is developed in §§8–9.

As indicated in §1, a value of this type of approach to higher-order deformation theory comes from the fact that it is well suited for computing the interaction of higher-order obstructions with the cohomology of the ambient manifold Z/Z'. To motivate this assertion, suppose that we are given a smooth family Z/Z' of deformations of Z_0, and let

$$T_{Z',0} \xrightarrow{\ v\ } H^0\left(N_{Y_0\backslash Z_0}\right)$$

be the obstruction map to deforming Y_0 to first order with $Z_{0.}$. Recall the easy first-order fact that the following diagram is commutative:

$$\begin{array}{ccc}
\wedge^r H^0\left(N_{Y_0\backslash Z_0}\right) \otimes T_{Z',0} \otimes H^{d-1}\left(\Omega_{Z_0}^{d+r+1}\right) & \xrightarrow{v} & H^0\left(\wedge^r N_{Y_0\backslash Z_0}\right) \otimes H^0\left(N_{Y_0\backslash Z_0}\right) \otimes H^{d-1}\left(\Omega_{Z_0}^{d+r+1}\right) \\
\downarrow \text{Kodaira-Spencer} & & \downarrow \text{contraction} \\
\wedge^r H^0\left(N_{Y_0\backslash Z_0}\right) \otimes H^d\left(\Omega_{Z_0}^{d+r}\right) & \longrightarrow & H^d\left(\omega_{Y_0}\right)
\end{array}$$

The contraction map across the bottom of this diagram is the infinitesimal Abel–Jacobi mapping. The diagram tells us that, at least when the natural mapping

$$\begin{array}{c} \wedge^r H^0\left(N_{Y_0\backslash Z_0}\right) \otimes H^{d-1}\left(\Omega_{Z_0}^{d+r+1}\right) \\ \big\downarrow \\ H^0\left(\wedge^r N_{Y_0\backslash Z_0}\right) \otimes H^{d-1}\left(\wedge^{r+1} N_{Y_0\backslash Z_0}^* \otimes \omega_{Y_0}\right) \\ \big\downarrow \\ H^{d-1}\left(N_{Y_0\backslash Z_0}^* \otimes \omega_{Y_0}\right) \xlongequal{\text{Serre duality}} H^1\left(N_{Y_0\backslash Z_0}\right)^* \end{array}$$

is surjective, the obstruction to moving Y_0 with Z_0 is completely computed from the infinitesimal Abel–Jacobi mapping for Y_0 in Z_0. A striking example is the case

$$d = r = 1 \quad n = 3 \quad \omega_{Z_0} = \mathcal{O}_{Z_0},$$

where the surjectivity is automatic. In this case, the above diagram tells us that the curve Y_0 deforms to first order in a direction in Z' if and only if the corresponding element of

$$H^1\left(\Omega_{Z_0}^2\right) = H^1\left(T_{Z_0}\right)$$

pairs to zero with $H^0\left(N_{Y_0\backslash Z_0}\right)$. Thus if Y_0 (infinitesimally) parametrizes some subspace $H^1\left(\Omega_{Z_0}^2\right)$, then Y_0 deforms (infinitesimally) over Z' in directions "normal" to that subspace.

If our higher-order operators $\left.\left(\left(p_{Y'}\right)_* D_{Y/Y''}^{Z/Z'}\right)\right|_{y'}$ pair nicely with relative cohomology of Z/Z', we should expect that higher-order obstruction deformation of Y/Y' in Z/Z' are computed from the higher derivatives of some generalized Abel–Jacobi mapping or normal function, at least when there is "sufficient" ambient cohomology around. This is indeed the case, as we show in §11.

3 Review of $\mathcal{D}$-modules

Let Y and X be complex analytic varieties and let

$$Y \xrightarrow{g} X$$

be a holomorphic mapping. Let $\mathcal{E}$ be a coherent sheaf on X and $\mathcal{F}$ a coherent sheaf on Y.

Definition 3.1. A (left) differential operator on $\mathcal{E}$ with values in $\mathcal{F}$ is a $\mathbb{C}$-linear map

$$g^{-1}\mathcal{E} \xrightarrow{T} \mathcal{F}$$

such that, for some $n \geqslant 0$ and for arbitrary sections $f_0, \ldots, f_n$ of $\mathcal{O}_X$, the $(n+1)$-fold commutator

$$[\ldots[[T, f_0], f_1]\ldots, f_n]$$

is zero, where

$$[T, f](e) = T(fe) - fT(e).$$

The minimum such n is called the *order* of the differential operator T, and we denote the set of all such operators as

$$\mathcal{D}(\mathcal{E}_X, \mathcal{F}_Y).$$

Let $\mathcal{D}(\mathcal{E}_X)$ denote $\mathcal{D}(\mathcal{E}_X, \mathcal{E}_X)$ for the identity map on X. Given $h : Z \to Y$ and a coherent sheaf $\mathcal{G}$ on Z, we have a natural composition

$$\mathcal{D}(\mathcal{F}_Y, \mathcal{G}_Z) \times h^{-1}\mathcal{D}(\mathcal{E}_X, \mathcal{F}_Y) \to \mathcal{D}(\mathcal{E}_X, \mathcal{G}_Z)$$

so that $\mathcal{D}(\mathcal{E}_X, \mathcal{F}_Y)$ has the natural structure of a right $g^{-1}\mathcal{D}(\mathcal{E}_X)$-module and a left $\mathcal{D}(\mathcal{F}_Y)$-module. (We will denote compositions of operators as ST, which means "do T first, then S.") Let $\mathcal{L}_X$ denote the sheaf of Lie algebras given by the sheaf of tangent vector fields on X.

For a complex manifold X of dimension n, let $\mathcal{D}_X = \mathcal{D}(\mathcal{O}_X, \mathcal{O}_X)$. The canonical bundle ω_X of X is a right $\mathcal{D}_X$-module as follows. Sections of $\mathcal{L}_X$ act on ω_X by Lie differentiation. Multiplying this action by (-1), we extend to a right action of the enveloping algebra $\mathcal{U}_X$ associated to the Lie algebra $\mathcal{L}_X$ on ω_X. But this enveloping algebra, considered as a ring of (left) differential operators on $\mathcal{O}_X$, maps to $\mathcal{D}_X$. One checks by direct computation in local coordinates that the kernel of this map acts trivially (on the right) on ω_X.

This last construction lets one pass from left $\mathcal{D}_X$-modules to right $\mathcal{D}_X$-modules and back again. One defines this correspondence on $\mathcal{L}_X$-modules as follows:

Left to right:

$$\begin{array}{rcl} \mathcal{E} & \to & \mathcal{E} \otimes_{\mathcal{O}_X} \omega_X \\ (f \otimes \omega) \cdot \xi & := & -\xi(f) \otimes \omega + f \otimes (\omega \cdot \xi) \end{array}$$

Right to left:

$$\begin{array}{rcl} \mathcal{F} & \to & \mathfrak{Hom}_{\mathcal{O}_X}(\omega_X, \mathcal{F}) \\ (\xi(\lambda))(\omega) & := & \lambda(\omega \cdot \xi) - \lambda(\omega) \cdot \xi \end{array}$$

and extend the action as above to $\mathcal{U}_X \to \mathcal{D}_X$. (See [Bor87, VI, 3.3].)

One relates the structure sheaf $\mathcal{O}_X$ of holomorphic functions to the constant sheaf $\mathbb{C}_X$ on X by viewing $\mathbb{C}_X$ as the "sheaf of solutions" associated to the (left) $\mathcal{D}_X$-module $\mathcal{O}_X$, that is, the set of section of $\mathcal{O}_X$ annihilated by the distinguished ideal of $\mathcal{D}_X$ generated by the sheaf $\mathcal{L}_X$ of vector fields on X. The corresponding functor from the category of bounded $\mathcal{D}_X$-modules to the bounded derived category of sheaves is called the "de Rham functor" (see [Bor87]). In the case of $\mathcal{O}_X$ we make the construction explicitly below. In what follows, notice that an easy computation in local coordinates shows that exterior differentiation is a global section of $\mathcal{D}\left(\Omega_X^k, \Omega_X^{k+1}\right)$ of degree one as a differential operator.

Lemma 3.2. *Via the natural map*

$$\mathcal{D}(\mathcal{O}_X, \Omega_X^*) = \underset{\mathcal{O}_X}{\mathfrak{Hom}}(\mathcal{O}_X, \Omega_X^*) \otimes_{\mathcal{O}_X} \mathcal{D}_X = \Omega_X^* \otimes_{\mathcal{O}_X} \mathcal{D}_X \to \omega_X \otimes_{\mathcal{O}_X} \mathcal{D}_X \to \omega_X$$

the sequence

$$0 \to \mathcal{D}(\mathcal{O}_X, \mathcal{O}_X) \xrightarrow{\partial\circ} \cdots \xrightarrow{\partial\circ} \mathcal{D}\left(\mathcal{O}_X, \Omega_X^{n-1}\right) \xrightarrow{\partial\circ} \mathcal{D}(\mathcal{O}_X, \omega_X) \to \omega_X \to 0$$

becomes a locally free resolution of ω_X as a right $\mathcal{D}_X$-module. In the same way, via

$$\mathcal{D}(\omega_X, \omega_X) = \mathcal{D}_X \otimes_{\mathcal{O}_X} \mathfrak{Hom}(\omega_X, \omega_X) = \mathcal{D}_X \otimes_{\mathcal{O}_X} \mathcal{O}_X \to \mathcal{O}_X$$

we have that

$$0 \to \mathcal{D}(\omega_X, \mathcal{O}_X) \xrightarrow{\partial\circ} \cdots \xrightarrow{\partial\circ} \mathcal{D}\left(\omega_X, \Omega_X^{n-1}\right) \xrightarrow{\partial\circ} \mathcal{D}(\omega_X, \omega_X) \to \mathcal{O}_X \to 0$$

becomes a locally free resolution of $\mathcal{O}_X$ as a left $\mathcal{D}_X$-module.

Proof. Filter negatively by degree of forms and positively by order of operator and use the Koszul lemma from elementary homological algebra, or see [Bor87, VI, 3.5]. For the second case use that Lie differentiation commutes with exterior differentiation, and define the left action of a vector field ξ on $\lambda \in \mathcal{D}(\omega_X, \Omega_X^p)$ by

$$(\lambda \cdot \xi)(\eta) = \lambda(\eta \cdot \xi).$$

One therefore obtains the action of the deRham functor $\omega_X \overset{L}{\otimes}_{\mathcal{D}_X}$ on $\mathcal{O}_X$ by acting on the above resolution of ω_X with $\otimes_{\mathcal{D}_X} \mathcal{O}_X$. But this computation gives back the deRham complex

$$0 \to \mathcal{O}_X \xrightarrow{\partial} \Omega_X^1 \xrightarrow{\partial} \cdots \xrightarrow{\partial} \omega_X \to 0.$$

So, by the Poincaré lemma,

$$\mathcal{H}\left(\omega_X \overset{L}{\otimes}_{\mathcal{D}_X} \mathcal{O}_X\right) = \mathbb{C}_X.$$

If $W \xrightarrow{i} X$ is the inclusion of W as a closed submanifold of X, then $\mathcal{D}(\mathcal{O}_X, \mathcal{O}_W)$ is a left $\mathcal{D}_W$-module and a right $\mathcal{D}_X$-module. In the (standard) notation of [Bor87], $\mathcal{D}(\mathcal{O}_X, \mathcal{O}_W) = \mathcal{D}_{W \to X}$. One reverses the sides on which the operators act by using the left-switching described above, thereby constructing a left $\mathcal{D}_X$-module and right $\mathcal{D}_W$-module:

$$\begin{aligned}\mathcal{D}_{X \leftarrow W} &= \underset{\mathcal{O}_X}{\mathfrak{Hom}}(\omega_X, \mathcal{D}(\mathcal{O}_X, \mathcal{O}_W)) \otimes_{\mathcal{O}_W} \omega_W \\ &= \mathcal{D}(\omega_X, \mathcal{O}_W) \otimes_{\mathcal{O}_W} \omega_W \\ &= \mathcal{D}(\omega_X, \omega_W)\end{aligned}$$

Now let $\mathcal{E}$ be a left $\mathcal{D}_X$-module which is a locally free $\mathcal{O}_X$-module of finite rank. By Kashiwara's theorem ([Bor87, VI, Theorem 7.13]) we can compute the local cohomology sheaf $R\mathcal{H}^0_W(\mathcal{E})$:

$$\begin{aligned}(i_+ \circ i^!)\,\mathcal{E} &= \mathcal{D}_{X \leftarrow W} \overset{L}{\otimes}_{\mathcal{D}_W} \mathcal{D}_{W \to X} \overset{L}{\otimes}_{\mathcal{D}_X} \mathcal{E} \\ &= \left(\mathcal{D}(\omega_X, \omega_W) \overset{L}{\otimes}_{\mathcal{D}_W} \mathcal{D}(\mathcal{O}_X, \mathcal{O}_W)\right) \overset{L}{\otimes}_{\mathcal{D}_X} \mathcal{E} \\ &= \left(\mathcal{D}(\omega_X, \omega_W) \overset{L}{\otimes}_{\mathcal{D}_W} (\mathcal{O}_W \otimes_{\mathcal{O}_X} \mathcal{D}(\mathcal{O}_X, \mathcal{O}_X))\right) \overset{L}{\otimes}_{\mathcal{D}_X} \mathcal{E} \\ &= \left(\mathcal{D}(\omega_X, \omega_W) \overset{L}{\otimes}_{\mathcal{D}_W} \mathcal{O}_W\right) \otimes_{\mathcal{O}_X} \mathcal{D}(\mathcal{O}_X, \mathcal{O}_X) \overset{L}{\otimes}_{\mathcal{D}_X} \mathcal{E} \\ &= \left(\mathcal{D}(\omega_X, \omega_W) \overset{L}{\otimes}_{\mathcal{D}_W} \mathcal{O}_W\right) \otimes_{\mathcal{O}_X} \mathcal{E} \\ &\underset{\text{Lemma 3.2}}{=} (\mathcal{D}(\omega_X, \omega_W) \otimes_{\mathcal{D}_W} \mathcal{D}(\omega_W, \Omega^*_W)) \otimes_{\mathcal{O}_X} \mathcal{E} \\ &= \mathcal{D}(\omega_X, \Omega^*_W) \otimes_{\mathcal{O}_X} \mathcal{E} \\ &= \mathcal{D}(\omega_X \otimes_{\mathcal{O}_X} \mathcal{E}^\vee, \Omega^*_W)\end{aligned}$$

(Here "$\vee$" denotes the dual as an $\mathcal{O}_X$-module. To check twists, let $W = X$. Via Lemma 3.2 we obtain, in the derived category, the tautology

$$\Omega^p_X = \mathcal{D}\left(\Omega^{n-p}_X, \Omega^*_W\right) = \mathfrak{Hom}\left(\Omega^{n-p}_X, \omega_X\right)$$

where the composed map is just given by exterior product in that case.)

Hence in the case of the left $\mathcal{D}_X$-module $\mathcal{E} = \mathcal{O}_X$ we obtain the isomorphism

$$R\mathcal{H}^0_W(\mathcal{O}_X) = \mathcal{D}(\omega_X, \Omega^*_W)$$

of right $\mathcal{D}(\omega_X)$-modules. Thus we have

$$\begin{aligned}R\mathcal{H}^0_W(\Omega^p_X) &= R\mathcal{H}^0_W(\mathcal{O}_X) \otimes_{\mathcal{O}_X} \Omega^p_X \\ &= \mathcal{D}(\omega_X, \Omega^*_W) \otimes_{\mathcal{O}_X} \Omega^p_X \\ &= \mathcal{D}((\wedge^p T_X) \otimes_{\mathcal{O}_X} \omega_X, \Omega^*_W) \\ &= \mathcal{D}\left(\Omega^{n-p}_X, \Omega^*_W\right).\end{aligned}$$

We will need to check that the isomorphism (10) commutes with the right action induced by exterior differentiation $\circ\partial_X$, that is, that the diagram

$$\begin{array}{ccc} R\mathcal{H}^0_W\left(\Omega^p_X\right) & = & \mathcal{D}\left(\Omega^{n-p}_X, \Omega^*_W\right) \\ \downarrow{\scriptstyle \circ\partial} & & \downarrow{\scriptstyle \circ\partial} \\ R\mathcal{H}^0_W\left(\Omega^{p+1}_X\right) & = & \mathcal{D}\left(\Omega^{n-p-1}_X, \Omega^*_W\right) \end{array}$$

is commutative. Since this is a local computation we may fix a set of local coordinates $(x_1, \ldots, x_n)$. For an ordered set of indices K, let K^c denote the increasing order on the complementary set of indices. Putting

$$dx_1 \wedge \ldots \wedge dx_n \leftrightarrow 1$$

equates $\mathcal{O}_X$ with ω_X as right $\mathcal{D}(\omega_X) = \mathcal{D}_X$-modules (with the adjoint action) and putting

$$dx_K \leftrightarrow \{(\mathrm{sgn}\,(K, K^c))\, dx_{K^c} \longmapsto 1\}$$

equates Ω^p_X and $\mathfrak{Hom}\left(\Omega^{n-p}_X, \mathcal{O}_X\right)$ as left $\mathcal{O}_X$-modules. We then compute

$$\begin{aligned} \partial\left(f dx_K\right) &= \sum\nolimits_{k\in K^c} \frac{\partial f}{\partial x_k} dx_k \wedge dx_K \\ &\leftrightarrow \sum\nolimits_{k\in K^c} \left(\mathrm{sgn}\left(\{k, K\}, \{k, K\}^c\right)\right) \frac{\partial f}{\partial x_k} dx_{\{k,K\}^c} \\ &= \left\{\mathrm{sgn}\,(K, K^c)\, dx_{K^c} \longmapsto f\right\} \circ \partial \\ &= \mathrm{sgn}\,(K, K^c) \sum\nolimits_{k\in K^c} \left\{ dx_{K^c - \{k\}} \longmapsto \frac{\partial f}{\partial x_k} \right\}. \end{aligned}$$

Thus

$$\left(R\mathcal{H}^0_W\left(\Omega^*_X\right), \partial\right) = \left(\mathcal{D}\left(\Omega^{n-*}_X, \Omega^*_W\right), \circ\partial\right). \tag{10}$$

Formula (10) is a bit removed from the classical treatment of local cohomology in algebraic geometry, but a local, non-canonical correspondence is not hard to see. Namely, if we have W realized as a complete intersection of hypersurfaces

$$W = Y_1 \cap \ldots \cap Y_c$$

then for the locally free $\mathcal{O}_X$-module Ω^p_X, local cohomology is computed from the sheaf

$$\frac{\Omega^p_X\left(\infty Y_1 + \ldots + \infty Y_c\right)}{\sum_j \Omega^p_X\left(\infty Y_1 + \ldots + \infty Y_{j-1} + \infty Y_{j+1} + \ldots + \infty Y_c\right)}.$$

On the other hand, if we fix a holomorphic projection

$$X \xrightarrow{p} W$$

as left inverse of the inclusion i, there is a well-defined residue operator

$$\frac{\omega_X(\infty Y_1 + \ldots + \infty Y_c)}{\sum_j \omega_X(\infty Y_1 + \ldots + \infty Y_{j-1} + \infty Y_{j+1} + \ldots + \infty Y_c)} \xrightarrow{\text{res}} \omega_W$$

and so a well-defined map

$$\begin{array}{ccc} \dfrac{\Omega_X^p(\infty Y_1 + \ldots + \infty Y_c)}{\sum_j \Omega_X^p(\infty Y_1 + \ldots + \infty Y_{j-1} + \infty Y_{j+1} + \ldots + \infty Y_c)} & \longrightarrow & \mathcal{D}\left(\Omega_X^{n-p}, \omega_W\right) \\ \eta & \longmapsto & (\psi \mapsto res\,(\psi \wedge \eta)) \end{array}$$

which induces (10).

4 The sheaf of deRham normal differential operators

4.1 Definition of $\mathcal{D}\left(\Omega_{Z/Z'}, \Omega_{Y/Y'}\right)$

Throughout this paper, we assume that we have a commutative diagram

$$\begin{array}{ccc} Y & \xrightarrow{h} & Z \\ \downarrow{\scriptstyle p_{Y'}} & & \downarrow{\scriptstyle p_{Z'}} \\ Y' & \xrightarrow{h'} & Z' \end{array} \qquad (11)$$

of holomorphic mappings of complex analytic manifolds. We assume that vertical maps are (surjective) smooth projections with connected fibers.

Definition 4.1. We call a diagram (11) relatively immersive if

$$Y \to Y' \times_{Z'} Z$$

is a closed immersion.

So we have a fibered product:

$$\begin{array}{ccc} X := X' \times_{Z'} Z & \xrightarrow{j} & Z \\ \downarrow{\scriptstyle p_{X'}} & & \downarrow{\scriptstyle p_{Z'}} \\ X' = Y' \times Z' & \xrightarrow{j'} & Z' \end{array}$$

We then have

$$\begin{array}{ccccc} Y & \xrightarrow{i} & X & \xrightarrow{j} & Z \\ \downarrow{\scriptstyle p_{Y'}} & & \downarrow{\scriptstyle p_{X'}} & & \downarrow{\scriptstyle p_{Z'}} \\ Y' & \xrightarrow{i'} & X' & \xrightarrow{j'} & Z' \end{array}$$

If (11) is relatively immersive, then i is a closed immersion and j is a smooth projection.

Let

$$\begin{aligned} n &= \operatorname{rel\,dim} Z/Z' \\ d &= \operatorname{rel\,dim} Y/Y' \\ c &= n-d\,. \end{aligned}$$

We will work throughout with sheaves of relative Kähler differentials. For example, on Z these are given by the formulas

$$\begin{aligned} \Omega^0_{Z/Z'} &= \mathcal{O}_Z \\ \Omega^q_{Z/Z'} &= \frac{\Omega^q_Z}{p^*_{Z'}(\Omega^1_{Z'}) \wedge \Omega^{q-1}_Z} \qquad q>0 \end{aligned}$$

We want to apply our discussion of sheaves of differential operators in §3 to the case

$$\mathcal{E}_X = \Omega^*_{X/X'} = j^*\Omega^*_{Z/Z'} \qquad\qquad \mathcal{F}_Y = \Omega^*_{Y/Y'}$$

We define the bicomplex

$$\mathcal{D}\left(\Omega_{Z/Z'}, \Omega_{Y/Y'}\right)^{p,q} = \left\{\left(\lambda^{p,q} \in \mathcal{D}\left(\Omega^{n-p}_{X/X'}, \Omega^q_{Y/Y'}\right)\right) : \lambda^{p,q} \text{ is } \mathcal{O}_Y\text{-linear}\right\}$$

with total differential

$$\left(\circ\partial_{Z/Z'}\right) + (-1)^p\left(\partial_{Y/Y'}\circ\right)$$

made up of two summands of degree $(1,0)$ and $(0,1)$ respectively.

4.2 Structure of $\mathcal{D}\left(\Omega_{Z/Z'}, \Omega_{Y/Y'}\right)$

As in §3, for any complex manifold W, let $\mathcal{L}_W$ denote the sheaf of Lie algebras of holomorphic vector fields on W. The product structure on $X' = Y' \times Z'$ above distinguishes a sub-Lie-algebra of $\mathcal{L}_{X'}$, namely the subsheaf of the vector fields tangent to the fibers of the projection to Y'. We denote this subsheaf of Lie algebras as $\mathcal{L}_{Y',Z'}$. We define

$$\mathcal{L}_{Y',Z} \tag{12}$$

as the sub-Lie-algebra of $\mathcal{L}_X$ made up of elements which are liftings of elements of $\mathcal{L}_{Y',Z'}$, that is, vector fields ξ for which there exists a vector field ξ' in $\mathcal{L}_{Y',Z'}$ such that

$$\xi\left(g\circ p_X\right)=\xi\left(g\right)\qquad\qquad g\in\mathcal{O}_X\,.$$

Then $\mathcal{L}_{Y',Z}$ acts on the right on the complex $\mathcal{D}\left(\Omega_{Z/Z'},\Omega_{Y/Y'}\right)$. $\mathcal{L}_{Y',Z}$ has a subsheaf $\mathcal{L}^0_{Y',Z}$ of vector fields which are lifts of the zero vector field, that is, vector fields pointing along the fibers of $p_{X'}$. Notice that $\mathcal{L}_{Y',Z}$ is an $\mathcal{O}_{X'}$-module but that the submodule $\mathcal{L}^0_{Y',Z}$ is (compatibly) an $\mathcal{O}_X$-module. We will abbreviate these two facts by referring to the "$\mathcal{O}_{X/X'}$-module structure" of $\mathcal{L}_{Y',Z}$. Notice that

$$\left[\mathcal{L}^0_{Y',Z},\mathcal{L}_{Y',Z}\right]\subseteq\mathcal{L}^0_{Y',Z}\tag{13}$$

and, near $x\in X$,

$$\left(\mathcal{L}_{Y',Z}\right)_x\cong\left(\mathcal{L}^0_{Y',Z}\right)_x\oplus p_{X'}^{-1}\left(j'\left(\mathcal{L}_{Z'}\right)\right)_x$$

Lie and exterior differentiation commute. Thus the complex $\mathcal{D}\left(\Omega_{Z/Z'},\Omega_{Y/Y'}\right)^{*,*}$ is a right $\mathcal{L}_{Y',Z}$-module. In fact, let $\mathcal{U}$, resp. $\mathcal{U}^0$, denote the $p_{X'}^{-1}\mathcal{O}_{X'}$-enveloping algebras associated to the Lie algebra $\mathcal{L}_{Y',Z}$, respectively $\mathcal{L}^0_{Y',Z}$. Similarly let $\mathcal{U}'$ denote the $\mathcal{O}_{X'}$-enveloping algebra associated to $\mathcal{L}_{Y',Z'}$. Then the complex $\mathcal{D}\left(\Omega_{Z/Z'},\Omega_{Y/Y'}\right)^{*,*}$ becomes a right $\mathcal{U}$-module.

4.3 Structure of $\mathcal{D}\left(\Omega_{Z/Z'},\Omega_{Y/Y'}\right)_{\mathcal{J}}$

If $\mathcal{J}'$ is a sheaf of ideals on X' which annihilates $i'\left(Y'\right)$ and we define $\mathcal{J}=p_{X'}^{*}\mathcal{J}'$, then we can define the complex

$$\mathcal{D}\left(\Omega_{Z/Z'},\Omega_{Y/Y'}\right)^{p,q}_{\mathcal{J}}=\left\{\lambda\in\mathcal{D}\left(\Omega_{Z/Z'},\Omega_{Y/Y'}\right)^{p,q}:\lambda\left(g\eta\right)=0,\forall g\in\mathcal{J},\eta\in\Omega^{n-p}_{X/X'}\right\}$$

Also let

$$\begin{aligned}\Omega^*_{X_{\mathcal{J}}/X'_{\mathcal{J}'}}&=\frac{\Omega^*_{X/X'}}{p_X^*\mathcal{J}\cdot\Omega^*_{X/X'}}\\ \mathcal{U}'_{\mathcal{J}'}&=\left\{u'\in\mathcal{U}':u'\left(g\right)|_{Y'}=0,\forall g\in\mathcal{J}'\right\}\\ \mathcal{U}_{\mathcal{J}}&=\left\{u\in\mathcal{U}:u\left(\eta\right)|_{Y}=0,\eta\in\mathcal{J}\cdot\Omega^*_{X/X'}\right\}\end{aligned}\tag{14}$$

where "$|_Y$" denotes the pullback i^*. Let ρ_Y denote the pullback via h composed with restriction of forms to Y. Then

$$\rho_Y\cdot\mathcal{U}_{\mathcal{J}}\subseteq\mathcal{D}\left(\Omega_{X_{\mathcal{J}}/X_{\mathcal{J}'}},\Omega_{Y/Y'}\right)^{p,n-p}=\mathcal{D}\left(\Omega_{X/X'},\Omega_{Y/Y'}\right)^{p,n-p}_{\mathcal{J}},\forall p.$$

4.4 Structure of $\mathcal{D}^{\perp}\left(\Omega^{p}_{Z/Z'}, \omega_{Y/Y'}\right)$

Next, we let $\mathcal{D}_{Y/Y'}$ be the algebra of $\mathcal{O}_{Y'}$-linear differential operators on Y. As we saw in §3, $\mathcal{D}_{Y/Y'}$ acts on the right on the relative dualizing sheaf $\omega_{Y/Y'}$ and we can form

$$\mathcal{D}^{\perp}\left(\Omega^{p}_{Z/Z'}, \omega_{Y/Y'}\right) = \omega_{Y/Y'} \otimes_{\mathcal{D}_{Y/Y'}} \mathcal{D}\left(\Omega^{p}_{Z/Z'}, \mathcal{O}_{Y}\right) \tag{15}$$

To compute the derived functors of (15), let $\mathcal{D}_{Y/Y'}\left(\mathcal{O}_{Y}, \Omega^{*}_{Y/Y'}\right)$ denote the sheaf of $\mathcal{O}_{Y'}$-linear elements of $\mathcal{D}\left(\mathcal{O}_{Y}, \Omega^{*}_{Y/Y'}\right)$ and form:

$$0 \to \mathcal{D}_{Y/Y'}(\mathcal{O}_{Y}) \xrightarrow{\partial_{\circ}} \ldots \xrightarrow{\partial_{\circ}} \mathcal{D}_{Y/Y'}\left(\mathcal{O}_{Y}, \Omega^{d-1}_{Y/Y'}\right) \xrightarrow{\partial_{\circ}} \mathcal{D}_{Y/Y'}(\mathcal{O}_{Y}, \omega_{Y/Y'}) \to \omega_{Y/Y'} \to 0$$

By Lemma 3.2, this is a (locally) free resolution of $\omega_{Y/Y'}$ as a right $\mathcal{D}_{Y/Y'}$-module, so we have the equality

$$\begin{aligned} \omega_{Y/Y'} \overset{L}{\otimes}_{\mathcal{D}_{Y/Y'}} \mathcal{D}\left(\Omega^{n-p}_{Z/Z'}, \mathcal{O}_{Y/Y'}\right) &= \mathcal{H}\Big(\mathcal{D}\left(\mathcal{O}_{Y/Y'}, \Omega^{*}_{Y/Y'}\right) \\ &\qquad \otimes_{\mathcal{D}_{Y/Y'}} \mathcal{D}\left(\Omega^{n-p}_{Z/Z'}, \mathcal{O}_{Y/Y'}\right)\Big) \\ &= \mathcal{H}\left(\mathcal{D}\left(\Omega_{Z/Z'}, \Omega_{Y/Y'}\right)^{p,*}\right) \end{aligned} \tag{16}$$

of right $\mathcal{U}$-modules. Similarly we let

$$\mathcal{D}_{Z/Z'}\left(\Omega^{n-*}_{Z/Z'}, \Omega^{*}_{Y/Y'}\right)$$

denote the $\mathcal{O}_{Z'}$-linear elements of $\mathcal{D}\left(\Omega^{n-*}_{Z/Z'}, \Omega^{*}_{Y/Y'}\right)$ and form the right $\mathcal{U}^{0}$-module

$$\mathcal{D}^{\perp}_{Z/Z'}\left(\Omega^{n-*}_{Z/Z'}, \omega_{Y/Y'}\right)$$

as above.

4.5 Cohomology of $\mathcal{D}\left(\Omega_{Z/Z'}, \Omega_{Y/Y'}\right)^{*,*}$

In this section we assume that

$$\begin{array}{ccc} Y & \xrightarrow{h} & Z \\ \downarrow{\scriptstyle p_{Y'}} & & \downarrow{\scriptstyle p_{Z'}} \\ Y' & \xrightarrow{h'} & Z' \end{array}$$

in (11) is relatively immersive. We can then explicitly compute the cohomology of the complex $\mathcal{D}\left(\Omega_{Z/Z'}, \Omega_{Y/Y'}\right)^{*,*}$. Referring to (16), the E_1-term in the "p-filtration" spectral sequence associated to the bicomplex

$$\sum\nolimits_{p,q} \mathcal{D}\left(\Omega_{Z/Z'}, \Omega_{Y/Y'}\right)^{p,q}$$

is exactly

$$\omega_{Y/Y'} \overset{L}{\otimes}_{\mathcal{D}_{Y/Y'}} \mathcal{D}\left(\Omega^{n-*}_{Z/Z'}, \mathcal{O}_{Y/Y'}\right)$$

Since (11) is relatively immersive, $\mathcal{D}\left(\Omega^{n-*}_{Z/Z'}, \mathcal{O}_{Y/Y'}\right)$ is a locally free $\mathcal{D}_{Y/Y'}$-module:

$$\begin{aligned} E_1^{p,d} &= \mathcal{D}^{\perp}\left(\Omega^{n-p}_{Z/Z'}, \omega_{Y/Y'}\right) \\ E_1^{p,q} &= 0, \quad q \neq d. \end{aligned}$$

For the immersion

$$Y \xrightarrow{\ i\ } Y' \times Z$$

we let $\mathcal{H}^{*}_{Y \backslash Y' \times Z}$ denote the derived functors of the functor which assigns to a sheaf on $X = Y' \times Z$ the subsheaf of the pullback consisting of sections supported on the submanifold Y. Now, by the assumption that i is locally a closed immersion,

$$\mathcal{H}^{k}_{Y \backslash Y' \times Z} = 0$$

for $k \neq n + c$. Also, by (10), we have

$$\mathcal{D}^{\perp}\left(\Omega^{n-p}_{Z/Z'}, \omega_{Y/Y'}\right) \cong \mathcal{H}^{n+c}_{Y \backslash Y' \times Z}\left(\Omega^{p}_{Z/Z'}\right). \tag{17}$$

We next filter $\mathcal{U}$ and $\mathcal{D}\left(\Omega^{n-*}_{Z/Z'}, \omega_{Y/Y'}\right)$ respectively by assigning to

$$\begin{aligned} & P \in \mathcal{U} \\ & Q \in \mathcal{D}\left(\Omega^{n-*}_{Z/Z'}, \omega_{Y/Y'}\right) \end{aligned}$$

an order r equal to the order of the operator

$$(f \longmapsto P(f)), \ f \in \mathcal{O}_{Z'},$$

respectively, the maximal order over all $\omega \in \Omega^{n-*}_{Z/Z'}$ of the operator

$$(f \longmapsto Q(f\omega)), \ f \in \mathcal{O}_{Z'}.$$

We let

$$\begin{aligned} {}_{[r]}\mathcal{U} &= \text{image}\left\{P \in \mathcal{U} : \deg P \leqslant r\right\} \\ {}_{[r]}\mathcal{D}^{\perp}\left(\Omega^{n-*}_{Z/Z'}, \omega_{Y/Y'}\right) &= \text{image}\left\{P \in \mathcal{D}\left(\Omega^{n-*}_{Z/Z'}, \omega_{Y/Y'}\right) : \deg P \leqslant r\right\}. \end{aligned} \tag{18}$$

The r-th graded quotient of the respective filtrations are

$$\begin{array}{c}\mathcal{U}^0 \otimes S^r T_{Z'} \\ \mathcal{D}^{\perp}_{Z/Z'}\left(\Omega^{n-*}_{Z/Z'}, \omega_{Y/Y'}\right) \otimes S^r T_{Z'}\end{array} \tag{19}$$

where S^r denotes the r-th symmetric product. Thus we have a filtration:

$$E_2^{pq} = \mathcal{H}^p\left(\mathcal{D}^{\perp}\left(\Omega^{n-*}_{Z/Z'}, \omega_{Y/Y'}\right), \circ\partial\right) \supseteq \ldots \supseteq \mathcal{H}^p_r \supseteq \ldots \supseteq \mathcal{H}^p_0 \tag{20}$$

where

$$\begin{aligned}\frac{\mathcal{H}^p_r}{\mathcal{H}^p_{r-1}} &= \mathcal{H}^p\left(\mathcal{D}^{\perp}_{Z/Z'}\left(\Omega^{n-*}_{Z/Z'}, \omega_{Y/Y'}\right) \otimes S^r T_{Z'}, \circ\partial\right) \\ &\cong \mathcal{H}^{c+p}_{Y \backslash Y' \times_{Z'} Z}\left(p_{Z'}^{-1} S^r T_{Z'}\right) \\ &= \begin{cases} 0 \text{ if } p \neq c \\ i^* S^r T_{Z'} \text{ if } p = c \end{cases}\end{aligned}$$

4.6 Definition of $\mathcal{D}^{Z/Z'}_{Y/Y'}$

Returning now to the general case (11), we truncate our original complex

$$\begin{aligned}\mathcal{D}\left(\Omega^{\leqslant d}_{Z/Z'}, \Omega_{Y/Y'}\right)^{p,q} &= \mathcal{D}\left(\Omega^{n-p}_{Z/Z'}, \Omega^q_{Y/Y'}\right) \text{ for } p \geqslant c \\ \mathcal{D}\left(\Omega^{\leqslant d}_{Z/Z'}, \Omega_{Y/Y'}\right)^{p,q} &= 0 \text{ for } p < c\end{aligned}$$

and we define our main object of study in this paper,

$$\mathcal{D}^{Z/Z'}_{Y/Y'} = \mathcal{H}^n\left(\mathcal{D}\left(\Omega^{\leqslant d}_{Z/Z'}, \Omega_{Y/Y'}\right)^{*,*}, (\circ\partial) + (-1)^p (\partial\circ)\right) \tag{21}$$

$\mathcal{D}^{Z/Z'}_{Y/Y'}$ is clearly a right $\mathcal{U}$-module. Notice that the map

$$\Omega^{\leqslant d}_{Z/Z'} \xrightarrow{\rho_Y} \Omega_{Y/Y'}$$

given by pulling back differential forms, is closed but not exact, and that we have a natural map

$$\mathcal{D}^{\perp}\left(\Omega^{d+1}_{Z/Z'}, \omega_{Y/Y'}\right) = \omega_{Y/Y'} \overset{L_0}{\otimes}_{\mathcal{D}_{Y/Y'}} \mathcal{D}\left(\Omega^{d+1}_{Z/Z'}, \mathcal{O}_Y\right) \overset{\circ\partial}{\to} \mathcal{D}^{Z/Z'}_{Y/Y'}.$$

4.7 $\mathcal{D}^{Z/Z'}_{Y/Y'}$ as local cohomology

Notice also that the relative deRham complex

$$0 \to \mathcal{O}_Z \overset{\partial}{\to} \Omega^1_{Z/Z'} \overset{\partial}{\to} \ldots \overset{\partial}{\to} \omega_{Z/Z'} \to 0$$

is a resolution of $p_{Z'}^{-1}\mathcal{O}_{Z'}$. So if (11) is relatively immersive, we have isomorphisms

$$\mathcal{H}\left(\mathcal{D}\left(\Omega_{Z/Z'},\Omega_{Y/Y'}\right)^{*,*},(\circ\partial)+(-1)^{p}(\partial\circ)\right)=\mathcal{H}_{Y/Y'}^{*}\left(\mathbb{C}_{Y'\times Z'}\right).$$

Thus, by §4.5, we have

$$\mathcal{D}_{Y/Y'}^{Z/Z'}=\ker\left(\mathcal{D}^{\perp}\left(\Omega_{Z/Z'}^{d},\omega_{Y/Y'}\right)\xrightarrow{\partial_d}\mathcal{D}^{\perp}\left(\Omega_{Z/Z'}^{d-1},\omega_{Y/Y'}\right)\right).$$

And using the increasing filtration ${}_{[r]}\mathcal{D}_{Y/Y'}^{Z/Z'}$ of $\mathcal{D}_{Y/Y'}^{Z/Z'}$ given by (18) and using (19) we have

$$\frac{{}_{[r]}\mathcal{D}_{Y/Y'}^{Z/Z'}}{{}_{[r-1]}\mathcal{D}_{Y/Y'}^{Z/Z'}}=\mathcal{H}\left(\mathcal{D}_{Z/Z'}\left(\Omega_{Z/Z'}^{\leqslant d},\Omega_{Y/Y'}\right)^{*,*}\right)\otimes S^{r}T_{Z'}. \tag{22}$$

Also $\mathcal{D}_{Y/Y'}^{Z/Z'}$ has a distinguished $\mathcal{U}$-submodule given as the image of the mapping

$$\mathcal{H}\left(\mathcal{D}\left(\Omega_{Z/Z'}^{>d}[1],\Omega_{Y/Y'}\right)^{*,*}\right)\to\mathcal{H}\left(\mathcal{D}\left(\Omega_{Z/Z'}^{\leqslant d},\Omega_{Y/Y'}\right)^{*,*}\right)$$

induced by the morphism of complexes

$$\Omega_{Z/Z'}^{\leqslant d}\xrightarrow{\partial_d}\Omega_{Z/Z'}^{>d}[1].$$

In fact (20) tells us that

$$\mathcal{H}\left(\mathcal{D}\left(\Omega_{Z/Z'}^{>d}[1],\Omega_{Y/Y'}\right)^{*,*}\right)=\mathcal{D}^{\perp}\left(\Omega_{Z/Z'}^{d+1},\omega_{Y/Y'}\right)\circ\partial_d \tag{23}$$

and that the sequence

$$0\to\mathcal{D}^{\perp}\left(\Omega_{Z/Z'}^{d+1},\omega_{Y/Y'}\right)\circ\partial_d\to\mathcal{D}_{Y/Y'}^{Z/Z'}\to p_{Y'}^{-1}\mathcal{D}^{\perp}\left(\Omega_{Z'}^{d'},\omega_{Y'}\right)\to 0 \tag{24}$$

is exact. Also

$${}_{[0]}\mathcal{D}_{Y/Y'}^{Z/Z'}=\mathcal{O}_{Y'}\cdot\rho\oplus\mathcal{D}_{Z/Z'}^{\perp}\left(\Omega_{Z/Z'}^{d+1},\omega_{Y/Y'}\right)\circ\partial_d \tag{25}$$

where ρ is the class given by the restriction element ρ_Y above.

4.8 Functoriality of $\mathcal{D}_{Y/Y'}^{Z/Z'}$

Returning to the general case, one easily checks that $\mathcal{D}_{Y/Y'}^{Z/Z'}$ satisfies the basic functorial properties:

Any commutative diagram of morphisms

$$\begin{array}{ccc} Z_1 & \longrightarrow & Z_2 \\ \downarrow & & \downarrow \\ Z_1' & \longrightarrow & Z_2' \end{array}$$

with the vertical maps smooth and surjective induces a natural map

$$\mathcal{D}_{Y/Y'}^{Z_1/Z_1'} \to \mathcal{D}_{Y/Y'}^{Z_2/Z_2'}.$$

Any fibered product

$$\begin{array}{ccc} Y_1 & \longrightarrow & Y_2 \\ \downarrow & & \downarrow \\ Y_1' & \longrightarrow & Y_2' \end{array}$$

(with the vertical maps smooth and surjective) induces a natural map

$$\mathcal{D}_{Y_2/Y_2'}^{Z/Z'} \to \mathcal{D}_{Y_1/Y_1'}^{Z/Z'}.$$

If $Z = Z'$ and so necessarily $Y = Y'$, we shall occasionally make the abbreviation

$$\mathcal{D}_Y^Z := \mathcal{D}_{Y/Y}^{Z/Z}.$$

as in equation (24).

Finally, referring to the decomposition (25), the basic formula

$$L_\varsigma = \langle \zeta \,| \partial\, () \rangle + \partial \, \langle \zeta \mid \, \rangle$$

for Lie differentiation shows that any element of $\mathcal{L}_{Z/Z'}^0$ (i.e., the sheaf of vector fields on Z which annihilate functions from Z') multiplies $\mathcal{O}_Y \cdot \rho$ into $\mathcal{D}_{Z/Z'}^{\perp}\left(\Omega_{Z/Z'}^{d+1}, \omega_{Y/Y'}\right) \circ \partial_d$. A more subtle point is that, for a general element $\xi \in \mathcal{L}_{Z/Z'}$,

$$\rho \cdot \xi = \rho \cdot (\partial \, \langle \xi \mid \, \rangle + \langle \xi \,| \partial \, \rangle) \notin \mathcal{D}^{\perp}\left(\Omega_{Z/Z'}^{d+1}, \omega_{Y/Y'}\right) \circ \partial_d.$$

This is because, unless $\xi \in \mathcal{L}_{Z/Z'}^0$, the component operations in the expression

$$\partial \, \langle \xi \mid \, \rangle + \langle \xi \,| \partial \, \rangle$$

cannot be split apart as the sum of well-defined operations on relative differentials. For example $\langle \xi \mid \, \rangle$ is not well-defined on relative differentials unless $\xi \in \mathcal{L}_{Z/Z'}^0$.

5 Distinguished filtrations, cyclicity

Throughout this section we continue with the situation of §4.1 and we assume that h in (11) is a relative immersion. We now consider the standard increasing filtration $\mathcal{U}_{[m]}$ of $\mathcal{U}$ given by assigning degree one to elements of $\mathcal{L}_{Y',Z}$ (and degree zero to elements of $p_{X'}^{-1}\mathcal{O}_{X'}$). This filtration induces increasing filtration

$$\left(\mathcal{D}_{Y/Y'}^{Z/Z'}\right)_{[m]}$$

on

$$\rho \cdot \mathcal{U} \subseteq \mathcal{D}_{Y/Y'}^{Z/Z'}.$$

Referring to (12), we can define a "relative normal bundle" as the image of

$$\frac{i^*\mathcal{L}_{Y',Z}}{\mathcal{L}_{Y',Y}}$$

in $\mathcal{D}_{Y/Y'}^{Z/Z'}$ via the map

$$\zeta \mapsto \rho_Y \cdot L_\zeta$$

This image $N_{Y\backslash Z}$ is the second summand of a splitting

$$\left(\mathcal{D}_{Y/Y'}^{Z/Z'}\right)_{[1]} = \mathcal{O}_Y \cdot \rho \cdot \oplus N_{Y\backslash Z}$$

Notice that $N_{Y\backslash Z}$ is not the usual normal sheaf $N_{Y\backslash Y'\times Z}$, but is rather the $\mathcal{O}_{Y/Y'}$-module given by normal vector fields in X which are liftings of tangent vector fields on Z, that is, we have an exact sequence

$$0 \to N_{Y\backslash Y'\times_{Z'}Z} \to N_{Y\backslash Z} \to p_{Y'}^{-1}h'^*T_{Z'} \to 0$$

with the sub-object being an $\mathcal{O}_Y$-module and the quotient an $\mathcal{O}_{Y'}$-module.

Lemma 5.1. *With reference to the filtration* ${}_{[r]}\mathcal{D}_{Y/Y'}^{Z/Z'}$ *of* $\mathcal{D}_{Y/Y'}^{Z/Z'}$ *defined in §4.7,* ${}_{[0]}\mathcal{D}_{Y/Y'}^{Z/Z'}$ *is the* $\mathcal{U}^0$*-submodule of* $\mathcal{D}_{Z/Z'}^{\perp}\left(\Omega_{Z/Z'}^d, \omega_{Y/Y'}\right)$ *generated by* $\rho\cdot$ ${}_{[0]}\mathcal{D}_{Y/Y'}^{Z/Z'}$ *has an increasing filtration induced by the natural filtration on* $\mathcal{U}^0$*. For* $m > 0$*, the* m*-th graded quotient of this filtration is*

$$S_{\mathcal{O}_Y}^m N_{Y\backslash Y'\times_{Z'}Z}.$$

Proof. Since all computations will be $\mathcal{O}_{Y'}$-linear, we can and will assume throughout this proof that Y' is a point. We let $\mathcal{L}_{Z/Z'}^0$ denote the vector fields on Z which annihilate functions from Z' and let $\mathcal{U}_{Z/Z'}^0$ denote the $\mathbb{C}$-enveloping algebra associated to $\mathcal{L}_{Z/Z'}^0$. We now consider the $\left(\mathcal{O}_{Z'}, \mathcal{L}_{Z/Z'}^0\right)$-enveloping algebra

$$\mathcal{U}_{\mathcal{O}_Z}\left(\mathcal{L}_{Z/Z'}^0\right)$$

obtained by defining a Lie algebra structure on

$$\mathcal{O}_Z + \mathcal{L}^0_{Z/Z'}$$

by the rule

$$[f+\lambda, g+\mu] = (\lambda(g) - \mu(f)) + [\lambda, \mu],$$

letting $\mathcal{U}\left(\mathcal{O}_Z + \mathcal{L}^0_{Z/Z'}\right)$ denote its universal enveloping algebra over $\mathbb{C}$ and defining

$$\mathcal{U}_{\mathcal{O}_Z}\left(\mathcal{L}^0_{Z/Z'}\right) = \frac{\left(\mathcal{O}_Z + \mathcal{L}^0_{Z/Z'}\right)\cdot \mathcal{U}\left(\mathcal{O}_Z + \mathcal{L}^0_{Z/Z'}\right)}{\langle f\otimes(g+\lambda) = (fg + f\lambda)\rangle}.$$

We consider the standard resolution of $\mathcal{O}_Z$ as a $\mathcal{U}_{\mathcal{O}_Z}\left(\mathcal{L}^0_{Z/Z'}\right)$-module and the standard identification of complexes

$$\underset{\mathcal{U}_{\mathcal{O}_Z}\left(\mathcal{L}^0_{Z/Z'}\right)}{\mathfrak{Hom}}\left(\mathcal{U}_{\mathcal{O}_Z}\left(\mathcal{L}^0_{Z/Z'}\right)\otimes_{\mathcal{O}_Z}\wedge^p\mathcal{L}^0_{Z/Z'},\ \mathcal{O}_Z\right), D$$

where D is the coboundary induced from the bar resolution. Since

$$\rho\cdot\partial = 0,$$

the standard formulas for Lie differentiation show that we have a morphism of complexes:

$$\begin{array}{ccc}
\downarrow & & \downarrow \\
\wedge^2\mathcal{L}^0_{Z/Z'}\otimes_{\mathcal{O}_Z}\mathcal{U}_{\mathcal{O}_Z}\left(\mathcal{L}^0_{Z/Z'}\right) & \longrightarrow & \mathcal{D}^{\perp}_{Z/Z'}\left(\Omega^{d+2}_{Z/Z'}, \omega_{Y/Y'}\right) \\
\downarrow & & \downarrow \\
\mathcal{L}^0_{Z/Z'}\otimes_{\mathcal{O}_Z}\mathcal{U}_{\mathcal{O}_Z}\left(\mathcal{L}^0_{Z/Z'}\right) & \longrightarrow & \mathcal{D}^{\perp}_{Z/Z'}\left(\Omega^{d+1}_{Z/Z'}, \omega_{Y/Y'}\right) \\
\downarrow & & \downarrow \\
\mathcal{U}_{\mathcal{O}_Z}\left(\mathcal{L}^0_{Z/Z'}\right) & \longrightarrow & \mathcal{D}^{\perp}_{Z/Z'}\left(\Omega^{d}_{Z/Z'}, \omega_{Y/Y'}\right) \\
\downarrow & & \\
\mathcal{O}_Z & &
\end{array} \tag{26}$$

(Here we change from left to right action of $\mathcal{U}_{\mathcal{O}_Z}\left(\mathcal{L}^0_{Z/Z'}\right)$ in the standard way, that is, by extending the (-1)-involution on $\mathcal{L}^0_{Z/Z'}$ multiplicatively.)

The horizontal maps in (26) are given by Lie differentiation, followed by contraction with elements of $\wedge^* \mathcal{L}^0_{Z/Z'}$, followed by multiplication on the left by the restriction class ρ. It is an easy calculation in local coordinates (see (50) in the appendix) to see that the horizontal maps are surjective. Thus, using the Poincaré–Birkoff–Witt Theorem we have increasing filtrations

$$\mathcal{D}^{\perp}_{Z/Z'}\left(\Omega^{d+k}_{Z/Z'}, \omega_{Y/Y'}\right)_{[m]} \subseteq \mathcal{D}^{\perp}_{Z/Z'}\left(\Omega^{d+k}_{Z/Z'}, \omega_{Y/Y'}\right)$$

and surjections

$$\wedge^k \mathcal{L}^0_{Z/Z'} \otimes_{\mathcal{O}_Z} S^m_{\mathcal{O}_Z}\left(\mathcal{L}^0_{Z/Z'}\right) \to \frac{\mathcal{D}^{\perp}_{Z/Z'}\left(\Omega^{d+k}_{Z/Z'}, \omega_{Y/Y'}\right)_{[m]}}{\mathcal{D}^{\perp}_{Z/Z'}\left(\Omega^{d+k}_{Z/Z'}, \omega_{Y/Y'}\right)_{[m-1]}}$$

Using commutativity, skew commutativity, and the fact that

$$\rho \cdot \langle \xi \mid \rangle = 0$$

if ξ points along the fibers of $p_{Y'}$, we see that this last map factors through

$$\wedge^k \mathcal{L}^0_{Z/Z'} \otimes_{\mathcal{O}_Y} S^m_{\mathcal{O}_Z} N_{Y \backslash Y' \times_{Z'} Z} \longrightarrow \frac{\mathcal{D}^{\perp}_{Z/Z'}\left(\Omega^{d+k}_{Z/Z'}, \omega_{Y/Y'}\right)_{[m]}}{\mathcal{D}^{\perp}_{Z/Z'}\left(\Omega^{d+k}_{Z/Z'}, \omega_{Y/Y'}\right)_{[m-1]}} \tag{27}$$

which must therefore be surjective.

Now the filtration $\left\{\left({}_{[0]}\mathcal{D}^{Z/Z'}_{Y/Y'}\right)_{[m]}\right\}$ induced on ${}_{[0]}\mathcal{D}^{Z/Z'}_{Y/Y'}$ from

$$\mathcal{D}^{\perp}_{Z/Z'}\left(\Omega^{d+1}_{Z/Z'}, \omega_{Y/Y'}\right)_{[m-1]} \circ \partial \subseteq {}_{[0]}\mathcal{D}^{Z/Z'}_{Y/Y'}$$

(with a degree shift of one) is the same one as that induced from the standard filtration on the preimage of $\mathcal{D}^{Z/Z'}_{Y/Y'}$ in $\mathcal{U}_{\mathcal{O}_Z}\left(\mathcal{L}^0_{Z/Z'}\right)$ given by number of Lie differentiations. On the other hand, let $\left\{\left(\mathcal{U}^0_{Z/Z'}\right)_{[m]}\right\}$ denote the filtration on $\mathcal{U}^0_{Z/Z'}$ given by number of Lie differentiations. Using (25) and the commutative diagram

$$\begin{array}{ccccc}
\mathcal{U}^0_{Z/Z'} & \longrightarrow & {}_{[0]}\mathcal{D}^{Z/Z'}_{Y/Y'} = \ker(\circ\partial) & \longrightarrow & \mathcal{O}_{Y'} = \mathbb{C} \\
\downarrow & & \downarrow & & \downarrow \\
\mathcal{U}_{\mathcal{O}_Z}\left(\mathcal{L}^0_{Z/Z'}\right) & \longrightarrow & \mathcal{D}^{\perp}_{Z/Z'}\left(\Omega^{d}_{Z/Z'}, \omega_{Y/Y'}\right) & \longrightarrow & \mathcal{O}_Y
\end{array}$$

we see that $\left\{\left(\mathcal{U}^0_{Z/Z'}\right)_{[m]}\right\}$ is subordinate to the filtration on $\mathcal{U}^0_{Z/Z'}$ induced from the standard one on $\mathcal{U}_{\mathcal{O}_Z}\left(\mathcal{L}^0_{Z/Z'}\right)$. Also, if $m > 0$, $\left(\mathcal{U}^0_{Z/Z'}\right)_{[m]}$ maps onto the m-th graded quotient of the latter filtration since it surjects to $S^m_{\mathcal{O}_Y} N_{Y\backslash Y'\times_{Z'}Z}$. Therefore by (27) the map

$$\begin{array}{ccc} \mathcal{U}^0_{Z/Z'} & \longrightarrow & \mathcal{D}^{Z/Z'}_{Y/Y'} \\ u & \longmapsto & \rho\cdot u \end{array}$$

must also be surjective. Thus the only thing left to check is that the natural surjection

$$S^m_{\mathcal{O}_Y} N_{Y\backslash Y'\times_{Z'}Z} \to \frac{\left({}_{[0]}\mathcal{D}^{Z/Z'}_{Y/Y'}\right)_{[m]}}{\left({}_{[0]}\mathcal{D}^{Z/Z'}_{Y/Y'}\right)_{[m-1]}}$$

is actually injective. But this is immediately seen by writing this map in local coordinates. (See the appendix.)

Lemma 5.2. *i)* $\mathcal{D}^{Z/Z'}_{Y/Y'}$ *is the* $\mathcal{U}$*-submodule of* $\mathcal{D}^{\perp}_{Z/Z'}\left(\Omega^d_{Z/Z'}, \omega_{Y/Y'}\right)$ *generated by* ρ*. The natural filtration* $\left\{\mathcal{U}_{[m]}\right\}$ *on* $\mathcal{U}$ *induces an increasing filtration on* $\mathcal{D}^{Z/Z'}_{Y/Y'}$ *such that*

$$Gr_0\left(\mathcal{D}^{Z/Z'}_{Y/Y'}\right) = \mathcal{O}_{Y'}\cdot\rho$$

and, if $m > 0$,

$$Gr_m\left(\mathcal{D}^{Z/Z'}_{Y/Y'}\right)$$

is itself filtered with associated gradeds being given by

$$S_{m,r} = S^{m-r}_{\mathcal{O}_Y} N_{Y\backslash Y'\times_{Z'}Z} \otimes_{\mathcal{O}_{Y'}} S^r_{\mathcal{O}_{Z'}} T_{Z'}$$

ii) Let Y_0 *denote a fiber of* $p_{Y'}$*. Suppose we are given a projection*

$$Z \xrightarrow{\ \nu\ } Y_0$$

locally around a point $y \in Y_0$ *(where, abusing notation, we temporarily denote the neighborhood of* $y \in Z$ *again by the letter* Z*). Let* $\{\xi_k\}_{k=1,\dots,c+\dim Z'}$ *be a commuting set of vector fields in* $\mathcal{L}_{Y',Z'}$ *which locally frame the fibers of* ν*. Then* $\mathcal{D}^{Z/Z'}_{Y/Y'}$ *is in fact locally generated over* $\mathcal{O}_{Y'}\cdot\rho$ *by the subalgebra of* $\mathcal{U}$ *generated by the vector fields in the set* $\{\mathcal{O}_{Y_0}\cdot\xi_k\}$.

Proof. By (22) we have an isomorphism

$$\frac{{}_{[r]}\mathcal{D}^{Z/Z'}_{Y/Y'}}{{}_{[r-1]}\mathcal{D}^{Z/Z'}_{Y/Y'}} = {}_{[0]}\mathcal{D}^{Z/Z'}_{Y/Y'} \otimes_{\mathcal{O}_{Z'}} S^r_{\mathcal{O}_{Z'}} T_{Z'}$$

So by Lemma 5.1 we have inductively that $\mathcal{U}^0 \cdot \mathcal{U}_{[r]}$ generates${}_{[r]}\mathcal{D}^{Z/Z'}_{Y/Y'}$.

Besides $\rho_Y \cdot \mathcal{U}$ we have another useful subset of $\mathcal{D}\left(\Omega_{Z/Z'}, \Omega_{Y/Y'}\right)$ which surjects to $\mathcal{D}^{Z/Z'}_{Y/Y'}$. One checks easily, from the fact that ρ is ∂-closed (see the appendix) and the basic expression

$$L_\varsigma = \langle \zeta \,| \partial\,() \rangle + \partial \langle \zeta \mid \rangle,$$

that, for $\zeta_i \in \mathcal{L}_{Y',Z}$,

$$\rho \cdot L_{\varsigma_1} \cdot \ldots \cdot L_{\varsigma_m} = \rho \cdot (\langle \zeta_1 \mid \rangle \cdot \partial \cdot \ldots \cdot \langle \zeta_m \mid \rangle \cdot \partial)$$

as elements of $\mathcal{D}^{Z/Z'}_{Y/Y'}$, where $\langle \zeta \mid \rangle$ is the contraction operator with the vector field ζ.

Another fact that will be important in what follows is:

Lemma 5.3. *i) Referring to §4.3, let $\mathcal{J}'$ be a sheaf of ideals on X' annihilating the image of Y'. Suppose that the natural map*

$$\mathcal{U}'_{\mathcal{J}'} \to \underset{\mathbb{C}}{\mathfrak{Hom}}\left(\left(\mathcal{O}_{X'}/\mathcal{J}'\right)_{y'}, \mathbb{C}\right)$$

is surjective for all $y' \in Y'$. Then for

$$\left(\mathcal{D}^{X/X'}_{Y/Y'}\right)_{\mathcal{J}} = \mathcal{H}\left(\mathcal{D}\left(\Omega^{\leqslant d}_{X_{\mathcal{J}}/X_{\mathcal{J}'}}, \Omega_{Y/Y'}\right)^{*,*}\right) \subseteq \mathcal{D}^{X/X'}_{Y/Y'}$$

we have

$$\rho \cdot \mathcal{U}_{\mathcal{J}} = \left(\mathcal{D}^{X/X'}_{Y/Y'}\right)_{\mathcal{J}}.$$

ii) Fix a foliation of a neighborhood of y in X so that leaves project isomorphically to a neighborhood of $h'(p_{Y'}(y))$ in Z' and project to points on Y'. This foliation determines a unique lifting of $\mathcal{L}_{Y',Z'}$ into $\mathcal{L}_{Y',Z}$ and of enveloping algebras

$$\begin{array}{ccc} \mathcal{U}'_{y'} & \longrightarrow & \mathcal{U}_y \\ u' & \longmapsto & u = u(u'). \end{array}$$

Let $\tilde{\mathcal{J}}' \subseteq \mathcal{J}'$ be a sub-ideal such that

$$\frac{\mathcal{J}'}{\tilde{\mathcal{J}}'} = \mathbb{C},$$

and let $x' \in \mathcal{J}'$ and $u' \in \mathcal{U}'$ be such that

$$u'(x') = 1.$$

Then

$$\rho_Y \cdot u(u')(x' \cdot \eta) = \rho_Y(\eta), \quad \eta \in \Omega^p_{X/X'}.$$

Proof. The foliation gives a local product structure

$$X \xrightarrow{(p_{X'},\pi)} X' \times Z_0.$$

Furthermore Z_0 itself is locally a product

$$Z_0 \xrightarrow{(\alpha,\beta)} Y_0 \times S.$$

Let V_S be the locally defined subalgebra of $\mathcal{U}$ generated by the lifting of a maximal commuting set of vector fields on S via the local product structure (see Lemma 5.2(ii)). V_S then corresponds to a uniquely chosen set of local coordinates on S such that, if we write differential forms in terms of these coordinates on S and arbitraily chosen coordinates on Y_0, the action of V_S via Lie differentiation is just given by the action of V_S on the coefficient functions. We use these coordinates as part of a local coordinate system on

$$X = X' \times Y_0 \times S$$

given by the product structure and compute $\mathcal{D}^{Z/Z'}_{Y/Y'}$ locally as in the appendix. By Lemma 5.2(ii)

$$\beta^{-1}V_S + u(\mathcal{U}')_y$$

generates $\mathcal{D}^{Z/Z'}_{Y/Y'}$. An easy computation in local coordinates shows that the action of $u(\mathcal{U}')_y$ on

$$\pi^*\Omega^p_{Z_0} = \pi^*\left(\sum\nolimits_s \alpha^*\Omega^{p-s}_{Y_0} \otimes \beta^*\Omega^s_S\right)$$

is simply given by the action of $\mathcal{U}'_{y'}$ on coefficient functions and this action commutes with the action of $(\beta \circ \pi)^{-1} V_S$. Both i) and ii) of the lemma then follow easily.

From (24) we also have

$$\mathcal{H}\left(\mathcal{D}^\perp\left(\Omega^{\geq d}_{Z/Z'}, \omega_{Y/Y'}\right), \circ\partial_d\right) = \mathcal{D}^\perp\left(\Omega^{d+1}_{Z/Z'}, \omega_{Y/Y'}\right) \circ \partial_d \subseteq \mathcal{D}^{Z/Z'}_{Y/Y'},$$

and from §4.5 and Lemmas (5.1) and (5.2) we have

Corollary 5.4. *The "Koszul complex"*

$$\begin{array}{l} \ldots \to \rho \cdot \left(\wedge^{r+1} \mathcal{L}^0_{Y',Z}\right) \otimes \mathcal{U} \overset{\circ\partial}{\to} \rho \cdot \left(\wedge^{r} \mathcal{L}^0_{Y',Z}\right) \otimes \mathcal{U} \overset{\circ\partial}{\to} \ldots \\ \ldots \overset{\circ\partial}{\to} \rho \cdot \mathcal{L}^0_{Y',Z} \otimes \mathcal{U} \overset{\circ\partial}{\to} \mathcal{D}^{\perp}\left(\Omega^{d+1}_{Z/Z'}, \omega_{Y/Y'}\right) \circ \partial_d \end{array}$$

is a resolution of $\mathcal{D}^{\perp}\left(\Omega^{d+1}_{Z/Z'}, \omega_{Y/Y'}\right) \circ \partial_d$ *as a (right)* $\mathcal{U}$*-module and the map*

$$\rho \cdot \left(\wedge^{\geqslant -1} \mathcal{L}^0_{Y',Z}\right) \otimes \mathcal{U} \to \mathcal{D}^{\perp}\left(\Omega^{\geqslant d}_{Z/Z'}, \omega_{Y/Y'}\right)$$

is an isomorphism. (Of course this is not a $\mathcal{U}$*-free resolution because of the relations implicit in the factor "* $\rho\cdot$ *" on the left-hand-side.)*

Proof. The only thing we have to check is that we have not changed anything in the proof of Lemma 5.1 by replacement of $\mathcal{U}_{\mathcal{O}_Z}\left(\mathcal{L}^0_{Z/Z'}\right)$ by the slightly smaller $\mathcal{U}$. But as above we still get surjectivity on each graded quotient and therefore surjectivity.

6 Canonical morphisms to $\mathcal{D}^{Z/Z'}_{Y/Y'}$ associated to deformations

As we stated at the outset, given the situation of §4.1 we are going to make a correspondence between certain homomorphisms into $\mathcal{D}^{Z/Z'}_{Y/Y'}$ and deformations of Y/Y'. In one direction that correspondence is essentially contained in the next lemma.

Lemma 6.1. *Suppose we have the commutative diagram*

$$\begin{array}{ccccc} Y & \xrightarrow{i} & X & \xrightarrow{j} & Z \\ \downarrow{\scriptstyle p_{Y'}} & & \downarrow{\scriptstyle p_{X'}} & & \downarrow{\scriptstyle p_{Z'}} \\ Y' & \xrightarrow{i'} & X' & \xrightarrow{j'} & Z' \end{array}$$

as in §4.1. Further suppose that the left-hand rectangle is a fibered product. Then, whenever $y' = p_{Y'}(y)$*, the above diagram induces a natural map*

$$\mathcal{U}'_{y'} \to \left(\mathcal{D}^{Z/Z'}_{Y/Y'}\right)_y$$

Proof. As in Lemma 5.3 fix a foliation of a neighborhood of y in X so that leaves project isomorphically to a neighborhood of $h'(p_{Y'}(y))$ in Z'

and project to points on Y'. This foliation determines a unique lifting of $\mathcal{L}_{Y',Z'}$ into $\mathcal{L}_{Y',Z}$ and of enveloping algebras

$$\begin{array}{ccc} \mathcal{U}'_{y'} & \longrightarrow & \mathcal{U}_y \\ u' & \longmapsto & u = u(u') \end{array}$$

and so induces a map:

$$\mathcal{U}'_{y'} \to \left(\mathcal{D}^{Z/Z'}_{Y/Y'}\right)_y$$

To see that the maps are independent of the foliation, notice that, since the maps from fibers of Y/Y' to fibers of Z/Z' are local isomorphisms, any operator in the right ideal generated by a Lie differentiation by an element of $\mathcal{L}^0_{Y,Z}$ annihilates ρ. Independence then comes from (13) and the fact that any two liftings of an element of $\mathcal{L}_{Y',Z'}$ differ by an element of $\mathcal{L}^0_{Y',Z}$.

Thus we have

Corollary 6.2. *Suppose with respect to (11) we have a commutative diagram*

$$\begin{array}{ccccc} Y & \longrightarrow & W & \longrightarrow & X \\ \downarrow{\scriptstyle p_{Y'}} & & \downarrow{\scriptstyle q_{X'}} & & \downarrow{\scriptstyle p_{X'}} \\ Y' & \longrightarrow & X' & \longrightarrow & X' \end{array}$$

for which the left-hand rectangle is a fibered product. Then there is a canonical induced morphism

$$\mathcal{U}'_{y'} \to \left(\mathcal{D}^{Z/Z'}_{Y/Y'}\right)_y$$

Proof. We apply Lemma 6.1 to the case in which we replace Z with W. Then use §4.8.

Later we will also need

Lemma 6.3. *i) Let $\mathcal{J}'$ be a sheaf of ideals on X' which annihilates the image of Y'. Suppose that the natural map*

$$\left(\mathcal{U}'_{\mathcal{J}'}\right)_{y'} \to \operatorname*{Hom}_{\mathbb{C}}\left((\mathcal{O}_{X'}/\mathcal{J}')_{y'}, \mathbf{C}\right)$$

is surjective for all $y' \in Y'$. Suppose in addition that we have a commutative diagram

$$\begin{array}{ccccc} Y & \longrightarrow & W_J & \longrightarrow & X \\ \downarrow{\scriptstyle p_{Y'}} & & \downarrow{\scriptstyle q_{X'}} & & \downarrow{\scriptstyle p_{X'}} \\ Y' & \longrightarrow & X'_{\mathcal{J}'} & \longrightarrow & X' \end{array}$$

and further suppose that the left-hand rectangle is a fibered product. Then there is a canonical induced morphism

$$\left(\mathcal{U}'_{\mathcal{J}'}\right)_{y'} \to \left(\mathcal{D}^{Z/Z'}_{Y/Y'}\right)_{y}.$$

ii) Suppose that $\tilde{\mathcal{J}}' \subseteq \mathcal{J}'$ is a sub-ideal such that

$$\frac{\mathcal{J}'}{\tilde{\mathcal{J}}'} = \mathbb{C}.$$

Suppose that we have two extensions

$$\begin{array}{ccccc} Y & \longrightarrow & W^i & \longrightarrow & X \\ \downarrow{\scriptstyle p_{Y'}} & & \downarrow{\scriptstyle q^i_{X'}} & & \downarrow{\scriptstyle p_{X'}} \\ Y' & \longrightarrow & X' & = & X' \end{array}$$

for $i = 0, 1$, which agree over $X'_{\mathcal{J}'}$. Suppose we have foliations of X as above which are compatible with W^0 and W^1 respectively and which agree over $X'_{\mathcal{J}'}$. Let

$$u^0, u^1 \in \mathcal{U}$$

be the liftings of

$$u' \in \mathcal{U}'_{\tilde{\mathcal{J}}'}$$

given in Lemma 5.3(ii) by the respective foliations. Then there is a vector field

$$\chi \in \mathcal{L}^0_{Y',Z}$$

such that, over $X'_{\mathcal{J}'}$

$$\rho \cdot \left(u^1 - u^0\right) = \rho \cdot L_{\chi}.$$

Proof. i) Apply Corollary 6.2 locally near y in Y to any family which extends

$$q_{X'}^{-1} X_{\mathcal{J}'} \to X_{\mathcal{J}'}$$

to a family over X'. The map given by Corollary 6.2 restricts to a map on $\left(\mathcal{U}'_{\mathcal{J}'}\right)_{y'}$. By Lemma 5.3 the map does not depend on the choice of extension.

ii) With respect to the family W^0, we write everything in the local coordinates chosen as in the proofs of Lemma 5.3. Suppose that, for S as in the proof of Lemma 5.3 with coordinates $t_1, \ldots, t_c$, the extension

$$W^0 \overset{q_X}{\to} X$$

is given by the conditions $t_k = 0$. Let $x' \in \tilde{\mathcal{J}}'$ be a generator of $\frac{\mathcal{J}'}{\tilde{\mathcal{J}}'}$. The extension

$$W^1 \overset{q_X}{\to} X'$$

is given over $X_{\tilde{\jmath}'}$ by the equations $\tilde{t}_k = 0$, where $\tilde{t}_k = t_k + a_k(y)\,x'$. Then we have

$$\begin{aligned}\left(u^1 - u^0\right)(t_k) &= u^1(t_k) \\ &= u^1\left(\tilde{t}_k - a_k(y)\,x'\right) \\ &= -u^1\left(a_k(y)\,x'\right) \\ &= -a_k(y)\cdot u'(x')\,.\end{aligned}$$

Also $\left(u^1 - u^0\right)$ annihilates pull-backs of functions from $X' \times Y_0$ and all products $t_k \cdot t_{k'} \cdot g$. Thus

$$\rho\cdot\left(u^1 - u^0\right) = \rho\cdot L_{-\sum_k a_k(y)u'(x')\frac{\partial}{\partial t_k}} = \rho\cdot L_{-u'\sum_k a_k(y)x'\frac{\partial}{\partial t_k}}$$

7 Dolbeault resolution of $\mathcal{D}^{Z/Z'}_{Y/Y'}$

Again in this section we assume that (11) is a relative immersion. From (23)–(24) we have pairings

$$\begin{array}{c} R\left(p_{Y'}\right)_*\mathcal{D}\left(\Omega^{\geq d}_{Z/Z'},\Omega^*_{Y/Y'}\right)\otimes R\left(p_{Z'}\right)_*\left(\Omega^{d+1}_{Z/Z'}\to\ldots\to\Omega^{n-1}_{Z/Z'}\to\omega_{Z/Z'}\right) \\ \downarrow \\ R\left(p_{Y'}\right)_*\left(\Omega^*_{Y/Y'}\right) \end{array} \tag{28}$$

and

$$\begin{array}{c} R\left(p_{Y'}\right)_*\left(\mathcal{D}^{Z/Z'}_{Y/Y'}\right)\otimes R\left(p_{Z'}\right)_*\left(\mathcal{O}_Z\to\Omega^1_{Z/Z'}\to\ldots\to\Omega^d_{Z/Z'}\right) \\ \downarrow \\ R\left(p_{Y'}\right)_*\left(\Omega^*_{Y/Y'}\right) \end{array} \tag{29}$$

Referring to (23), these pairings are related by the commutative diagram

$$\begin{array}{ccc}
R\left(p_{Y'}\right)_{*}\left(\mathcal{D}_{Y/Y'}^{Z/Z'}\right) \otimes R\left(p_{Z'}\right)_{*}\left(\Omega_{Z/Z'}^{\leqslant d}\right) & \searrow & \\
\uparrow{\scriptstyle (\circ\partial)\otimes 1} & & \\
R\left(p_{Y'}\right)_{*} \mathcal{D}\left(\Omega_{Z/Z'}^{>d}[1], \Omega_{Y/Y'}^{*}\right) \otimes R\left(p_{Z'}\right)_{*}\left(\Omega_{Z/Z'}^{\leqslant d}\right) & & R\left(p_{Y'}\right)_{*}\left(\Omega_{Y/Y'}^{*}\right) \\
\downarrow{\scriptstyle 1\otimes\partial} & \nearrow & \\
R\left(p_{Y'}\right)_{*} \mathcal{D}\left(\Omega_{Z/Z'}^{>d}[1], \Omega_{Y/Y'}^{*}\right) \otimes R\left(p_{Z'}\right)_{*}\left(\Omega_{Z/Z'}^{>d}\right)[1] & &
\end{array} \tag{30}$$

If, for example, $\mathcal{D}_{Y/Y'}^{Z/Z'}$ were a subsheaf of $\mathcal{D}\left(\Omega_{Z/Z'}^{\leqslant d}, \Omega_{Y/Y'}^{*}\right)$, we could compute its pairings at the level of complexes of sheaves, but in general we are not in this situation and the pairing is only defined in the derived category. To work at the level of complexes of sheaves it is useful to reformulate $\left(p_{Y'}\right)_{*}\left(\mathcal{D}_{Y/Y'}^{Z/Z'}\right)$ in terms of Dolbeault complexes.

Recall from (21) that

$$\mathcal{D}_{Y/Y'}^{Z/Z'} = \mathcal{H}\left(\sum\nolimits_{p \geqslant c} D\left(\Omega_{Z/Z'}^{n-p}, \Omega_{Y/Y'}^{q}\right), (\circ\partial) + (-1)^{p}(\partial\circ)\right)$$

We let $\mathcal{A}_{Z/Z'}^{p,r}$, resp. $\mathcal{A}_{Y/Y'}^{q,s}$ denote the sheaves of C^{∞}-differential forms of type (p,r) on Z/Z', resp. of type (q,s) on Y/Y'. Now let $\mathcal{F}$ be a locally free $\mathcal{O}_Z$-module, and let $\mathcal{D}_h^{\infty}\left(\mathcal{F}, \mathcal{A}_{Y/Y'}^{q,s}\right)$ denote the sheaf of $\mathbb{C}$-linear operators T from the sheaf of (holomorphic) sections of $\mathcal{F}$ to the sheaf of C^{∞}-sections $\mathcal{A}_{Y/Y'}^{q,s}$ such that, for some n and for arbitrary holomorphic functions $f_0, \ldots, f_n$,

$$[\ldots[[T, f_0], f_1] \ldots, f_n] = 0$$

We can consider $\mathcal{D}\left(\mathcal{F}, \Omega_{Y/Y'}^{*}\right)$ as a subsheaf of $\mathcal{D}_h^{\infty}\left(\mathcal{F}, \mathcal{A}_{Y/Y'}^{*,0}\right)$, and conclude that we have, over Y':

$$\begin{aligned}
\mathcal{O}_Y \otimes_{\mathcal{D}\left(\omega_{Y/Y'}\right)} \mathcal{D}\left(\mathcal{F}, \omega_{Y/Y'}\right) &\equiv \mathcal{H}\left(\mathcal{D}\left(\mathcal{F}, \Omega_{Y/Y'}^{*}\right), (\partial\circ)\right) \\
&\equiv \mathcal{H}\left(\mathcal{D}_h^{\infty}\left(\mathcal{F}, \mathcal{A}_{Y/Y'}^{*}\right), (d\circ)\right)
\end{aligned} \tag{31}$$

The first equivalence is derived as in §§4.4–4.5 or [Bor87, 13.2] and the second is derived from the quasi-isomorphism

$$\left(\Omega_{Y/Y'}^{*}, \partial\right) \longrightarrow \left(\mathcal{A}_{Y/Y'}^{*}, d\right).$$

(There is a small point here that the cohomology of the left-hand complex is the sheaf of holomorphic sections of a holomorphic vector bundle whereas the cohomology of the right-hand complex is the sheaf of C^∞-sections of a C^∞-vector bundle. Our assertion is that the underlying complex vector bundles are naturally isomorphic.)

Alternatively let

$$\mathcal{D}^\infty\left(\mathcal{F}, \mathcal{A}^{q,s}_{Y/Y'}\right)$$

denote the sheaf of C^∞-differential operators from the sheaf of C^∞-sections of $\mathcal{F}$ to the sheaf of C^∞-sections of $\mathcal{A}^{q,s}_{Y/Y'}$. We characterize $\mathcal{D}^\infty_h\left(\mathcal{F}, \mathcal{A}^{q,s}_{Y/Y'}\right)$ as the subsheaf of $\mathcal{D}^\infty\left(\mathcal{F}, \mathcal{A}^{q,s}_{Y/Y'}\right)$ consisting of those operators P such that, for every holomorphic section ζ of $\mathcal{F}$ and holomorphic function f on Z,

$$P\left(\bar{f}\cdot\zeta\right) = \bar{f}\cdot P(\zeta)$$

Thus $\mathcal{D}^\infty_h\left(\mathcal{F}, \mathcal{A}^{q,s}_{Y/Y'}\right)$ is a right $\overline{\mathcal{O}}_Z$-module. Tensoring over $\overline{\mathcal{O}}_Z$ with the exact sequence of locally free left $\overline{\mathcal{O}}_Z$-modules

$$0 \to \overline{\mathcal{D}_Z\left(\Omega^n_{Z/Z'}, \mathcal{O}_Z\right)} \overset{\circ\bar{\partial}}{\to} \cdots \overset{\circ\bar{\partial}}{\to} \overline{\mathcal{D}_Z\left(\Omega^1_{Z/Z'}, \mathcal{O}_Z\right)} \overset{\circ\bar{\partial}}{\to} \overline{\mathcal{D}_Z\left(\mathcal{O}_Z, \mathcal{O}_Z\right)} \to \overline{\mathcal{O}_Z} \to 0$$

we obtain

$$\begin{aligned}&\mathcal{H}\left(\mathcal{D}^\infty\left(\mathcal{F}, \mathcal{A}^{*,*}_{Y/Y'}\right), (d\circ)\right)\\ &\equiv \mathcal{H}\left(\sum\nolimits_{r,q,s}^\infty \mathcal{D}^\infty\left(\mathcal{A}^{0,n-r}_{Z/Z'}(\mathcal{F}), \mathcal{A}^{q,s}_{Y/Y'}\right), (\circ\bar{\partial}) + (-1)^r (d\circ)\right)\end{aligned} \tag{32}$$

We apply (31) and (33) to the case in which $\mathcal{F}$ is a graded quotient of $\left(\sum_{p\geqslant c}\Omega^{n-p}_{Z/Z'}, \partial\right)$. We obtain

$$\begin{aligned}&\mathcal{D}^\perp\left(\Omega^{n-p}_{Z/Z'}, \omega_{Y/Y'}\right)\\ &= \mathcal{H}\left(\sum\nolimits_{r,q,s} \mathcal{D}^\infty\left(\mathcal{A}^{n-p,n-r}_{Z/Z'}, \mathcal{A}^{q,s}_{Y/Y'}\right), (\circ\bar{\partial}) + (-1)^{p+r} (d\circ)\right)\end{aligned} \tag{33}$$

The standard spectral sequence argument then allows us to conclude that the natural inclusion of complexes gives a quasi-isomorphism:

$$\mathcal{D}^{Z/Z'}_{Y/Y'} = \mathcal{H}\left(\sum\nolimits_{p\geqslant c}\sum\nolimits_{q,r,s} \mathcal{D}^\infty\left(\mathcal{A}^{n-p,n-r}_{Z/Z'}, \mathcal{A}^{q,s}_{Y/Y'}\right), (\circ d) + (-1)^{p+r} (d\circ)\right). \tag{34}$$

By the acyclicity of the sheaves in (34) we have

$$\begin{aligned}&R\left(p_{Y'}\right)_*\left(\mathcal{D}^\perp\left(\Omega^{n-p}_{Z/Z'}, \omega_{Y/Y'}\right)\right)\\ &= \mathcal{H}\left(\sum\nolimits_{r,q,s}^\infty \mathcal{D}^\infty\left(\mathcal{A}^{n-p,n-r}_{Z/Z'}, \mathcal{A}^{q,s}_{Y/Y'}\right), (\circ\bar{\partial}) + (-1)^{p+r} (d\circ)\right)\end{aligned}$$

where

$$\mathcal{A}^{*,*} = (p_{Y'})_* \mathcal{A}^{*,*}$$

and by the acyclicity of the sheaves in and (34) we have

$$\begin{aligned} & R(p_{Y'})_* \left(\mathcal{D}^{Z/Z'}_{Y/Y'}\right) \\ & = \mathcal{H}\left(\sum\nolimits_{p \geqslant c} \sum\nolimits_{q,r,s} \mathcal{D}^{\infty}\left(\mathcal{A}^{n-p,n-r}_{Z/Z'}, \mathcal{A}^{q,s}_{Y/Y'}\right), (\circ d) + (-1)^{p+r}(d\circ)\right). \end{aligned}$$

Thus for any fixed p and q the natural pairing of $\bar{\partial}$-complexes

$$\sum\nolimits_{r,s} \mathcal{D}^{\infty}\left(\mathcal{A}^{p,r}_{Z/Z'}, \mathcal{A}^{q,s}_{Y/Y'}\right) \otimes \sum\nolimits_{r} \mathcal{A}^{p,r}_{Z/Z'} \to \sum\nolimits_{s} \mathcal{A}^{q,s}_{Y/Y'} \tag{35}$$

obtained by "letting differential operators operate" induces the pairings (28) and (29) on Dolbeault cohomology. Thus, using Corollary 5.4, the pairings (28) and (29) are induced from (35) and the map

$$(p_{Y'})_* \left(\mathcal{A}^{0,*}_{Z/Z'} \otimes \wedge^j (\mathcal{L}_{Y',Z}) \otimes \mathcal{U}\right) \longrightarrow \mathcal{D}^{\infty}\left(\mathcal{A}^{*+j,*}_{Z/Z'}, \mathcal{A}^{*,*}_{Y/Y'}\right).$$

8 The Gauss-Manin connection

If we are in the situation of Corollary 6.2, that is. We have a geometric extension

$$\begin{array}{ccccc} Y & \longrightarrow & W & \longrightarrow & Z \\ \downarrow{\scriptstyle p_{Y'}} & & \downarrow{\scriptstyle q_{X'}} & & \downarrow{\scriptstyle p_{Z'}} \\ Y' & \xrightarrow{i'} & X' & \xrightarrow{j'} & Z' \end{array} \tag{36}$$

of Y/Y', then (29) and Corollary 6.2 imply that we have an induced pairing

$$\mathcal{U}' \otimes j'^* R(p_{Z'})_* \left(\mathcal{O}_Z \to \Omega^1_{Z/Z'} \to \ldots \to \Omega^d_{Z/Z'}\right) \to R(p_{Y'})_* \left(\Omega^*_{Y/Y'}\right).$$

We next wish to see that this pairing is nothing more than the pull-back map

$$\begin{aligned} & R(p_{Z'})_* \left(\mathcal{O}_Z \to \Omega^1_{Z/Z'} \to \ldots \to \Omega^d_{Z/Z'}\right) \\ \to \quad & R(q_{X'})_* \left(\mathcal{O}_W \to \Omega^1_{W/X'} \to \ldots \to \Omega^d_{W/X'}\right) \\ = \quad & R(q_{X'})_* (\mathbb{C}_W) \end{aligned}$$

followed by the Gauss–Manin connection on the family W/X'.

Definition 8.1. Let $\xi' \in \mathcal{L}_{Y',Z'}$. By a C^∞-lifting of type $(1,0)$ of ξ' to X/X' we mean a C^∞-vector field ξ on X such that

i) $[\xi, \bar{\xi}] = 0$,

ii) $\xi(f' \circ p_{X'}) = \xi'(f')$ for every C^∞-function f' on X'.

Given (36) one constructs a C^∞-projection of a tubular neighborhood of Y/Y' in W/X' compatible with the projection $X' \to Y'$ and such that the fibers are complex holomorphic polydisks isomorphic to Z'. Any such projection induces a unique C^∞-lifting of each $\xi' \in \mathcal{L}_{Y',Z'}$ to a vector field tangent to the fibers of the projection and so a homomorphism of sheaves of Lie algebras

$$\mathcal{L}_{Y',Z'} \to \mathcal{A}^0_X \otimes \mathcal{L}_{Y',Z}$$

and therefore an induced mapping

$$\mathcal{U}' \to \mathcal{D}^\infty\left(\mathcal{A}^{\leqslant d,*}_{Z/Z'}, \mathcal{A}^{*,*}_{Y/Y'}\right). \tag{37}$$

As in §4.2 Lie differentiation commutes with the action of $\circ d$ so that $\mathcal{U}'$ acts on the complex $\mathcal{D}^\infty\left(\mathcal{A}^{\leqslant d,*}_{Z/Z'}, \mathcal{A}^{*,*}_{Y/Y'}\right)$ and the mapping (37) is simply the image of that action on the distinguished generator ρ. Under the standard Cech-to-Dolbeault isomorphism, (37) is identified with the mapping given by Corollary 6.2. Thus we get that the pairing (29) is given by

$$\mathcal{U}' \otimes \mathcal{D}^\infty\left(\mathcal{A}^{\leqslant d,*}_{Z/Z'}, \mathcal{A}^{*,*}_{Y/Y'}\right) \longrightarrow \mathcal{U}' \otimes \mathcal{D}^\infty\left(\mathcal{A}^{*,*}_{W/X'}, \mathcal{A}^{*,*}_{Y/Y'}\right) \xrightarrow{\text{Gauss–Manin}} \mathcal{A}^{*,*}_{Y/Y'}$$

9 The deformation criterion

Suppose now that we have commutative diagram

$$\begin{array}{ccc} Y & \xrightarrow{h} & Z \\ \downarrow{\scriptstyle p_{Y'}} & & \downarrow{\scriptstyle p_{Z'}} \\ Y' & \xrightarrow{h'} & Z' \end{array} \tag{38}$$

as in (11) where $Y/Y' \to Z/Z'$ is relatively immersive. We wish to consider extensions of (38) to diagrams of the form

$$\begin{array}{ccccc} Y & \longrightarrow & W & \longrightarrow & X \\ \downarrow{\scriptstyle p_{Y'}} & & \downarrow{\scriptstyle q_{X'}} & & \downarrow{\scriptstyle p_{X'}} \\ Y' & \xrightarrow{i'} & X' & = & X' \end{array} \tag{39}$$

for which the left-hand rectangle is a fibered product. For example, if Y' is a one-point space, (39) amounts to giving a deformation of the submanifold Y in Z/Z'.

We first apply Corollary 6.2 to associate to (39) a family of natural morphisms

$$\mathcal{U}'_{y'} \xrightarrow{\varphi_{y',y}} \left(\mathcal{D}^{Z/Z'}_{Y/Y'}\right)_y$$

whenever $y \in Y_{y'} = p_{Y'}^{-1}(y')$, so to define a morphism

$$\mathcal{U}' \to (p_{Y'})_* \left(\mathcal{D}^{Z/Z'}_{Y/Y'}\right). \tag{40}$$

In the rest of this section, we shall be concerned with the opposite direction. Namely, given a diagram (38) and a morphism (40):

1. What are the conditions on the morphism such that it is induced by a diagram (39)?
2. Does the morphism (40) completely determine the diagram (39) that induces it?

To state our result precisely, we need two definitions.

Definition 9.1. If $y' = p_{Y'}(y)$, an $\mathcal{O}_{X/X'}$-linear homomorphism of Lie algebras of tangent vector fields

$$\psi : (\mathcal{L}_{Y',Z'})_{y'} \to (\mathcal{L}_{Y',Z})_y$$

is called a lifting homomorphism if, for all $\xi' \in (\mathcal{L}_{Y',Z'})_{y'}$:

i) $\psi(g \cdot \xi') = (g \circ p_{X'}) \cdot \psi(\xi')$ for all $g \in \mathcal{O}_{X'}$,

ii) $\psi(\xi')(g \circ p_{X'}) = \xi'(g)$ for all $g \in \mathcal{O}_{X'}$.

Definition 9.2. Given $y \in Y$, $y' \in Y'$ with $y' = p_{Y'}(y)$, a mapping

$$\mathcal{U}'_{y'} \xrightarrow{\varphi_{y',y}} \left(\mathcal{D}^{Z/Z'}_{Y/Y'}\right)_y$$

is called almost-multiplicative if there exists a lifting homomorphism

$$\psi : (\mathcal{L}_{Y',Z'})_{y'} \to (\mathcal{L}_{Y',Z})_y$$

such that, for the induced homomorphism

$$\mathcal{U}'_{y'} \xrightarrow{\psi} \mathcal{U}_y$$

and for all $u' \in \mathcal{U}'$,

$$\varphi_{y'}(u') = \rho \cdot \psi(u')$$

where ρ is the distinguished generator of $\mathcal{D}^{Z/Z'}_{Y/Y'}$. (See (25).)

We call a map

$$\mathcal{U}'_{y'} \xrightarrow{\varphi_{y'}} \left((p_{Y'})_* \left(\mathcal{D}^{Z/Z'}_{Y/Y'} \right) \right)_{y'}$$

almost multiplicative if the induced maps

$$\mathcal{U}'_{y'} \xrightarrow{\varphi_{y',y}} \left(\mathcal{D}^{Z/Z'}_{Y/Y'} \right)_{y}$$

are almost multiplicative whenever $y \in Y$, $y' \in Y'$ with $y' = p_{Y'}(y)$. In short, almost multiplicative homomorphisms are those induced locally by "cotransversals," that is, by local foliations X transverse to the fibers of Y/Y'. (The important point here is that the existence of ψ is required only locally; a global lifting of $\mathcal{L}_{Y',Z'}$ to $(p_{Y'})_* (\mathcal{L}_{Y',Z})$ is usually not possible.)

Notice that the mapping (40) derived from (39) and Corollary 6.2 is, by construction, almost multiplicative. By Lemma 5.2(ii), these almost multiplicative maps are determined by

i) choosing a commuting set of elements of $\mathcal{L}_{Y',Z'}$ which locally frame the tangent space of Z' along Y',

ii) lifting these to a commuting family of vector fields on X.

The chief result we wish to prove in §9 can now be stated as follows.

Theorem 9.3. *There exists a natural one-to-one correspondence between formal extensions of the diagram (38) to diagrams (39) and almost multiplicative homomorphisms*

$$\mathcal{U}' \xrightarrow{\varphi} (p_{Y'})_* \left(\mathcal{D}^{Z/Z'}_{Y/Y'} \right).$$

Here "formal" means that W is a formal neighborhood of $i(Y)$ in X.

Proof. As already mentioned, the mapping

$$\mathcal{U}' \xrightarrow{\varphi} (p_{Y'})_* \left(\mathcal{D}^{Z/Z'}_{Y/Y'} \right)$$

associated to a diagram (39) is almost multiplicative by construction. Also, since all morphisms are left $\mathcal{O}_{Y'}$-linear, it suffices to establish the result in the case in which $Y' = \{0\}$ is a point, which we will therefore assume throughout the proof. So to complete the proof of the theorem it will suffice to establish the following assertion: The mapping

$$\mathcal{U}'_0 \xrightarrow{\varphi_y} \left(\mathcal{D}^{Z/Z'}_{Y/Y'} \right)_{y} \tag{41}$$

uniquely determines a (formal) local family

$$W \to Z'$$

fitting into (39).

As is standard in local deformation theory, we work inductively over increasing large (infinitesimal) neighborhoods of $\{0\}$ in Z'. We can extend the notion of almost multiplicative morphism to the restriction of φ_y to a scheme containing $\{0\}$ and lying in Z' as follows:

Let $\mathcal{J}_1'$ be the ideal defining $\{0\}$ in Z', and suppose we have a filtration

$$\mathcal{J}_{s+1}' \subseteq \cdots \subseteq \mathcal{J}_1' \subseteq \mathcal{J}_0' = \mathcal{O}_{Z'}$$

by ideals on Z' such that successive quotients

$$J_r = \frac{\mathcal{J}_r'}{\mathcal{J}_{r+1}'}$$

are dimension-one complex vector spaces generated by

$$x_r' \in \mathcal{J}_r'.$$

Then the set $\{x_j'\}_{j=0,\dots,r}$ gives each quotient $\mathcal{O}_r = \frac{\mathcal{O}_{Z'}}{J_{r+1}}$ the structure of a $\mathbb{C}$-vector space. Let

$$\mathcal{O}_r \xrightarrow{D_r} J_r = \mathbb{C}$$

be the natural projection homomorphism associated with this basis. It is immediate to see that D_r is a differential operator of order $\leqslant r$ with values in $\mathbb{C}$. Thus D_r comes from an element $u_r' \in \mathcal{U}_0'$. In fact, if we let

$$\mathcal{U}_{[r]}' = \{u' \in \mathcal{U}' : u'(g)(0) = 0 \ for \ all \ g \in \mathcal{J}_{r+1}'\}$$

we have

$$u_r' \in \mathcal{U}_{[r]}' \subseteq \mathcal{U}'.$$

Letting $Z_{[r]}'$ denote the scheme defined in Z' by $\mathcal{J}_{r+1}'$, and referring to (14), our definition for "infinitesimal almost-multiplicativity over $Z_{[r]}'$" is obtained from Definition 9.2 by replacing the condition

$$\varphi_y(u') = \rho \cdot \psi(u')$$

for $u \in \mathcal{U}'$ with the same condition restricted to $u' \in \mathcal{U}_{[r]}'$.

If we have constructed

$$W_{[s]} \xrightarrow{q} Z_{[s]}'$$

over $Z_{[s]}'$, then by Lemma 6.3(i) we have a uniquely determined almost multiplicative map

$$\mathcal{U}_{[s]}' \xrightarrow{\varphi_y^s} \left(\mathcal{D}_{Y/Y'}^{Z/Z'}\right)_y$$

We must show a natural bijective correspondence between extensions of φ_y^s to an almost multiplicative map

$$\mathcal{U}'_{[s+1]} \xrightarrow{\varphi_y^{s+1}} \left(\mathcal{D}_{Y/Y'}^{Z/Z'}\right)_y$$

and extensions of $W_{[s]} \xrightarrow{q} Z'_{[s]}$ to a family over $Z'_{[s+1]}$.

We let Y_0 denote the fiber of $p_{Y'}$. Let

$$\mathcal{J} = \text{ideal sheaf of } Y_0 \text{ in } Z$$

whose restriction gives the ideal sheaf $\mathcal{J}_0$ of Y_0 in Z_0. At $y \in Y_0$ let

$$\{z_k \in \mathcal{J}_y\}_k$$

restrict to a free set of generators of $\mathcal{J}_0$. For $y_0 \in Y_0$, abuse notation slightly and let Y_0 denote the local germ of Y_0 at y_0. Standard deformation theory shows that the set of local extensions of a (flat) family

$$\begin{array}{ccccc} Y_0 & \longrightarrow & W_{[s]} & \longrightarrow & Z \\ \Big\downarrow{\scriptstyle p_{Y'}} & & \Big\downarrow{\scriptstyle q} & & \Big\downarrow{\scriptstyle p_{Z'}} \\ \{0\} & \rightarrowtail & Z'_{[s]} & \longrightarrow & Z' \end{array}$$

over $Z'_{[s]}$ to a (flat) family over $Z'_{[s+1]}$ is a principal homogeneous space for

$$\mathrm{Hom}\left(\frac{\mathcal{J}_0}{\mathcal{J}_0^2}, p_{Y'}^* \frac{\mathcal{J}'_{s+1}}{\mathcal{J}'_{s+2}}\right)_y = \mathrm{Hom}\left(\frac{\mathcal{J}_0}{\mathcal{J}_0^2}, p_{Y'}^* J_{s+1}\right)_y .$$

(See [Kol96, Chapter I].) More precisely, fix a local product structure

$$Z = Y_0 \times S \times Z'$$

such that $p_{Z'}$ is given by projection onto the last factor. Suppose that functions

$$z_k = w_k + a_{k,0}(y)\, x'_0 + \ldots + a_{k,s}(y)\, x'_s$$

are a free set of local generators of the ideal of $W_{[s]}$ in

$$Z_{[s]} = Z'_{[s]} \times_{Z'} Z .$$

(We use the same notation for functions on Z' and their pullbacks to Z.) Fix reference extensions of z_k to $Z_{[s+1]}$, and continue to call them z_k. The possible choices of extensions of $W_{[s]}$ to $W_{[s+1]}$ are then uniquely determined by the functions

$$\tilde{z}_k = z_k + a_k(y)\, x'_{s+1} . \tag{42}$$

Thus extensions are in one-to-one correspondence with elements of

$$\operatorname{Hom}\left(\frac{\mathcal{J}_0}{\mathcal{J}_0^2}, p_{Y'}^*\mathcal{J}'_{s+1}\right)_y = \left(N_{Y_0\setminus Z_0}\right)_y .$$

So we need only show that the almost multiplicative extension

$$\mathcal{U}'_{[s+1]} \xrightarrow{\varphi_y^{s+1}} \left(\mathcal{D}_{Y/Y'}^{Z/Z'}\right)_y$$

induced via Lemma 6.3(i) by a chosen extension $W_{[s+1]}$ determines the element of

$$\left(N_{Y_0\setminus Z_0}\right)_y$$

that defines $W_{[s+1]}$ with respect to the reference extension. But by Lemma 6.3(ii) we conclude that

$$\rho\cdot\tilde{\psi}\left(u'_{s+1}\right) - \rho\cdot\psi\left(u'_{s+1}\right) = \rho\cdot L_{-\sum_k a_k(y)\frac{\partial}{\partial w_k}} . \tag{43}$$

Thus we read off from (43) the exact element of

$$\left(N_{Y_0\setminus Z_0}\right)_y$$

that determined the extension of the deformation given by the equations $\tilde{z}_k = 0$ in (42). So the extension

$$\mathcal{U}'_{[s+1]} \xrightarrow{\varphi_y^{s+1}} \left(\mathcal{D}_{Y/Y'}^{Z/Z'}\right)_y$$

determines the extension

$$W_{[s+1]} \xrightarrow{q} Z'_{[s+1]} .$$

But Lemma 6.3(i) tell us that the extension

$$W_{[s+1]} \xrightarrow{q} Z'_{[s+1]}$$

determines the extension

$$\mathcal{U}'_{[s+1]} \xrightarrow{\varphi_y^{s+1}} \left(\mathcal{D}_{Y/Y'}^{Z/Z'}\right)_y$$

Thus the proof of assertion (41) is complete, which in turn completes the proof of Theorem 9.3.

Inside complex projective varieties, formal subvarieties give rise to actual geometric subvarieties, so we have

Theorem 9.4. *Suppose that we have a diagram (38) with the same hypotheses as given there. Suppose, in addition, $p_{Y'}$ is proper and Z/Z' is relatively quasi-projective. Given an almost multiplicative morphism*

$$\mathcal{U}' \xrightarrow{\varphi} (p_{Y'})_* \left(\mathcal{D}^{Z/Z'}_{Y/Y'}\right)$$

there is a geometric deformation

$$\begin{array}{ccccc} Y & \rightarrowtail & W & \rightarrowtail & X \\ \downarrow{\scriptstyle p_{Y'}} & & \downarrow{\scriptstyle q_{X'}} & & \downarrow{\scriptstyle p_{X'}} \\ Y' & \xrightarrow{i'} & X' & = & X' \end{array}$$

such that the associated

$$\mathcal{U}' \xrightarrow{\tilde{\varphi}} (p_{Y'})_* \left(\mathcal{D}^{Z/Z'}_{Y/Y'}\right)$$

approximates the morphism φ to arbitrary pregiven order.

Theorem 9.3 gives a formal family

$$\begin{array}{ccccc} Y & \rightarrowtail & \widehat{W} & \rightarrowtail & X \\ \downarrow{\scriptstyle p_{Y'}} & & \downarrow{\scriptstyle q_{X'}} & & \downarrow{\scriptstyle p_{X'}} \\ Y' & \xrightarrow{i} & \widehat{X}' & = & X' \end{array}$$

projective over $\widehat{X}$. So the formal subvariety $\widehat{W}$ can be approximated to arbitrary order by a W which is projective over X [Art68, Theorem 1.4], so over a sufficiently small analytic neighborhood of Y in X.

From the proof of Theorem 9.3 and Lemma 6.3(ii) we have

Corollary 9.5. *i) Let*

$$\tilde{\mathcal{J}}' \subseteq \mathcal{J}'$$

be sheaves of ideals on X' contained in the ideal of Y' such that

$$\frac{\mathcal{J}'}{\tilde{\mathcal{J}}'} = \mathcal{O}_{Y'}.$$

If $\mathfrak{Hom}$ denotes almost multiplicative homomorphisms of sheaves and φ is an almost multiplicative homomorphism giving the family

$$W_{\mathcal{J}'} \xrightarrow{q_{\mathcal{J}'}} X_{\mathcal{J}'}$$

the obstruction to extending the family to

$$W_{\tilde{\mathcal{J}}'} \overset{q_{\mathcal{J}'}}{\rightarrow} X_{\tilde{\mathcal{J}}'}$$

is the element of

$$\underset{\mathcal{O}_{Y'}}{\mathfrak{Hom}}\left(\left(\frac{\mathcal{J}'}{\tilde{\mathcal{J}}'}\right)^{*}, R^{1}\left(p_{Y'}\right)_{*} N_{Y\backslash Y'\times_{Z'}Z}\right) = R^{1}\left(p_{Y'}\right)_{*} N_{Y\backslash Y'\times_{Z'}Z}$$

which measures the obstruction to lifting φ under the homomorphism

$$\mathfrak{Hom}\left(\mathcal{U}'_{\tilde{\mathcal{J}}'}, \left(p_{Y'}\right)_{*}\left(\mathcal{D}^{Z/Z'}_{Y/Y'}\right)\right) \rightarrow \mathfrak{Hom}\left(\mathcal{U}'_{\mathcal{J}'}, \left(p_{Y'}\right)_{*}\left(\mathcal{D}^{Z/Z'}_{Y/Y'}\right)\right).$$

ii) The map

$$\begin{array}{rcl} N_{Y\backslash Y'\times_{Z'}Z} & \longrightarrow & \mathcal{D}^{\perp}\left(\Omega^{d+1}_{Z/Z'}, \omega_{Y/Y'}\right) \\ \xi & \longmapsto & \langle \xi \mid \rangle \end{array}$$

induces a natural morphism

$$R^{1}\left(p_{Y'}\right)_{*} N_{Y\backslash Y'\times_{Z'}Z} \rightarrow R^{1}\left(p_{Y'}\right)_{*}\left(\mathcal{D}^{\perp}\left(\Omega^{d+1}_{Z/Z'}, \omega_{Y/Y'}\right)\circ \partial_{d}\right)$$

so that from i) we obtain an obstruction element

$$v \in R^{1}\left(p_{Y'}\right)_{*}\left(\mathcal{D}^{\perp}\left(\Omega^{d+1}_{Z/Z'}, \omega_{Y/Y'}\right)\circ \partial_{d}\right).$$

Proof. i) By Lemma 6.3(ii), the sheaf sequence

$$\begin{aligned} 0 \rightarrow \underset{\mathcal{O}_{Y'}}{\mathfrak{Hom}}\left(p^{*}_{Y'}\left(\frac{\mathcal{J}'}{\tilde{\mathcal{J}}'}\right)^{*}, N_{Y\backslash Y'\times_{Z'}Z}\right) & \rightarrow \mathfrak{Hom}\left(\mathcal{U}'_{\tilde{\mathcal{J}}'}, \mathcal{D}^{Z/Z'}_{Y/Y'}\right) \\ & \rightarrow \mathfrak{Hom}\left(\mathcal{U}'_{\mathcal{J}'}, \mathcal{D}^{Z/Z'}_{Y/Y'}\right) \rightarrow 0 \end{aligned}$$

is exact. The obstruction element is computed from a covering $\{V_a\}$ of Z by coordinate disks as the Cech cochain

$$\chi_{ab} = -\sum\nolimits_{k} a^{ab}_{k}(y)\frac{\partial}{\partial v^{ab}_{k}} \in \Gamma_{\mathcal{U}_a\cap\mathcal{U}_b}\mathcal{L}^{0}_{Y',Z}$$

constructed as in the proof of Theorem 9.3.

ii) The second assertion then follows from the exact sequence (24).

10 Some immediate consequences

Corollary 10.1. *A compact submanifold Y of dimension d has a geometric deformation in a projective manifold Z of dimension n if and only if there is an almost multiplicative homomorphism*

$$\mathcal{D}' \xrightarrow{\varphi} H^0\left(Y; \mathcal{D}_{Y/\{0\}}^{Z/\{0\}}\right)$$

where $\mathcal{D}'$ is the ring of constant-coefficient differential operators in one variable and Δ is the unit disc. Thus, referring to (17), a necessary condition for Y to have a deformation in Z is that

$$\overset{\lim}{\longleftarrow}_s H^d\left(\Omega_Z^d / \mathcal{I}_{Y,Z}^s \Omega_Z^d\right)$$

be infinite-dimensional.

Proof. The inverse limit $\overset{\lim}{\leftarrow} H^d\left(\Omega_Z^d / \mathcal{I}_{Y,Z}^s \Omega_Z^d\right)$ is the dual of

$$H_Y^c\left(\Omega_Z^c\right) \cong H^0\left(\mathcal{D}^\perp\left(\Omega_Z^d, \omega_Z\right)\right) .$$

(See [Bor87, Theorem 7.13].)

Again suppose $Y = \{0\}$. For a geometric family (39), it is easy to see, by induction on the order of neighborhood of $0 \in Z'$, that

$$\mathcal{U}'_{\{0\}} \xrightarrow{\varphi} H^0\left(D_{Y_0/\{0\}}^{W/X'}\right)$$

is in fact an isomorphism since that is trivially true for the associated graded objects with respect to the filtration in Lemma 5.2. It is natural to ask whether the composition

$$\mathcal{D}_{\{0\}}^{X'} \xrightarrow{\varphi} H^0\left(\mathcal{D}_{Y_0/\{0\}}^{Z/Z'}\right)$$

associated to a diagram (39) is an isomorphism if the associated first-order map

$$T_{X',0} \xrightarrow{\varphi} H^0\left(N_{Y_0 \backslash Z}\right)$$

is. This is not in general true. The simplest example of this is perhaps the classical logarithmic transform

$$Z = Z_0 = \frac{\mathbb{C} \times E}{\{(t,e) = (-t, e+h)\}}, \quad Z' = \{0\}$$

where E is an elliptic curve, h is a non-trivial half-period, and Y_0 is given by $t = 0$. Then

$$H^0\left(N_{Y_0 \backslash Z_0}\right) = H^1\left(N_{Y_0 \backslash Z_0}\right) = 0$$

but the local isomorphisms

$$\begin{array}{ccccc} \left(\mathcal{D}^{\mathbb{C}\times E/\mathbb{C}}_{\{0\}\times E/\{0\}}\right)_{(0,e)} & \xrightarrow{\approx} & \left(\mathcal{D}^{\mathbb{C}\times E/\mathbb{C}}_{\{0\}\times E/\{0\}}\right)_{(0,e+h)} & \xrightarrow{\approx} & \left(\mathcal{D}^{X_0/\{0\}}_{Y/\{0\}}\right)_{(0,e)} \\ \rho\cdot\frac{\partial}{\partial t} & \longmapsto & \rho\cdot\left(-\frac{\partial}{\partial t}\right) & \longmapsto & \rho\cdot\xi \end{array}$$

imply that there is a non-trivial second-order element of

$$H^0\left(\mathcal{D}^{Z/\{0\}}_{Y_0/\{0\}}\right)$$

given everywhere locally by $\rho\cdot\xi\cdot\xi$.

However, there are situations in which, for example, the map

$$\varphi:\left(\mathcal{D}^{X'/\{0\}}_{\{0\}/\{0\}}\right)\to(p_{Y'})_*\left(\mathcal{D}^{X'\times X_0/\{0\}}_{\{0\}/\{0\}}\right)=H^0\left(\mathcal{D}^{X_0/\{0\}}_{\{0\}/\{0\}}\right)$$

induced by a geometric family (39) is in fact an isomorphism.

Lemma 10.2. *For a geometric family (39), suppose that the natural maps*

$$S^mT_{X',0}\xrightarrow{\varphi_m}H^0\left(S^mN_{Y_0\backslash Z_0}\right)$$

are isomorphisms for each m, that is

$$T_{X',0}\overset{\cong}{\to}H^0\left(N_{Y_0\backslash Z_0}\right)$$

and

$$S^mH^0\left(N_{Y_0\backslash Z_0}\right)=H^0\left(S^mN_{Y_0\backslash Z_0}\right).$$

(This is the case if, for example, $N_{Y_0\backslash Z_0}=\left(\mathcal{O}_{Y_0}\right)^{c'}$.) Then the natural map

$$\mathcal{D}^{X'}_{\{0\}}\xrightarrow{\varphi}H^0\left(\mathcal{D}^{Z_0/\{0\}}_{Y_0/\{0\}}\right)$$

is an isomorphism.

Proof. As in the proof of Theorem 9.3, we proceed by induction on the order of the neighborhood of 0 in Z'. Using Lemma 5.1, the lemma is then an immediate consequence of the commutative diagram with exact rows:

$$\begin{array}{ccccccc} 0 \to & \left(\mathcal{D}^{X'}_{\{0\}}\right)^s & \longrightarrow & \left(\mathcal{D}^{X'}_{\{0\}}\right)^{s+1} & \longrightarrow & S^{s+1}T_{X',0} & \to 0 \\ & \| & & \downarrow & & \| & \\ 0 \to & H^0\left(\left(\mathcal{D}^{Z_0/\{0\}}_{Y_0/\{0\}}\right)^s\right) & \longrightarrow & H^0\left(\left(\mathcal{D}^{Z_0/\{0\}}_{Y_0/\{0\}}\right)^{s+1}\right) & \longrightarrow & S^{s+1}N_{Y_0\backslash Z_0} & \end{array}$$

We now turn to some more general consequences of Theorem 9.3. First, suppose that in (39) we have a proper smooth family:

$$\begin{array}{ccc} W & \longrightarrow & Z \\ \Big\downarrow{\scriptstyle q_{X'}} & & \Big\downarrow{\scriptstyle p_{X'}} \\ X' & \xrightarrow{j'} & Z' \end{array}$$

We would like to have a criterion to decide when the fibers $W_{x'}$ of $q_{X'}$ are of constant modulus (that is, mutually isomorphic) when x' moves along $\{y'\} \times Z'$. Referring to §4.6 we let

$$\mathcal{Z}_{Y/Y'}^{Z/Z'} = \ker\left(\mathcal{D}\left(\Omega_{Z/Z'}^{\leqslant d}, \Omega_{Y/Y'}^{*}\right) \overset{(\circ\partial)\pm(\partial\circ)}{\longrightarrow} \mathcal{D}\left(\Omega_{Z/Z'}^{\leqslant d}, \Omega_{Y/Y'}^{*}\right)\right).$$

Theorem 10.3. *The family*

$$\begin{array}{ccccc} Y & \rightarrowtail & W & \rightarrowtail & X' \times_{Z'} Z \\ \Big\downarrow{\scriptstyle p_{Y'}} & & \Big\downarrow{\scriptstyle q_{X'}} & & \Big\downarrow{\scriptstyle p_{X'}} \\ Y' & \xrightarrow{i'} & X' & = & X' \end{array}$$

in (39) is expressible as a family

$$\begin{array}{ccccc} Y & \rightarrowtail & X' \times_{Z'} Y & \longrightarrow & X' \times_{Z'} Z \\ \Big\downarrow{\scriptstyle p_{Y'}} & & \Big\downarrow{\scriptstyle q_{X'}} & & \Big\downarrow{\scriptstyle p_{X'}} \\ Y' & \xrightarrow{i'} & X' & = & X' \end{array}$$

if and only if the associated almost multiplicative morphism

$$\mathcal{U}' \xrightarrow{\varphi} (p_{Y'})_* \left(\mathcal{D}_{Y/Y'}^{Z/Z'}\right)$$

lifts to an almost multiplicative map

$$\mathcal{U}' \xrightarrow{\tilde{\varphi}} (p_{Y'})_* \left(\mathcal{Z}_{Y/Y'}^{Z/Z'}\right).$$

(Here the definition of almost multiplicative is extended in the obvious way for a map to the right $\mathcal{U}$-module $\mathcal{Z}_{Y/Y'}^{Z/Z'}$.)

Proof. Suppose first that we have a lifting

$$\mathcal{U}' \xrightarrow{\tilde{\varphi}} (p_{Y'})_* \left(\mathcal{Z}^{Z/Z'}_{Y/Y'}\right).$$

Replace Z in (38) and (39) with $Y \times Z$ and the map

$$Y \xrightarrow{h} Z$$

with

$$\begin{array}{ccc} Y & \xrightarrow{\tilde{h}} & Y \times Z \\ y & \longmapsto & (y, h(y)) . \end{array}$$

The decomposition

$$\Omega^{\leqslant d}_{Y \times Z} = \sum\nolimits_{i+j \leqslant d} p^*_Y \Omega^i_Y \otimes p^*_Z \Omega^j_Z$$

induces a natural map

$$\mathcal{D}\left(\Omega^{\leqslant d}_Z, \Omega^*_Y\right) \to \mathcal{D}\left(\Omega^{\leqslant d}_{Y \times Z}, \Omega^*_Y\right)$$

which in turn induces

$$(p_{Y'})_* \left(\mathcal{Z}^{Z/Z'}_{Y/Y'}\right) \to (p_{Y'})_* \left(\mathcal{D}^{Y \times Z/Z'}_{Y/Y'}\right)$$

To see that the composition

$$\widehat{\varphi} : \mathcal{U}' \to (p_{Y'})_* \left(\mathcal{Z}^{Z/Z'}_{Y/Y'}\right) \to (p_{Y'})_* \left(\mathcal{D}^{Y \times Z/Z'}_{Y/Y'}\right)$$

is almost multiplicative, notice that it is given by lifting vector fields so that they always are subordinate to the foliation of $Y \times X$ with leaves $\{point\} \times X$. Now apply Theorem 9.3 to the almost multiplicative morphism $\widehat{\varphi}$. In the other direction, the product structure allows a globally consistent choice of the local vector fields $\psi(\xi')$ in Definition 9.2.

11 The obstruction pairing

Suppose again that we are given the relative immersion

$$\begin{array}{ccc} Y & \xrightarrow{h} & Z \\ \downarrow{\scriptstyle p_{Y'}} & & \downarrow{\scriptstyle p_{Z'}} \\ Y' & \xrightarrow{h'} & Z' \end{array}$$

and that $p_{Z'}$ is proper and relatively Kähler. Suppose now that we are in the situation of Corollary 9.5 where we have an extension

$$W_{\mathcal{J}'} \overset{q_{\mathcal{J}'}}{\to} X'$$

and $\tilde{\mathcal{J}}' \subseteq \mathcal{J}'$ with $\frac{\mathcal{J}'}{\tilde{\mathcal{J}}'} = \mathcal{O}_{Y'}$.

Suppose further that $\left(\frac{\mathcal{J}'}{\tilde{\mathcal{J}}'}\right)^{\vee}$ is spanned by

$$u' = v' \cdot \xi' \in \mathcal{U}'_{\tilde{\mathcal{J}}'} \tag{44}$$

for some $v' \in \mathcal{U}'_{\mathcal{J}'}$ and some vector field $\xi' \in \mathcal{L}_{Y',Z'}$. We can then compute a relation that must be satisfied by the obstruction element v given by Corollary 9.5(ii). This computation will be carried out in this final section of the paper.

We apply the pairing (30) to a Cech cochain

$$\left\{\chi_{ab} \in (p_{Y'})_* \left(N_{Y\backslash Y'\times_{Z'}Z}|_{V_a \cap V_b}\right)\right\}$$

giving the element v in Corollary 9.5, where, as in Lemma 5.3, we have an open cover $\{X_a\}$ of X such that for each X_a we have local decompositions

$$X_a := Y_0 \times S \times Y' \times Z'$$

which locally simultaneously trivialize the families Y/Y' and Z/Z'. On $X_a \cap X_b$, the product structure associated to each of the two trivializations determines a lifting of

$$\xi' \in \mathcal{L}_{Y',Z'}.$$

The difference of the two liftings gives

$$\theta_{ab}(\xi') \in (p_{X'})_* \left(\mathcal{L}^0_{Y',Z}|_{X_a \cap X_b}\right).$$

We denote by

$$\mathcal{U}' \xrightarrow{u_a} \mathcal{U}|_{X_a}$$

the (multiplicative) lifting morphism corresponding to the trivialization of X_a. By Corollary 9.5

$$\begin{aligned}
\rho \cdot L_{\chi_{ab}} &= \rho \cdot (u_b(u') - u_a(u')) \\
&= \rho \cdot (u_b(v' \cdot \xi') - u_a(v' \cdot \xi')) \\
&= \rho \cdot (u_b(v') \cdot u_b(\xi') - u_a(v') \cdot u_b(\xi')) \\
&= \rho \cdot u(v') \cdot \theta_{ab}(\xi')
\end{aligned}$$

since

$$\rho \cdot u(v') = \rho \cdot u_a(v') = \rho \cdot u_b(v')$$

because liftings to $W_{\mathcal{J}'}$ are already well-defined, independently of local trivializations. Thus, from (30) we have

Theorem 11.1. *Under the decomposition assumption (44), the map*

$$R^{m}\left(p_{Z'}\right)_{*}\left(\Omega_{Z/Z'}^{\geq d}\right)\stackrel{\upsilon\cdot}{\to}R^{m+1}\left(p_{Y'}\right)_{*}\left(\Omega_{Y/Y'}^{*}\right)$$

given by the pairing (28) evaluated on the obstruction class υ is the same as the map

$$\begin{array}{ccccc} R^{m}\left(p_{Z'}\right)_{*}\left(\Omega_{Z/Z'}^{\geq d}\right) & \longrightarrow & R^{m+1}\left(p_{Z'}\right)_{*}\left(\Omega_{Z/Z'}^{d}\right) & \longrightarrow & R^{m+1}\left(p_{Y'}\right)_{*}\left(\Omega_{Y/Y'}^{*}\right) \\ \eta & \longmapsto & \langle\theta_{ab}\mid\rangle\cup\eta & \longmapsto & u\left(v'\right)\left(\langle\theta_{ab}\mid\rangle\cup\eta\right) \end{array}$$

where the first map in the composition is the Kodaira–Spencer map for the family Z/Z' and the second is given by the pairing (29) (induced from the Gauss–Manin connection—see §8).

Corollary 11.2. *Assume the factorization (44). Also assume that $p_{Y'}$ can be factored as*

$$Y\stackrel{\pi}{\to}V'\stackrel{\sigma}{\to}Y'$$

with π proper, smooth, and of relative dimension $d'\leqslant d$ so that the extension $q_{\mathcal{J}'}$ can be factored as

$$W'_{\mathcal{J}'}\stackrel{\pi}{\to}V'_{\mathcal{J}'}\stackrel{\sigma}{\to}X'_{\mathcal{J}'}.$$

Then a particular case of the obstruction map in Theorem 11.1 becomes

$$\begin{array}{c} R^{d-1}\left(p_{Z'}\right)_{*}\left(\Omega_{Z/Z'}^{d+1}\right) \\ \Big\downarrow \text{Kodaira–Spencer} \\ R^{d}\left(p_{Z'}\right)_{*}\left(\Omega_{Z/Z'}^{d}\right) \\ \Big\downarrow \text{pull-back} \\ R^{d}\left(q_{\mathcal{J}'}\right)_{*}\left(\omega_{W'_{\mathcal{J}'}/X'_{\mathcal{J}'}}\right) \\ \Big\downarrow \\ \sigma_{*}\left(\omega_{V'_{\mathcal{J}'}/X'_{\mathcal{J}'}}\otimes R^{d'}\left(\pi\right)_{*}\left(\omega_{W'_{\mathcal{J}'}/V'_{\mathcal{J}'}}\right)\right) \\ \Big\downarrow \text{Gauss–Manin} \\ \sigma_{*}\left(\omega_{V'/Y'}\otimes R^{d}\left(\pi\right)_{*}\left(\omega_{Y/V'}\right)\right) \\ \Big\downarrow \text{integration over fiber} \\ \sigma_{*}\omega_{V'/Y'} \end{array}$$

We first apply Corollary 11.2 in the case $d = d'$, that is, $p_{Y'}$ itself is proper.

Corollary 11.3. ([Ran93], [Kaw95]) *Suppose in Corollary 11.2 we have $d = d'$ and are given that the cohomology class*

$$\gamma := c(Y/Y') \in R^{2c}(p_{Z'})_*(\mathbb{C}_Z)$$

stays of type (c,c). (Recall that $c = n - d$ is the codimension of the fiber of Y/Y' in the corresponding fiber of Z/Z'.) Then the obstruction pairing

$$R^{d-1}(p_{Z'})_*\left(\Omega^{d+1}_{Z/Z'}\right) \to \mathcal{O}_{Y'}$$

vanishes.

Proof. For $\omega \in R^{d-1}(p_{Z'})_*\left(\Omega^{d+1}_{Z/Z'}\right)$ and $\xi' \in \mathcal{L}_{Y',Z'}$,

$$(\xi'(\omega)) \cdot \gamma|_{W_{\mathcal{J}'}} = \langle \theta_{ab} \mid \rangle \cup \omega|_{W_{\mathcal{J}'}} \in R^d\left(q_{W_{\mathcal{J}'}}\right)_*\left(\Omega^d_{W_{\mathcal{J}'}/X'_{\mathcal{J}'}}\right) = \mathcal{O}_{X'_{\mathcal{J}'}}$$

is zero because of the assumption that γ is everywhere of type (c,c). Now apply Corollary 11.2.

Finally we wish to generalize this application of Corollary 11.2. Suppose as above that we have an extension

$$\begin{array}{ccccc} Y & \longrightarrow & W_{\mathcal{J}'} & \rightarrowtail & X \\ \downarrow p_{Y'} & & \downarrow q_{\mathcal{J}'} & & \downarrow p_{X'} \\ Y' & \longrightarrow & X_{\mathcal{J}'} & \rightarrowtail & X' \end{array}$$

and wish to examine the obstruction pairing to extending over $X_{\tilde{\mathcal{J}}'}$. (In particular we continue to assume the factorization (44).) In general, we have an Abel–Jacobi map in homology:

$$\begin{array}{ccc} L_{d-d'}\sigma_*\left(\mathbb{Q}_{V_{\mathcal{J}'}}\right) = \left(R^{d-d'}\sigma_*\left(\mathbb{Q}_{V_{\mathcal{J}'}}\right)\right)^\vee & \longrightarrow & \left(j'^{-1}\left(R^{d+d'}(p_{Z'})_*(\mathbb{Q}_Z)\right)^\vee\right)\Big|_{X'_{\mathcal{J}'}} \\ \gamma & \longmapsto & \pi^{-1}(\gamma) \end{array} \tag{45}$$

The image of (45) must lie in (the dual of) a Hodge sub-structure

$$A_{\mathcal{J}'} \subseteq \left(j'^{-1}\left(R^{d+d'}(p_{Z'})_*(\mathbb{Q}_Z)\right)\right)\Big|_{X'_{\mathcal{J}'}} \tag{46}$$

of type

$$(d,d') + \ldots + (d',d)$$

Theorem 11.4. *Assume that there is a rational sub-Hodge structure of*

$$R^{d+d'}(p_{Z'})_*(\mathbb{Q}_Z)$$

of type

$$(d,d')+\ldots+(d',d)$$

extending the one given by the image of the mapping (46) (via the Gauss–Manin connection). Then the image of the obstruction mapping in Corollary 11.2 lies in the subspace of $\sigma_*(\omega_{V'/Y'})$ *given by the (relatively) exact forms, that is, this subspace maps to zero under the natural map*

$$\sigma_*(\omega_{V'/Y'}) \rightarrow R^{d-d'}\sigma_*(\mathbb{C}_{V'})$$

Proof. For $\omega \in R^{d-1}(p_{Z'})_*\left(\Omega^{d+1}_{Z/Z'}\right)$, we have by Corollary 11.3 that

$$\int_{\pi^{-1}(\gamma)} \langle v\,|\omega\rangle = \int_{\pi^{-1}(\gamma)} u(v')\,\xi'(\omega) = v'\int_{\pi^{-1}(\gamma)} \xi'(\omega)$$

But the hypothesis that the sub-Hodge structure extends over Z implies that

$$\int_{\pi^{-1}(\gamma)} \xi'(\omega) \equiv 0$$

for all $\gamma \in \left(R^{d-d'}\sigma_*(\mathbb{Q}_{V'})\right)^{\vee}$, $\omega \in j^{-1}R^{d-1}(p_{Z'})_*\left(\Omega^{d+1}_{Z/Z'}\right)$ and $\xi' \in \mathcal{L}_{Y',Z'}$.

Unfortunately, we do not, as yet, know any good applications of Theorem 11.4 to deformation theory. There are situations, however, such as the case

$$n=3, \quad d=2, \quad d'=1$$
$$\omega_{Z/Z'} = \mathcal{O}_{Z/Z'}$$

in which the map in Corollary 11.2 seems to carry most of the obstruction theory (see [Cle98] and [CK98]). So hopefully the above formalism will eventually yield generalizations of variational Hodge-conjecture results like Theorem 4.13 of [Cle98].

Appendix A Local coordinates

Let Y_0 be a d-dimensional complex submanifold of the complex n-manifold Z_0. For any $y \in Y_0$, choose a complete set of local coordinates near y in Z_0 made up as follows:

x_k such that $x_k = 0$ for all k define Y_0 in Z_0,

y_j such that $y_j = 0$ for all j define y in Y_0.

For any tuple of non-negative integers K indexed by the set of indices $\{k\}$, we have associated the differential operator ∂_x^K which give the composition of differentiation with respect to each of the coordinates x_k the number of times given by the k-th integer in K. Then the elements of the stalk of $\mathcal{D}^{\perp}_{Z_0/\{0\}}\left(\Omega^p_{Z_0/\{0\}}, \omega_{Y_0/\{0\}}\right)$ can be uniquely written as

$$\sum\nolimits_{J,K,K',|J|+|K|=p} \left\{a_{K,J,K'}(y)\left((dx_K dy_J)^* \, dy\right)\partial_x^{K'}\right\} \tag{47}$$

where $J =$ subset of $\{1, \dots, d\}$, ordered by $<$, $K =$ subset of $\{1, \dots, c\}$, ordered
by $<$,

$$\begin{aligned} dy_J &= dx_{j_1} \wedge \dots \wedge dx_{j_{|J|}} \\ dx_K &= dx_{k_1} \wedge \dots \wedge dx_{k_{|K|}} \end{aligned}$$

and

$$dy = dy_1 \wedge \dots \wedge y_d.$$

For example, the restriction map ρ given above becomes $dy^* \cdot dy$. Notice next that right multiplication by $\partial_x^{K'}$ is identical to right multiplication by the operator $L_{K'}$ given by the composition of (commuting) Lie differentiations with respect to the corresponding $|K'|$ vector fields $\frac{\partial}{\partial x_{k'}}$.

Right multiplication of (47) by $\partial = \partial_{Z_0}$ is given by composition with

$$\sum (dx_k \wedge)\frac{\partial}{\partial x_k} + \sum (dy_j \wedge)\frac{\partial}{\partial y_j},$$

and this action commutes with right multiplication by $\partial_x^{K'}$, so that it suffices to study its action on elements

$$a_{J,K}(y)((dx_K \wedge dy_J)^* \cdot dy)\,. \tag{48}$$

An easy computation gives that multiplying (48) on the right by ∂ gives

$$\begin{aligned} &\left\{(-1)^{|K|-1} \sum_{j\in J} \tfrac{\operatorname{sgn}(j(J-\{j\}))\partial a_{K,J}(y)}{\partial y_j}\left(\left(dx_K dy_{J-\{j\}}\right)^* dy\right)\right\} \\ &+\left\{\sum_{k\in K} \operatorname{sgn}\left(k\left(K-\{k\}\right)\right) a_{J,K}(y)\left(\left(dx_{K-\{k\}} dy_J\right)^* dy\right)\tfrac{\partial}{\partial x_k}\right\}. \end{aligned} \tag{49}$$

Thus, if ∂ is to annihilate a sum of elements (49), $a_{J,K}(y) = 0$ whenever K can be written as $K' \cup \{k\}$, that is, whenever K is not empty. So we are left to compute $\partial(a(y)(dy^* \cdot dy))$, which gives

$$\sum_{j=1,\dots,d} \frac{\pm\partial a(y)}{\partial y_j}\left(\left(dy_{J-\{j\}}\right)^* dy\right)$$

Thus, to get $\partial(a(y)(dy^* \cdot dy)) = 0$, $a(y)$ must be constant in the y-coordinates.

To see that

$$\rho \cdot \wedge^r \mathcal{L}_{Z_0} \cdot \mathcal{U}_{Z_0} = \mathcal{D}^{\perp}_{Z_0/\{0\}} \left(\Omega^{d+r}_{Z_0/\{0\}}, \omega_{Y_0/\{0\}} \right) \tag{50}$$

where $\mathcal{U}_{Z_0}$ is the enveloping algebra of the Lie algebra $\mathcal{L}_{Z_0}$, we need only show how to generate elements

$$(dx_K dy_J)^* \, dy$$

by Lie differentiations and contractions. But this is immediate from the fact that, for example,

$$L_{y_j \frac{\partial}{\partial x_k}}$$

removes dx_k and replaces it with dy_j.

References

[Art68] M. Artin, *On the solutions of analytic equations*, Invent. Math. **51** (1968), 277–291.

[Bor87] A. Borel, *Algebraic D-modules*, Perspectives in Mathematics, Academic Press, Orlando, FL, 1987.

[CK98] H. Clemens and H. Kley, *Counting curves which move with threefolds*, math.AG/9807109, Math Archive, July 1998, Preprint.

[Cle77] H. Clemens, *Degeneration of Kähler manifolds*, Duke Math. Journal **44** (1977), 215–290.

[Cle98] H. Clemens, *Cohomology and obstructions*, math. AG/9809127, Math Archive, September 1998, Preprint.

[GM90] W. Goldman and J. Millson, *The homotopy invariance of the Kuranishi space*, Ill. J. Math. **34**:2 (1990), 337–367.

[Kaw95] Y. Kawamata, *Unobstructed deformations II*, J. Alg. Geom. **4** (1995), 277–279.

[Kol96] J. Kollár, *Rational curves on algebraic varieties*, Springer-Verlag, Berlin, 1996.

[Ran93] Z. Ran, *Hodge theory and the Hilbert scheme*, J. Diff. Geom. **37** (1993), 191–198.

Herb Clemens
Mathematics Department
University of Utah
Salt Lake City, UT 84112
`clemens@math.utah.edu`
`burchard@pobox.com`

Notes on the Construction of the Moduli Space of Curves

Dan Edidin

The purpose of these notes is to discuss the problem of moduli for curves of genus $g \geq 3$[1] and outline the construction of the (coarse) moduli scheme of stable curves due to Gieseker. The notes are broken into four parts.

In Section 1 we discuss the general problem of constructing a moduli "space" of curves. We will also state results about its properties, some of which will be discussed in the sequel.

We begin Section 2 by recalling from [DM] (see also [Vi]) the definition of a groupoid, and define the moduli groupoid of curves, as well as the quotient groupoid of a scheme by a group. We then discuss morphisms and fiber products in the 2-category of groupoids. Once this is in place we state the geometric conditions required for a groupoid to be a stack, and prove that the quotient groupoid of a scheme by a group is a stack. After discussing properties of morphisms of stacks, we define a Deligne–Mumford stack and prove that if a group acts on a scheme so that the stabilizers of geometric points are finite and reduced then the quotient stack is Deligne–Mumford. In the last part of Section 2 we talk about some basic algebro-geometric properties of Deligne–Mumford stacks.

In Section 3 the notion of a stable curve is introduced, and we define the groupoid of stable curves. The groupoid of smooth curves is a sub-groupoid. We then prove that the groupoid of stable curves of genus $g \geq 3$ is equivalent to the quotient groupoid of a Hilbert scheme by the action of the projective

---

The author was partially supported by an NSA grant during the preparation of this work.

[1]Because every curve of genus 1 and 2 has non-trivial automorphisms, the problem of moduli is more subtle in this case than for curves of higher genus.

linear group with finite, reduced stabilizers at geometric points. By the results of the previous section we can conclude the the groupoid of stable curves is a Deligne–Mumford stack defined over $\operatorname{Spec}\mathbb{Z}$ (as is the groupoid of smooth curves). We also discuss the results of [DM] on the irreducibility of the moduli stack.

We begin Section 4 by defining the moduli space of a Deligne–Mumford stack, proving that a geometric quotient of a scheme by a group is the moduli space of the quotient stack. We then discuss the method of geometric invariant theory for constructing geometric quotients for actions of reductive groups. Finally, we briefly outline Gieseker's approach to constructing the coarse moduli space over an algebraically closed field as the quotient of Hilbert schemes of pluricanonically embedded stable curves.

Acknowledgments: These notes are based on lectures the author gave at the Weizmann Institute, Rehovot, Israel in July 1994. It is a pleasure to thank Amnon Yekutieli and Victor Vinnikov for the invitation, and for many discussions on the material in these notes. Special thanks to Angelo Vistoli for a careful reading and many critical comments. Thanks also to Alessio Corti, Andrew Kresch and Ravi Vakil for useful discussions.

1 Basics

Definition 1.1. Let S be a scheme. By a smooth curve of genus g over S we mean a proper, flat, family $C \to S$ whose geometric fibers are smooth, connected 1-dimensional schemes of genus g.

Remark 1.1. By the genus of a smooth, connected curve C over an algebraically closed field, we mean $\dim H^0(C,\omega_C) = \dim H^1(C,\mathcal{O}_C) = g$, where ω_C is the sheaf of regular 1-forms on C. If the ground field is $\mathbb{C}$, then C is a smooth, compact Riemann surface, and the algebraic definition of genus is the same as the topological one.

The basic problem of moduli is to classify curves of genus g. As a start, it is desirable to construct a space $\mathcal{M}_g$ whose geometric points represent all possible isomorphism classes of smooth curves. In the language of complex varieties, we are looking for a space that parametrizes all possible complex structures on a fixed surface of genus g.

However, as modern (post-Grothendieck) algebraic geometers we would like $\mathcal{M}_g$ to have further functorial properties. In particular given a scheme S, a curve $C \to S$ should correspond to a morphism of S to $\mathcal{M}_g$ (when S is the spectrum of an algebraically closed field this is exactly the condition of the previous paragraph).

In the language of functors, we are trying to find a scheme $\mathcal{M}_g$ which represents the functor $\mathcal{F}_{\mathcal{M}_g}$: Schemes $\to$ Sets which assigns to a scheme S the set of isomorphism classes of smooth curves of genus g over S.

Unfortunately, such a moduli space cannot exist because some curves have non-trivial automorphisms. As a result, it is possible to construct non-trivial families $C \to B$ where each fiber has the same isomorphism class. Since the image of B under the corresponding map to the moduli space is a point, if the moduli space represented the functor $\mathcal{F}_{\mathcal{M}_g}$ then $C \to B$ would be isomorphic to the trivial product family.

Given a curve X and a non-trivial (finite) group G of automorphisms of X we construct a non-constant family $C \to B$ where each fiber is isomorphic to X as follows. Let B' be a scheme with a free G action, and let $B = B'/G$ be the quotient. Let $C' = B' \times X$. Then G acts on C' by acting as it does on B' on the first factor, and by automorphism on the second factor. The quotient C'/G is a family of curves over B. Each fiber is still isomorphic to X, but C will not in general be isomorphic to $X \times B$.

There is however, a *coarse* moduli scheme of smooth curves. By this we mean

Definition 1.2. [GIT, Definition 5.6, p. 99] There is a scheme $\mathcal{M}_g$ and a natural transformation of functors $\phi : \mathcal{F}_{\mathcal{M}_g} \to \mathrm{Hom}\,(\cdot, \mathcal{M}_g)$ such that

1. For any algebraically closed field k, the map

$$\phi : \mathcal{F}_{\mathcal{M}_g}(\Omega) \to \mathrm{Hom}\,(\Omega, \mathcal{M}_g)$$

is a bijection, where $\Omega = \mathrm{Spec}\, k$.

2. Given any scheme M and a transformation $\psi : \mathcal{F}_{\mathcal{M}_g} \to \mathrm{Hom}\,(\cdot, M)$ there is a unique transformation $\chi : \mathrm{Hom}\,(\cdot, \mathcal{M}_g) \to \mathrm{Hom}\,(\cdot, M)$ such that $\psi = \chi \circ \phi$.

The existence of ϕ means that given a family of curves $C \to B$ there is an induced map to $\mathcal{M}_g$. We do not require however, that a map to moduli gives a family of curves (as we have already seen a non-constant family with iso-trivial fibers). However, condition 1 says that giving a curve over an algebraically closed field is equivalent to giving a map of that field into $\mathcal{M}_g$.

Condition 2 is imposed so that the moduli space is a universal object.

In his book on geometric invariant theory Mumford proved the following theorem.

Theorem 1.1. [GIT, Chapter 5]) *Given an algebraically closed field k there is a coarse moduli scheme $\mathcal{M}_g$ of dimension $3g-3$ defined over* $\mathrm{Spec}\, k$, *which is quasi-projective and irreducible.*

The proof of this theorem will be subsumed in our general discussion of the construction of $\overline{\mathcal{M}}_g$, the moduli space of stable curves.

A natural question to ask at this point, is whether $\mathcal{M}_g$ is complete (and thus projective). The answer is no. It is quite easy to construct curves $C \to \mathrm{Spec}\, \mathcal{O}$ with $\mathcal{O}$ a D.V.R. which has function field K, where C is smooth, but the fiber over the residue field is singular. Since C is smooth, the restriction $C_K \to \mathrm{Spec}\, K$ is a smooth curve over $\mathrm{Spec}\, K$, so there is a map $\mathrm{Spec}\, K \to \mathcal{M}_g$. The existence of such a family does not prove anything, since we must show that we cannot replace the special fiber by a smooth curve. The total space $\tilde{C}$ of a family with modified special fiber is birational to C. Since C and $\tilde{C}$ are surfaces (being curves over 1-dimensional rings) there must be a sequence of birational transformations (centered in the special fiber) taking one to the other. It appears therefore, that it suffices to construct a family $C \to \operatorname{Spec} \mathcal{O}$ such that no birational modification of C centered in the special fiber will make it smooth. Unfortunately, because $\mathcal{M}_g$ is only a coarse moduli scheme, the existence of such a family does not prove that $\mathcal{M}_g$ is incomplete. The reason is that there may be a map $\operatorname{Spec} \mathcal{O} \to \mathcal{M}_g$ extending the original map $\operatorname{Spec} K \to \mathcal{M}_g$ without there being a family of smooth curves $\tilde{C} \to \operatorname{Spec} \mathcal{O}$ extending $C_K \to \operatorname{Spec} K$. However, we will see when we discuss the valuative criterion of properness for Deligne–Mumford stacks that it suffices to show that for every finite extension $K \subset K'$ we cannot complete the induced family of smooth curves $C_{K'} \to \operatorname{Spec} K'$ to a family of smooth curves $C' \to \operatorname{Spec} \mathcal{O}'$, where $\mathcal{O}'$ is the integral closure of $\mathcal{O}$ in K'.

Example 1.1. The following family shows that $\mathcal{M}_3$ is not complete. It can be easily generalized to higher genera. Consider the family $x^4 + xyz^2 + y^4 + t(z^4 + z^3x + z^3y + z^2y^2)$ of quartics in $\mathbf{P}^2 \times \operatorname{Spec} \mathcal{O}$ where $\mathcal{O}$ is a D.V.R. with uniformizing parameter t. The total space of this family is smooth, but over the closed point the fiber is a quartic with a node at the point $(0 : 0 : 1) \in \mathbf{P}^2$. Moreover, even after base change, any blow-up centered at the singular point of the special fiber contains a (-2) curve, so there is no modification that gives us a family of smooth curves.

Since $\mathcal{M}_g$ is not complete, a natural question is to ask whether it is affine. The answer again is no. This follows from the fact that $\mathcal{M}_g$ has a projective compactification in the moduli space of abelian varieties, such that boundary has codimension 2. In particular, this means that there are complete curves in $\mathcal{M}_g$. On the other hand, Diaz proved the following theorem ([Di]).

Theorem 1.2. *Any complete subvariety of $\mathcal{M}_g$ has dimension less than $g - 1$.*

It is not known how close this bound is to being sharp.

Finally we state a spectacular theorem proved by Harris-Mumford, Harris and Eisenbud-Harris.

Theorem 1.3. *For $g > 23$, $\mathcal{M}_g$ is of general type.*

Remark 1.2. The importance of this theorem is that until its proof, there was some belief that $\mathcal{M}_g$ was rational, or at least unirational. The reason is that for $g = 3, 4, 5$ it is very easy to show that $\mathcal{M}_g$ is unirational and not much more difficult to show that it is in fact rational. For $g \leq 10$, the unirationality of $\mathcal{M}_g$ had been affirmed by the Italian school. (For a nice discussion of the rationality of moduli spaces of curves of low genus, see

Dolgachev's article [Do].)

2 Stacks

Let S be a scheme, and let $\mathcal{S} = (Sch/S)$ be the category of schemes over S.

2.1 Groupoids

Definition 2.1. A category over S is a category F together with a covariant functor $p_F : F \to \mathcal{S}$. If B is an object of $\mathcal{S}$ we say X lies over B if $p_F(X) = B$.

Definition 2.2. (see also [Vi, Definition 7.1]) If (F, p_F) is a category over S, then it is a groupoid over S if the following conditions hold:

(1) If $f : B' \to B$ is a morphism in $\mathcal{S}$, and X is an object of F lying over B, then there is an object X' over B' and a morphism $\phi : X' \to X$ such that $p_F(\phi) = f$.

(2) Let X, X', X'' be objects of F lying over B, B', B'' respectively. If $\phi : X' \to X$ and $\psi : X'' \to X$ are morphisms in F, and $h : B' \to B''$ is a morphism such that $p_F(\psi) \circ h = p_F(\phi)$ then there is a unique morphism $\chi : X' \to X''$ such that $\psi \circ \chi = \phi$ and $p_F(\chi) = h$.

Remark 2.1. Condition (2) implies that a morphism $\phi : X' \to X$ of objects over B' and B respectively is an isomorphism if and only if $p_F(\phi) : B' \to B$ is an isomorphism. (To see that $p_F(\phi)$ being an isomorphism is sufficient to ensure that ϕ is an isomorphism, apply condition (2) where one of the maps is $p_F(\phi)$ and the other the identity, and lift $p_F(\phi)^{-1} : B \to B'$ to $\phi^{-1} : X \to X'$. The other direction is trivial.) Define $F(B)$ to be the subcategory consisting of all objects X such that $p_F(X) = B$ and morphisms f such $p_F(f) = id_B$. Then $F(B)$ is a groupoid; i.e. a category where all morphisms are isomorphisms. This is the reason we say that F is a groupoid over S.

Condition (2) also implies that the object X' over B' in condition (1) is unique up to canonical isomorphism. This object will be called the pullback of X via f and denoted f^*X. Moreover, if $X \xrightarrow{s} X'$ is a morphism in $F(B)$ then there is a canonical morphism $f^*X \xrightarrow{f^*s} f^*X'$. In other words, given a morphism $B' \xrightarrow{f} B$ of S-schemes, there is an induced functor $f^* : F(B) \to F(B')$. Note that f^* is actually a *covariant* functor.

Example 2.1. If $F : \mathcal{S} \to$ Sets is a contravariant functor, then we can associate a groupoid (also called F) whose objects are pairs (B, β) where B is an object of $\mathcal{S}$ and $\beta \in F(B)$. A morphism $(B', \beta') \to (B, \beta)$ is an S-morphism $f : B' \to B$ such that $F(f)(\beta) = \beta'$. In this case $F(B)$ in the groupoid sense is just the set $F(B)$ in the functor sense; i.e. all morphisms in the groupoid $F(B)$ are the identities.

In particular, if X is any S-scheme then its functor of points gives a groupoid $\underline{X}$. Objects are X-schemes, i.e. morphisms $B \xrightarrow{\phi} X$, and a morphism from $B' \xrightarrow{\phi'} X'$ to $B \xrightarrow{\phi} X$ is a morphism $B' \xrightarrow{f} B'$ such that $\psi f = \psi'$. The functor $p_{\underline{X}}$ simply forgets the X-structure, and views schemes and morphisms as being over S.

Example 2.2. If X/S is a scheme and G/S is a flat group scheme (of finite type) acting on the left on X then we define the quotient groupoid $[X/G]$ as follows. The sections (i.e. objects) of $[X/G]$ over B are G-principal bundles $E \to B$ together with a G-equivariant map $f : E \to X$. A morphism from $E' \to B'$ with equivariant map $f' : E' \to B'$ to $E \to B$ is a cartesian diagram

$$\begin{array}{ccc} E' & \xrightarrow{g} & E \\ \downarrow & & \downarrow \\ B' & \to & B \end{array}$$

where g is an equivariant map such that $gf = f'$. If the action is free and a quotient scheme X/G exists, then there is an equivalence of categories between $[X/G]$ and the groupoid $\underline{X/G}$.

Example 2.3. The central example in these notes is the moduli groupoid $F_{\mathcal{M}_g}$ defined over $\operatorname{Spec}\mathbb{Z}$. The objects of $F_{\mathcal{M}_g}$ are smooth curves as defined in Part 1. A morphism from $X' \to B'$ to $X \to B$ is a cartesian diagram

$$\begin{array}{ccc} X' & \to & X \\ \downarrow & & \downarrow \\ B' & \to & B \end{array}$$

The functor $F_{\mathcal{M}_g} \to \operatorname{Sch}\mathbb{Z}$ sends $X \to B$ to B. We will eventually prove that $F_{\mathcal{M}_g}$ is a quotient groupoid as in the previous example.

A related groupoid is the universal curve $F_{\mathcal{C}_g}$, also defined over $\operatorname{Spec}\mathbb{Z}$. Objects of $F_{\mathcal{C}_g}$ are smooth curves $X \to B$ together with a section $\sigma : B \to X$. A morphism is a cartesian diagram which is compatible with the sections.

Remark 2.2. [Warning] The groupoid we have just defined is not the groupoid associated to the moduli functor we defined in Part 1. The groupoid here is not a functor, since if X/B is a curve with non-trivial automorphisms $F(B)$ will not be a set because there are morphisms which are not identities. (A set is a groupoid where all the morphisms are identities.)

2.2 Morphisms of groupoids

Definition 2.3. (a) If (F_1, p_{F_1}) and (F_2, p_{F_2}) are groupoids over S then a morphism $F_1 \to F_2$ is a functor $p : F_1 \to F_2$ such that $pp_{F_2} = p_{F_1}$.

(b) A morphism p is called an isomorphism if it is an equivalence of categories.

Example 2.4. The functor $F_{\mathcal{C}_g} \to F_{\mathcal{M}_g}$ defined by forgetting the section is a morphism of groupoids.

Here are some more subtle examples:

Example 2.5. If $f : X \to Y$ is a morphism of schemes then it induces a morphism of groupoids $\underline{X} \xrightarrow{p} \underline{Y}$ in a fairly obvious way: Objects of $\underline{X}$ are X-schemes, which, via the morphism f, can be viewed as Y-schemes, i.e., objects of $\underline{Y}$. Thus, $p(B \xrightarrow{u} X) = B \xrightarrow{fu} Y$. If $s : B \to B'$ is a morphism of X-schemes, then $p(s)$ is the morphism s viewed now as a morphism of Y-schemes.

Conversely, if $p : \underline{X} \to \underline{Y}$ is a morphism of groupoids over $\mathcal{S}$ then, because p preserves the projection to $\mathcal{S}$, $p(X \xrightarrow{id} X) = X \xrightarrow{f} Y$ for some morphism f. Yoneda's lemma implies that p is induced by the morphism of schemes $f : X \to Y$.

More generally, if F is a groupoid and B is a scheme then giving a morphism $\underline{B} \xrightarrow{p} F$ is equivalent to giving an object X in $F(B)$. The object X is simply $p(B \xrightarrow{id} B)$.

Example 2.6. We can view the category $\mathcal{S}$ of S-schemes (which is trivially a groupoid over $\mathcal{S}$) as the groupoid $\underline{S}$. If F is a groupoid over $\mathcal{S}$ then functor p_F is a morphism of groupoids $F \to \underline{S}$.

Example 2.7. Let X/S be a scheme and G/S a group scheme acting on the left on X. Then we can define a morphism $p : \underline{X} \to [X/G]$ as follows: If $B \xrightarrow{s} X$ is an object of $\underline{X}(B)$ then $p(s)$ is the bundle $G \times B \to B$ where G acts by left translation on G and trivially on B. The equivariant map $G \times B \to X$ is given by the formula $(g, b) \mapsto gs(b)$. If $f : B' \to B$ is a morphism in $\underline{X}$ then $p(f)$ is the cartesian diagram

$$\begin{array}{ccc} G \times B' & \xrightarrow{id \times f} & G \times B \\ \downarrow & & \downarrow \\ B' & \xrightarrow{f} & B \end{array}$$

Remark 2.3. Because isomorphisms of stacks are defined as equivalences of categories some subtleties arise. In particular an isomorphism $p : F_1 \to F_2$ need not have an inverse; i.e. there need not be a functor $q : F_2 \to F_1$ such that $pq = id_{F_2}$ and $qp = id_{F_1}$. However, any equivalence of categories has a quasi-inverse; that is a functor $q : F_2 \to F_1$ such that pq (resp. qp) are naturally isomorphic to the identity functors 1_{F_2} (resp. 1_{F_2}). Moreover, if $p : F_2 \to F_1$ is a equivalence which is a morphism of $\mathcal{S}$-groupoids, then there is a quasi-inverse which is a morphism of groupoids.

The usual way to describe this point is to say that groupoids over $\mathcal{S}$ form a 2-category. This 2-category has objects, which are the groupoids; 1-morphisms, which are functors, and 2-morphisms, which are natural isomorphisms of functors. In other words, the category of groupoids contains extra information about *isomorphisms* between morphisms.

For the most part, the fact that groupoids (and thus stacks) form a 2-category will not require too much thought but there are a few situations where it is relevant. For example, as we discuss below, a cartesian diagram commutes only up to homotopy. A more geometric situation is the valuative criterion, where (see Theorem 2.2) we require two extensions of morpisms be isomorphic.

The following proposition shows that the notion of an isomorphism of groupoids is an extension of our notion of an isomorphism of schemes.

Proposition 2.1. *Let X and Y be S-schemes. Then there is an isomorphism $X \xrightarrow{f} Y$ as S-schemes if and only if there is an equivalence of $\mathcal{S}$ groupoids $\underline{X} \xrightarrow{p} \underline{Y}$.*

Proof. If f is an isomorphism, then the induced functor $\underline{X} \xrightarrow{p} \underline{Y}$ is in fact a strong equivalence; i.e. the functor $\underline{Y} \xrightarrow{q} \underline{X}$ induced by f^{-1} has the property that $pq = id_Y$ and $qp = id_X$.

Conversely, suppose that $\underline{X} \xrightarrow{p} \underline{Y}$ is an isomorphism of groupoids with quasi-inverse $\underline{Y} \xrightarrow{q} \underline{X}$. As we saw in Example 2.5 the functor p (resp. q) is induced by a morphism $f : X \to Y$ (resp. $g : Y \to X$). Then $qp(X \xrightarrow{id} X) = X \xrightarrow{gf} X$. Since q and p are equivalences $X \xrightarrow{id} X$ and $X \xrightarrow{gf} X$ are isomorphic as X-schemes. Hence $gf : X \to X$ is an automorphism. Likewise, $fg : Y \to Y$ is also an automorphism. Therefore, $f : X \to Y$ must be an isomorphism. □

Remark 2.4. If B is a scheme then from now on we will use the simpler notation $B \to F$ (resp. $F \to B$) to refer to a morphism $\underline{B} \to F$ (resp. $F \to \underline{B}$).

2.3 Fiber products and cartesian diagrams

Let F and G be groupoids over S. If $f : F \to G$ and $h : H \to G$ are morphisms of groupoids, then we define the fiber product $F \times_G H$ to be the following S-groupoid. Objects are triples (x, y, ψ) where $(x, y) \in F(B) \times H(B)$ and $\psi : g(x) \to h(y)$ is an isomorphism in $G(B)$. Here B is a fixed scheme in $\mathcal{S}$.

Suppose (x', y', ψ') is another object with $(x', y') \in F(B')$ and $\psi' : g(x) \to h(y)$ an isomorphism in $G(B')$. A morphism from (x', y', ψ') to (x, y, ψ) is a pair of morphisms $x' \xrightarrow{\alpha} x$, $y' \xrightarrow{\beta} y$ lying over the same morphism $B' \to B$, such that $\psi \circ f(\alpha) = g(\beta) \circ \psi'$.

By construction there are obvious functors $p : F \times_G H \to F$ and $q : F \times_G H \to H$. Note however, that the diagram

$$\begin{array}{ccc} F \times_G H & \to & H \\ \downarrow & & \downarrow \\ F & \to & G \end{array}$$

does not strictly commute, since $fp(x, y, \psi) = f(x)$ and $gq(x, y, \psi) = g(y)$. The objects $f(x)$ are $g(y)$ isomorphic but not necessarily equal. There is however a natural isomorphism between the functors fp and gq. Following the language of 2-categories we say that such a diagram is 2-commutative. More generally given a 2-commutative diagram of groupoids

$$\begin{array}{ccc} T & \to & H \\ \downarrow & & \downarrow \\ F & \to & G \end{array}$$

there is a morphism $T \to F \times_G H$ which is unique up to canonical isomorphism. If this morphism is an isomorphism then we say the diagram is *cartesian.*

Example 2.8. If X, Y, Z are schemes then $\underline{X} \times_{\underline{Z}} \underline{Y}$ is isomorphic $\underline{X \times_Y Z}$, so our notion of fiber product is an extension of the usual one for schemes.

Remark 2.5. Despite the subtleties, this notion is correct for considering base change. In particular, suppose that F is an S-groupoid and $T \to S$ a morphism of schemes. Then if $B \to T$ is a T-scheme, one can check that the groupoids $F(B)$ and $(F \times_S T)(B)$ are equivalent; i.e. $F \times_S T$ is the T-groupoid obtained by base change to T.

2.4 Definition of a stack

Let (F, p_F) be an S-groupoid. Let B be an S-scheme and let X and Y be any objects in $F(B)$. Define a contravariant functor $\underline{\mathrm{Iso}}_B(X,Y) : (\mathrm{Sch}/B) \to (\mathrm{Sets})$ by associating to any morphism $f : B' \to B$ the set of isomorphisms in $F(B')$ between f^*X and f^*Y. Let $B' \xrightarrow{f} B$ and $B'' \xrightarrow{g} B$ be B-schemes. If $h : B'' \to B'$ is a morphism of B-schemes (i.e. $g = fh$) then by construction of the pullback, there are canonical isomorphisms $\psi_X : g^*X \to h^*f^*X$ and $\psi_Y : g^*Y \to h^*f^*Y$. We define a map $\underline{\mathrm{Iso}}_B(X,Y)(B') \to \underline{\mathrm{Iso}}_B(X,Y)(B'')$ as follows: If $f^*X \xrightarrow{\phi} f^*Y$ is an isomorphism then (since $h^* : F(B') \to F(B'')$ is a functor) we obtain an isomorphism $h^*f^*X \xrightarrow{h^*\phi} h^*f^*Y$. The composite, $\psi_Y^{-1} \circ h^*\phi \circ \phi_X$ is the image of ϕ in $\underline{\mathrm{Iso}}_B(X,Y)(B'')$.

If $X = Y$ then $\underline{\mathrm{Iso}}_B(X,X)$ is the functor whose sections over B' mapping to B are the automorphisms of the pull-back of X to B'.

In the case of curves of genus $g \geq 2$, Deligne and Mumford proved that $\underline{\mathrm{Iso}}_B(X,Y)$ is represented by a scheme $\mathbf{Iso}_B(X,Y)$, because X/B and Y/B have canonical polarizations ([DM, p. 84]). When $X = Y$ then Deligne and Mumford prove directly that the $\mathbf{Iso}_B(X,X)$ is finite and unramified over B ([DM, Theorem 1.11]). Applying the theorem to $B = \operatorname{Spec} k$, where k is an algebraically closed field, this theorem proves that every curve has a finite automorphism group.

The scheme $\mathbf{Iso}_B(X,X)$ is naturally a group scheme over B. However, in general it will not be flat over B. For example, if X/B is a family of curves, the number of points in the fibers of $\mathbf{Iso}_B(X, X)$ over B will jump over the points $b \in B$ where the fiber X_b has non-trivial automorphisms.

Definition 2.4. A groupoid (F, p_F) over S is a stack

(1) If $\underline{\text{Iso}}_B(X, Y)$ is a sheaf in the étale topology for all B, X and Y.

(2) If $\{B_i \to B\}$ is a covering of B in the étale topology, and X_i is a collection of objects in $F(B_i)$ with isomorphisms

$$\phi_{ij} : X_{j|B_i \times_B B_j} \to X_{i|B_i \times_B B_j}$$

in $F(B_i \times_B B_j)$ satisfying the cocycle condition. Then there is an object $X \in F(X)$ with isomorphisms $X_{|B_i} \stackrel{\simeq}{\to} X_i$ inducing the isomorphisms ϕ_{ij} above.

Note that F is a groupoid associated to a functor. Then conditions (1) and (2) just assert that the functor is a sheaf in the étale topology. A functor represented by a scheme will be a stack, since condition (1) is trivially satisfied and (2) is equivalent to saying that the functor of points is a sheaf in the étale topology. The moduli functor we defined in Part 1 is not a stack, since it doesn't satisfy condition (2). In particular, as noted above, given a curve C with automorphism group G and $B' \to B$ a Galois cover with group G, there are two ways to descend the family $C \times B'/B'$ to a family over B, so a section of F over B is not determined by its pull-back to an étale cover.

However, the moduli groupoid defined above is a stack. We will not prove this here, but instead we will prove that the moduli groupoid is the quotient groupoid of a scheme by $\mathbf{PGL}\,(N+1)$.

Proposition 2.2. (cf. [Vi, Example 7.17]) *Let G/S be a smooth*[2] *affine group scheme over X. Then the groupoid $[X/G]$ defined above is a stack.*

Proof. Let e, e' be sections of $[X/G](B)$ corresponding to principal bundles $E \to B$ and $E' \to B$ and G-maps $f : E \to X$ and $f' : E' \to X$. Then $\underline{\text{Iso}}_B(e, e')$ is the étale sheaf which is the quotient of $(X \times_{X \times X} E \times_B E')$ by the free product action of G. Moreover, descent theory shows that this sheaf is in fact represented by a scheme.

When $E = E'$ and $f = f'$ then the isomorphisms correspond to elements of G which preserve f. In other words $\mathbf{Iso}_B(e, e)$ is the is the stabilizer of the $G-$map $f : E \to X$ (see [GIT, Definition 0.4] for the definition of stabilizer).

Since any principal bundle $E \to B$ is locally trivial in the étale topology it determines descent data as follows: Let $\{B_i \stackrel{p_i}{\to} B\}$ be an étale cover on which $E \to B$ is trivial. Then we have equivariant isomorphisms $\phi_i : p_i^* E \to G \times B_i$. If ϕ_{ij} is the pullback of ϕ_i to $B_i \times_B B_j$, then the ϕ_{ij}'s satisfy the cocycle condition; i.e. $\phi_{ij}\phi_{ik} = \phi_{jk}$.

---

[2]In fact the proposition holds for flat affine group schemes as well.

Descent theory gives us the opposite direction. Given principal bundles (not necessarily trivial) $E_i \to B_i$ and isomorphisms of $E_{i|B_i \times_B B_j} \to E_{j|B_i \times_B B_j}$ satisfying the cocycle condition, there is a principal bundle $E \to B$ such that $E_i = p_i^* E$. This is condition (2) in the definition of a stack. □

Example 2.9. If F, G, H are stacks over S then one can check that the fiber product $F \times_G H$ is also a stack.

2.5 *Representable morphisms*

Most of the material in this section is taken from [DM, Section 4].

Definition 2.5. A morphism $f : F \to G$ of stacks is said to be representable if for any map of a scheme $B \to G$ the fiber product $F \times_G B$ is isomorphic to a stack associated to a scheme.

Remark 2.6. The definition of representable morphism is the one given by Deligne–Mumford in [DM]. Artin, who considered a larger category of stacks, extended the definition to require that if $B \to F$ is a map of an algebraic space (which we define below) then the the fiber product $F \times_G B$ is also an algebraic space. For the stacks we consider here, the Deligne–Mumford definition suffices.

Example 2.10. Let G/S be a smooth affine group scheme acting on X. We saw that there is a projection morphism $X \to [X/G]$ which associates to any X-scheme $Z \xrightarrow{s} X$ the principal bundle $G \times Z$ with the equivariant map given by $(g, z) \mapsto gs(z)$. Suppose $B \to [X/G]$ is a morphism from a scheme corresponding to a principal bundle $E \to B$ with equivariant map $E \to X$. Let us consider the category $B \times_{[X/G]} X$. Its objects are triples $\{B_1, B_2, \psi\}$ where $B_1 \to B$, $B_2 \to X$ and ψ is an isomorphism between the images in $[X/G]$. Since B_1 and B_2 lie over the same object in the base category, there is a scheme B' such that $B_1 = B_2 = B'$. If $B' \xrightarrow{f} B$ is a morphism then the image of B' is the principal bundle $f^*E \to B'$ with equivariant map $f^*E \to E \to X$. Thus objects of $B \times_{[X/G]} X$ are triples $(B' \xrightarrow{f} B, B' \xrightarrow{s} X, \psi)$ where $\psi : G \times B' \xrightarrow{\sim} f^*E$ is a trivialization. The section $b \in B' \mapsto (1_G, b')$ gives a morphism $B' \to f^*E \to E$. Conversely, suppose we are given a morphism $B' \to E$. Let $f : B' \to E \to B$ be the composite morphism. The diagonal gives a canonical trivialization of the bundle $E \times_B E \to E$ and hence of $f^*E \to B'$. Thus we can factor the morphism $B' \to E$ as $B' \to f^*E \to E$. A similar analysis shows that morphisms in $B \times_{[X/G]} X$ correspond to morphisms of E-schemes, so $B \times_{[X/G]} X$ is isomorphic to $\underline{E}$. Therefore, the morphism $X \to [X/G]$ is representable.

Example 2.11. The morphism $F_{\mathcal{C}_g} \to F_{\mathcal{M}_g}$ forgetting the section is also representable. If $B \to F_{\mathcal{M}_g}$ corresponds to a smooth curve $C \to B$ then

$B \times_{F\mathcal{M}_g} F_{\mathcal{C}_g}$ is represented by the scheme C.

Let **P** be a property of morphisms of schemes that is stable under base change. Most properties of morphisms of schemes have this ability. For example, finite type, separated, proper, affine, flat, smooth (hence étale), etc.

Definition 2.6. [DM, Definition 4.3] A representable morphism of stacks $f : F \to G$ has property **P** if for all maps of scheme $B \to G$ the corresponding morphism of schemes $F \times_G B \to B$ has property **P**.

Example 2.12. The projection morphism $X \to [X/G]$ in Example 2.10 is smooth since for any $B \to [X/G]$ the corresponding map $E \to B$ is smooth because E is a principal bundle over E.

2.6 Definition of a Deligne–Mumford stack

Definition 2.7. A stack is Deligne–Mumford if

(1) The diagonal $\Delta_F : F \to F \times_S F$ is representable, quasi-compact and separated. If the diagonal is proper, then we say that the stack is *separated.*

(2) There is a scheme U and an étale surjective morphism $U \to F$. Such a morphism $U \to F$ is called an (étale) atlas.

Remark 2.7. The representability of the diagonal automatically implies that the functor $\underline{\mathrm{Iso}}_B(X, Y)$ is representable. The reason is that if X and Y are objects in $F(B)$, then $\underline{\mathrm{Iso}}_B(X, Y)$ is represented by the fiber product $(B \times B)_{F \times F} F$ where the map $B \times B \to F \times F$ is the product of the the maps $B \to F$ corresponding to the objects X and Y.

Remark 2.8. In [DM], such a stack is called an *algebraic* stack. To conform to current terminology we use the term Deligne–Mumford stack. A more general class of stacks was studied by Artin, and they are now called Artin stacks. The basic difference is that an Artin stack need only have a *smooth* atlas. An *algebraic space* is defined as an étale sheaf with an étale cover by a scheme. This is the same as a Deligne–Mumford stack where the diagonal is an embedding. An algebraic space is separated if the diagonal is a closed embedding.

Remark 2.9. Condition (1) above is equivalent to the following condition:

(1′) Every morphism $B \to F$ from a scheme is representable, so condition (2) makes sense.

(This fact is stated in [DM] and proved in [Vi, Prop 7.13]).

Condition (2) asserts the existence of a universal deformation space for deformations over Artin rings.

Remark 2.10. (Separated morphisms) The proof of the following lemma can be found in [L-MB].

Lemma 2.1. *Let $f : F \to G$ be a morphism of stacks satisfying property (1) above. Then the diagonal $\Delta_{F/G} : F \to F \times_G F$ is representable.*

As a consequence of the lemma we make the following extension of the notion of separated morphisms of schemes.

Definition 2.8. A morphism of Deligne–Mumford stacks is separated if $\Delta_{F/G}$ is proper.

Vistoli also proves the following proposition:

Proposition 2.3. [Vi, Prop 7.15] *The diagonal of a Deligne–Mumford stack is unramified.*

As a consequence of this proposition we can prove [Vi, p. 666]

Corollary 2.1. *If F is a Deligne–Mumford stack, B quasi-compact, and $X \in F(B)$, then X has only finitely many automorphisms.*

Remark 2.11. There are Artin stacks which are not Deligne–Mumford where each object has a finite automorphism group. In this case the diagonal is quasi-finite but ramified. Objects in the groupoid have *infinitesimal automorphisms.* This phenomenon only occurs in characteristic p, because all group schemes of finite type over a field of characteristic 0 are smooth.

Proof. Let $B \to F$ be a map corresponding to X, and let $B \to F \times_S F$ be the composition with diagonal. The pullback $B \times_{F \times_S F} F$ can be identified with scheme $\underline{\mathrm{Iso}}_B(X, X)$. Since F is a Deligne–Mumford stack the map $\underline{\mathrm{Iso}}_B(X, X)$ is unramified over X. Furthermore, since B is quasi-compact, the map $\underline{\mathrm{Iso}}_B(X, X) \to X$ can have only finitely many sections. Therefore, X has only finitely many automorphisms over B. □

The following theorem is stated (but not proved) in [DM, Theorem 4.21]. We give the proof below with a slight additional assumption.

Theorem 2.1. *Let F be a quasi-separated stack over a Noetherian scheme S. Assume that*

(1) The diagonal is representable and unramified,

(2) There exists a scheme U of finite type over S and a smooth surjective S-morphism $U \to F$,

Then F is a Deligne–Mumford stack.

Remark 2.12. This theorem says that condition (1) and the existence of a versal deformation space (condition (2)) are actually equivalent to the existence of a universal deformation space.

Remark 2.13. We give the proof below under the additional assumption that the residue fields of the closed points of S are perfect. In particular we prove the theorem for stacks of finite type over $\mathrm{Spec}\,\mathbb{Z}$. Using the theorem we will prove that the stack of stable curves is a Deligne–Mumford stack of finite type over $\mathrm{Spec}\,\mathbb{Z}$. By assumption F is of finite type over S, so it is relatively straightforward to reduce to the case that F is actually of finite type over $\operatorname{Spec}\mathbb{Z}$. Thus, the general statement can be reduced to the case we prove. However, we do not give the details here.

Proof. The only thing to prove is that F has an étale atlas of finite type over S. Let $u \in U$ be any closed point. Set $U_u = \delta^{-1}(u \times_S u) = u \times_F U$. Let $z \in U_u$ be a closed point which is separable (i.e. étale) over u. (The set of such closed points is dense in a smooth variety.) Since U_u is smooth, the point z is cut out by a regular sequence in the local ring of U_u at z.

The diagonal $\delta : F \to F \times_S F$ is unramified. Thus, the map $U_u \to u \times_S U$ obtained by pulling back the morphisms $u \times_S U \to F \times_S F$ along the diagonal is unramified. We assume U is of finite type and that the residue fields of S are perfect. Thus, $k(u)$ is a finite, hence separable, extension of the residue field of its image in S. Hence the morphism $u \times_S U \to U$ is unramified and so is the composition $U_u \to u \times_S U \to U$.

Let x be the image of z in U. By [EGA4, 18.4.8] there are étale neighborhoods W' and U' of x and z respectively and a closed immersion $W' \hookrightarrow U'$ such that the diagram commutes

$$\begin{array}{ccc} z' \in W' & \hookrightarrow & U' \\ {\scriptstyle \text{étale}} \downarrow & & \downarrow {\scriptstyle \text{étale}} \\ z \in U_u & \to & x \in U \end{array}$$

Let z' be any point lying over z. Let Z_u be the closed subscheme of U' defined by lifts to $\mathcal{O}'_U$ of the local equations for $z' \in W'$. By construction, Z_u intersects U' transversally at z'. We will show that the induced morphism $Z_u \to F$ is étale in a neighborhood of z.

By definition, this means that for every map of a scheme $B \to F$, the induced map of schemes $B \times_F Z_u \to B$ is étale in a neighborhood of $z' \times_F Z_u$. Since $U \to F$ is smooth and surjective, it suffices to check that the morphism is étale after a base change to U.

By construction, $Z_u \subset U'$ is cut out by a regular sequence in a neighborhood of $z' \in U'$ (since z' is a smooth point of W'). Thus $Z_u \times_F U \to U' \times_F U$ is a regular embedding in a neighborhood of $z' \times_F U$. Since $U' \times_F U \to U'$ is smooth, we can apply [EGA4, Theorem 17.12.1], and conclude that $Z_u \times_F U' \to U'$ is smooth in a neighborhood of z'. Moreover, the relative dimension of this morphism is 0. Therefore, $Z_u \to F$ is étale in a neighborhood of z'.

Since U is of finite type over S, the Z_u's are as well. The union of the Z_u's cover F (since their pullbacks via the morphism $U \to F$ cover U). Also, U is Noetherian because it is of finite type over a Noetherian scheme. Thus a finite number of the Z_u's will cover the F. (To see this, we can pullback via the map $U \to F$. The pullback of the Z_u's form an étale cover of U which is Noetherian.) □

The theorem has a useful corollary.

Corollary 2.2. *Let X/S be a Noetherian scheme of finite type and let G/S be a smooth affine group scheme (also of finite type over S) acting on X such that the stabilizers of geometric points are finite and reduced.*

(i) $[X/G]$ is a Deligne–Mumford stack. If the stabilizers are trivial, then $[X/G]$ is an algebraic space.

(ii) The stack is separated if and only if the action is proper.

Proof. The condition on the action ensures that $\underline{\mathrm{Iso}}_B(E,E)$ is unramified over E for any map $B \to [X/G]$ corresponding to the principal bundle $E \to B$. This in turn implies that the diagonal is also unramified, so condition (1) is satisfied. Furthermore, condition (2) is satisfied by the smooth map $X \to [X/G]$.

Suppose that $[X/G]$ is separated, i.e. the diagonal $[X/G] \to [X/G] \times [X/G]$ is proper and representable. One can check that

$$[X/G] \times_{[X/G]\times[X/G]} X \times X$$

is represented by the scheme $G \times X$ where the projection

$$[X/G] \times_{[X/G]\times[X/G]} X \times X \to X \times X$$

corresponds to the action map $G \times X \to X \times X$, so the action is proper.

Conversely, suppose that map $G \times X \to X \times X$ is proper. This implies that the diagonal is proper after base change to $X \times X$; i.e.

$$[X/G] \times_{[X/G]\times[X/G]} X \times X \to X \times X$$

is proper. Let $Z \to [X/G] \times [X/G]$ be any scheme and set

$$W = [X/G] \times_{[X/G]\times[X/G]} [X/G].$$

We will use descent to show that $W \to Z$ is proper.

If $Z' = Z \times_{[X/G]\times[X/G]} X \times X$ then the map $W' = W \times_Z Z' \to Z$ is proper. Since $X \to [X/G]$ is smooth and surjective (in particular it is faithfully flat) $Z' \to Z$ is as well. Descent theory for faithfully flat morphisms implies that the map $W \to Z$ is proper. Therefore $[X/G]$ is separated. □

Example 2.13. In order for a group action to be proper it must have have finite stabilizers. However, it is not difficult to construct examples of group actions which, despite having finite stabilizers, are not proper. In [GIT, Example 0.4] there is an example of an $\mathbf{SL}(2,\mathbb{C})$ action on a four-dimensional variety X which has trivial stabilizers but is not proper. The quotient $[X/\mathbf{SL}(2,\mathbb{C})]$ is a non-separated algebraic space.

2.7 Further properties of Deligne–Mumford stacks

From now on the term *stack* will mean a Deligne–Mumford stack, though we will often use the term *Deligne-Mumford stack* for emphasis.

Not all morphisms of stacks are representable, so we can not define algebro-geometric properties of these morphisms as we did for representable morphisms. However, if we consider morphisms of Deligne–Mumford stacks then we can define properties of morphisms as follows (see [DM, p. 100]).

Let **P** be a property of morphisms of schemes which at source and target is of a local nature for the étale topology. This means that for any family of commutative squares

$$\begin{array}{ccc} X_i & \stackrel{g_i}{\to} & X \\ f_i \downarrow & & f \downarrow \\ Y_i & \stackrel{h_i}{\to} & Y \end{array}$$

where the g_i (resp h_i) are étale and cover X (resp. Y), then f has property **P** if and only if f_i has property **P** for all i.

Examples of such properties are f flat, smooth, étale, unramified, locally of finite type, locally of finite presentation, etc.

Then if $f : F \to G$ is any morphism of Deligne–Mumford stacks we say that f has property **P** if there are étale atlases $U \to F$, $U' \to G$ and a compatible morphism $U \to U'$ with property **P**.

Likewise, if **P** is a property of schemes which is local in the étale topology (for example regular, normal, locally Noetherian, of characteristic p, reduced, Cohen–Macaulay, etc.) then a Deligne–Mumford stack F has property **P** if for one (and hence every) étale atlas $U \to F$, the scheme U has property **P**.

A stack F is quasi-compact if it has an étale atlas which is quasi-compact. A morphism $f : F \to G$ of stacks is quasi-compact if for any map of scheme $X \to G$ the fiber product $X \times_G F$ is a quasi-compact stack. We can now talk about morphisms of finite type; a morphism of finite type is a quasi-compact morphism which is locally of finite type. Similarly, a stack is Noetherian if it quasi-compact and locally Noetherian.

Definition 2.9. [DM, Definition 4.11] A morphism $f : F \to G$ is proper if it is separated, of finite type and locally over F there is a Deligne–Mumford stack $H \to F$ and a (representable) proper map $H \to G$ commuting with the projection to F and the original map $F \to G$.

$$\begin{array}{ccc} H & & \\ \downarrow & \searrow & \\ F & \longrightarrow & G \end{array}$$

Remark 2.15. When f is a scheme it is easy to show that there is a lift $\operatorname{Spec} V' \to F$ if and only if there is a lift $\operatorname{Spec} V \to F$.

As is the case with schemes, there are valuative criteria for separation and properness ([DM, Theorems 4.18–4.19]). The valuative criterion for separation is equivalent to the criterion for schemes, but we only construct an *isomorphism* between two extensions.

Theorem 2.2. *A morphism $f : F \to G$ is separated iff the following condition holds:*

For any complete discrete valuation ring V and fraction field K and any morphism $f : \operatorname{Spec} V \to G$ with lifts $g_1, g_2 : \operatorname{Spec} V \to F$ which are isomorphic when restricted to $\operatorname{Spec} K$, then the isomorphism can be extended to an isomorphism between g_1 and g_2.

Theorem 2.3. *A separated morphism $f : F \to G$ is proper if and only if for any complete discrete valuation ring V with field of fractions K and any map $\operatorname{Spec} V \to G$ which lifts over $\operatorname{Spec} K$ to a map to F, there is a finite separable extension K' of K such that the lift extends to all of $\operatorname{Spec} V'$ where V' is the integral closure of V in K'.*

Remark 2.16. When applied to schemes, Theorem 2.3 seems stronger than the usual valuative criterion for properness. However, this is not the case, as it is easy to show that if there is a lift $\operatorname{Spec} V' \to F$, then there is in fact a lift $\operatorname{Spec} V \to F$, as long as the image of V' is contained in an affine subscheme of F—which is always the case if F is a scheme.

Example 2.14. [Angelo Vistoli] Here is an example showing the necessity of passing to a cover.

Let $G = \{\pm 1\}$ act on $X = \mathbf{A}^1_{\mathbb{C}}$ by by left multiplication and set $Y = \mathbf{A}^1$. The double cover $f : X \to Y$ given by $z \mapsto z^2$ is clearly G-invariant (and in fact is a geometric quotient in the sense of [GIT]—see Section 4.1). In particular if $E \to B$ is a principal G-bundle with equivariant map to X, then composition $E \to X \to Y$ is G-invariant. Hence, since $E \to B$ is a categorical quotient in the sense of [GIT] and there is a unique map $B \to Y$ making the diagram

$$\begin{array}{ccc} E & \to & X \\ \downarrow & & \downarrow \\ B & \to & Y \end{array}$$

commute. Moreover,

$$\begin{array}{ccc} E' & \to & E \\ \downarrow & & \downarrow \\ B' & \to & B \end{array}$$

is a cartesian diagram corresponding to a morphism in $[X/G]$, then the morphism $B' \to B$ is actually a morphism of Y-schemes, and we obtain a morphism $\psi : [X/G] \to Y$.

It is easy to check that the double cover $f : X \to Y$ factors as $X \xrightarrow{p} [X/G] \xrightarrow{\psi} Y$ where $p : X \to [X/G]$ is the map defined in Example 2.7. Since G is finite, p is a finite surjective (and representable) morphism, and since f is finite, it follows that the non-representable morphism $[X/G] \to Y = \mathbf{A}^1$ is finite. We wish to test the valuative criterion for this morphism.

Let $R = \mathbb{C}[[t]]$ be the complete local ring of Y at 0, and let $\operatorname{Spec} R \to Y$ be the obvious morphism. The restriction of the double cover $X \to Y$ to the generic point is a non-trivial degree 2 Galois covering corresponding the extension $\mathbb{C}(t) \subset \mathbb{C}(u)$ where $t = u^2$. We can view this cover as a principal G-bundle, giving us a lift $\operatorname{Spec}\mathbb{C}(t) \to [X/G]$. However, this lift cannot be extended to all of $\operatorname{Spec} R$ since it has no non-trivial Galois covers. To obtain a lift we must first trivialize the bundle $\operatorname{Spec}\mathbb{C}(u) \to \operatorname{Spec}\mathbb{C}(t)$ by normalizing R in $\mathbb{C}(u)$.

2.8 The topology of stacks

Most of the topological properties of schemes make sense for Deligne–Mumford stacks. Thus in many ways we can think of them as spaces.

Remark 2.17. [Connectedness] If F_1 and F_2 are stacks over S, define the disjoint sum $F = F_1 \coprod F_2$ as follows: Objects are disjoint unions $X_1 \coprod X_2$ where X_1 is an object of F_1 and X_2 is an object of X_2. A morphism from $X_1' \coprod X_2' \to X_1 \coprod X_2$ is specified by giving morphisms $X_1' \to X_1$ $X_2' \to X_2$.

A stack is connected if it is not the disjoint sum of two non-void stacks.

Proposition 2.4. [DM, Proposition 4.14] *A Noetherian stack F over a field is connected if and only if there is a surjective morphism $X \to F$ from a connected scheme.*

Remark 2.18. [Open and closed substacks] An open substack $F \subset G$ is a full subcategory of G such that for any $x \in Obj(F)$, all objects in G isomorphic to x are also in F. Furthermore, the inclusion morphism $F \to G$ is represented by open immersions. In a similar way we can talk about closed (or locally closed) substacks.

In particular a stack is irreducible if it is not the disjoint union of two closed substacks.

A normal stack is irreducible if and only if it is connected. (Since being normal is an étale local property of schemes, a stack is normal iff it has a normal étale atlas).

The following theorem is crucial to the proof of irreducibility of $F_{\overline{\mathcal{M}}_g}$ over $\operatorname{Spec}\mathbb{Z}$.

Theorem 2.4. [DM, Theorem 4.17(iii)] *Let $f : F \to S$ be a proper flat morphism with geometrically normal fibers. Then the number of connected components of the geometric fibers is constant.*

3 Stable curves

In this section we discuss stable curves and the compactification of the moduli of curves to the moduli of stable curves.

Definition 3.1. [DM, Definition 1.1] A Deligne–Mumford stable (resp. semi-stable) curve of genus g over a scheme S is a proper flat family $C \to S$

whose geometric fibers are reduced, connected, 1-dimensional schemes C_s such that:

(1) C_s has only ordinary double points as singularities.

(2) If E is a non-singular rational component of C, then E meets the other components of C_s in more than two points (resp. in at least 2 points).

(3) C_s has arithmetic genus g; i.e. $\dim H^1(\mathcal{O}_{C_s}) = g$.

Remark 3.1. Clearly, a smooth curve of genus g is stable. Condition (2) ensures that stable curves have finite automorphism groups, so that we will be able to form a Deligne–Mumford stack out of the category of stable curves. We will not use the notion of semi-stable curves until we discuss geometric invariant theory in Section 4.

Denote by $F_{\overline{\mathcal{M}}_g}$ the groupoid over $\operatorname{Spec}\mathbb{Z}$ whose sections over a scheme B are families of stable curves $X \to B$. As is the case with smooth curves, we define a morphism from $X' \to B'$ to $X \to B$ as a cartesian diagram

$$\begin{array}{ccc} X' & \rightarrow & X \\ \downarrow & & \downarrow \\ B' & \rightarrow & B \end{array}$$

3.1 The stack of stable curves is a Deligne–Mumford stack

Let $\pi : C \to S$ be a stable curve. Since π is flat and its geometric fibers are local complete intersections, the morphism is a local complete intersection morphism. It follows from the theory of duality that there is a canonical invertible dualizing sheaf $\omega_{C/S}$ on C. If C/S is smooth, then this sheaf is the relative cotangent bundle. The key fact we need about this sheaf is a theorem of Deligne and Mumford [DM, p. 78].

Theorem 3.1. *Let $C \xrightarrow{\pi} S$ be a stable curve of genus $g \geq 2$. Then $\omega_{C/S}^{\otimes n}$ is relatively very ample for $n \geq 3$, and $\pi_*(\omega_{C/S}^{\otimes n})$ is locally free of rank $(2n-1)(g-1)$.*

Remark 3.2. When π is smooth, the theorem follows from the classical Riemann–Roch theorem for curves. The general case is proved by analyzing the locally free sheaf obtained by restricting $\omega_{C/S}$ to the geometric fibers of C/S. In particular, if $S = \operatorname{Spec} k$, with k algebraically closed, then $\omega_{C/S}$ can be described as follows. Let $f : C' \to C$ be the normalization of C (note C' need not be connected). Let $x_1, \ldots, x_n, y_1, \ldots, y_n$ be the points of C' such that the $z_i = f(x_i) = f(y_i)$ are the double points of C. Then $\omega_{C/S}$ can be identified with the sheaf of 1-forms η on C' regular except for simple poles at the x's and y's and with $Res_{x_i}(\eta) + Res_{y_i}(\eta) = 0$.

As a result, if $N = (2n-1)(g-1) - 1$, then every stable curve can be realized as a curve in $\mathbf{P}^N$ with Hilbert polynomial $P_{g,n}(t) = (2nt-1)(g-1)$. There is a subscheme (defined over $\operatorname{Spec}\mathbb{Z}$) $\overline{H}_{g,n} \subset Hilb_{\mathbf{P}^N}^{P_{g,n}}$ of the Hilbert scheme corresponding to n-canonically embedded stable curves. Likewise there is a subscheme $H_{g,n} \subset \overline{H}_{g,n}$ corresponding to n-canonically embedded smooth curves. A map $S \to \overline{H}_{g,n}$ corresponds to a stable curve $C \xrightarrow{\pi} S$ of genus g and an isomorphism of $\mathbf{P}(\pi_*(\omega_{C/S}))$ with $\mathbf{P}^N \times S$.

Now, $\mathbf{PGL}\,(N+1)$ naturally acts on $H_{g,n}$ and $\overline{H}_{g,n}$.

Theorem 3.2. $F_{\mathcal{M}_g} \simeq [H_{g,n}/\mathbf{PGL}\,(N+1)]$ *and*

$$F_{\overline{\mathcal{M}}_g} \simeq [\overline{H}_{g,n}/\mathbf{PGL}\,(N+1)].$$

Note that the theorem asserts that the quotient is independent of n.

Proof. We construct a functor $p : F_{\overline{\mathcal{M}}_g} \to [\overline{H}_{g,n}/\mathbf{PGL}\,(N+1)]$ which takes $F_{\mathcal{M}_g}$ to $[H_{g,n}/\mathbf{PGL}\,(N+1)]$ as follows: Given a family of stable curves $C \xrightarrow{\pi} B$, let $E \to B$ be the principal $\mathbf{PGL}\,(N+1)$ bundle associated to the projective bundle $\mathbf{P}(\pi_*(\omega_{C/B}^{\otimes n}))$. Let $\pi' : C \times_B E \to E$ be the pullback family. The pullback of this projective bundle to E is trivial and is isomorphic to $\mathbf{P}(\pi'_*(\omega_{C \times_B E/E}))$, so there is a map $E \to \overline{H}_{g,n}$ which is clearly $\mathbf{PGL}\,(N+1)$ invariant. If

$$\begin{array}{ccc} C' & \to & C \\ \pi' \downarrow & & \pi \downarrow \\ B' & \xrightarrow{\phi} & B \end{array}$$

is a morphism in $F_{\overline{\mathcal{M}}_g}$, then $\pi'_*(\omega_{C'/B'}) \simeq \phi^*\pi_*(\omega_{C/B})$, so we obtain a morphism of associated $\mathbf{PGL}\,(N+1)$ bundles

$$\begin{array}{ccc} E' & \to & E \\ \downarrow & & \downarrow \\ B' & \to & B \end{array}$$

The next step is to show that if C/B is a stable curve then any automorphism of C/B is induced by an automorphism of the projective bundle $\mathbf{P}(\pi_*(\omega_{C/B}))$. This is proved for smooth curves in [GIT, Proposition 5.2], and is easily generalized to stable curves because $\pi_*(\omega_{C/B})$ has the same properties as the smooth case. It then follows that our functor is fully faithful.

Now if $E \to B$ is an object of $[\overline{H}_{g,n}/\mathbf{PGL}\,(N+1)]$ then there is a family $C_E \xrightarrow{\pi_E} E$ of curves of genus g together with an isomorphism $\mathbf{P}(\pi_{E,*}(\omega_{C_E/E})) \simeq \mathbf{P}^N_E$, where $\mathbf{PGL}\,(N+1)$ acts by changing the polarization. The morphism $C_E \to E$ is equivariantly projective and the quotient $B = E/\mathbf{PGL}\,(N+1)$ is a scheme, so descent theory says that $C = C_E/\mathbf{PGL}\,(N+1)$ exists as a scheme. Then $C_E \simeq C \times_B E$, so the object $E \to B$ is isomorphic to of $p(C \to B)$. Therefore p is an equivalence of categories. □

Corollary 3.1. $F_{\mathcal{M}_g}$ *and* $F_{\overline{\mathcal{M}}_g}$ *are Deligne–Mumford stacks.*

Proof. We have just shown that $F_{\mathcal{M}_g}$ and $F_{\overline{\mathcal{M}}_g}$ are quotients of a scheme by a smooth group, so they have smooth atlases. Every stable curve defined over an algebraically closed field has a finite and reduced automorphism group, so the diagonal is unramified. Therefore, they are Deligne–Mumford stacks by Theorem 2.1. □

3.2 Properness of $F_{\overline{\mathcal{M}}_g}$

Given that $F_{\overline{\mathcal{M}}_g}$ is a Deligne–Mumford stack, the valuative criterion of properness and the following stable reduction theorem show that it is proper over $\operatorname{Spec}\mathbb{Z}$. Hence $F_{\overline{\mathcal{M}}_g}$ are $F_{\mathcal{M}_g}$ are separated.

Theorem 3.3. *Let B be the the spectrum of a DVR with function field K, and let $X \to B$ be a family of curves such that its restriction $X_K \to \operatorname{Spec} K$ is a smooth curve. Then there is a finite extension K'/K and a unique stable family $X' \to B'$ where B' is the normalization of B in K' such that the restriction $X' \to \operatorname{Spec} K'$ is isomorphic to $X_K \times_K K'$.*

Remark 3.3. [Remarks on the proof of Theorem 3.3] The properness of $F_{\overline{\mathcal{M}}_g}$ follows from the observation: To check the properness of a morphism using the valuative criterion, it suffices to consider maps where the image of $\operatorname{Spec} K$ is contained in a fixed dense open substack (see the discussion after the statement of [DM, Theorem 4.19]).

This theorem was originally proved (but not published) in characteristic zero by Mumford and Mayer ([GIT, Appendix D]). There is a relatively straightforward algorithmic version of this theorem in characteristic 0 which I learned from Joe Harris. Blow up the singular points of the special fiber of X/B until the total space of the family is smooth and the special fiber has only nodes as singularities. The modified special fiber will have a number of components with positive multiplicity coming from the exceptional divisors in the blowups. Next, do a base change of degree equal to the l.c.m. of the multiple components. After base change all components of the special fiber will have multiplicity 1. Then contract all (-1) and (-2) rational components in the total space. (That this can be done follows from the existence of minimal models for surfaces.) The special fiber is now stable. Furthermore, the total space of the new family is a minimal model for the surface. Since minimal models of surfaces are unique, the stable limit curve is unique.

This algorithmic proof fails in characteristic $p > 0$, because after blowing up some components of the special fiber may have multiplicity divisible by p. In this case, it will not be possible to make the component become reduced after base change.

Deligne and Mumford proved the stable reduction theorem in arbitrary characteristic using Neron models of the Jacobians of the curves ([DM]). Later Artin and Winters [AW] gave a direct geometric proof using the theory of curves on surfaces.

3.3 Irreducibility of $F_{\mathcal{M}_g}$ and $F_{\overline{\mathcal{M}}_g}$

Using the description of the moduli stacks as quotients of H_g and $\overline{H}_g$ we can deduce properties of the stacks from the corresponding properties of the Hilbert scheme. In particular, deformation theory shows that H_g and $\overline{H}_g$ are smooth over $\operatorname{Spec}\mathbb{Z}$ ([DM, Cor 1.7]). Since the map $H_g \to F_{\mathcal{M}_g}$ (resp. $\overline{H}_g \to F_{\overline{\mathcal{M}}_g}$) is smooth we see that $F_{\overline{\mathcal{M}}_g}$ is smooth.

Further analysis [DM, Cor 1.9] shows that the scheme $\overline{H}_g - H_g$ representing polarized, singular, stable curves is a divisor with normal crossings in $\overline{H}_g$. This property descends to the moduli stacks.

Theorem 3.4. [DM, Thm 5.2] *$F_{\overline{\mathcal{M}}_g}$ is smooth and proper over* $\operatorname{Spec}\mathbb{Z}$. *The complement $F_{\overline{\mathcal{M}}_g} - F_{\mathcal{M}_g}$ is a divisor with normal crossings in $F_{\overline{\mathcal{M}}_g}$.*

The main result of [DM] is the following theorem:

Theorem 3.5. [DM] *$F_{\overline{\mathcal{M}}_g}$ has irreducible geometric fibers over* $\operatorname{Spec}\mathbb{Z}$.

Remark 3.4. Deligne and Mumford gave two proofs of this theorem. In both cases they deduce the result from the classical characteristic 0 result stated below. We outline below their second proof, which uses Deligne–Mumford stacks.

Proposition 3.1. *$F_{\mathcal{M}_g} \times_{\operatorname{Spec}\mathbb{Z}} \operatorname{Spec}\mathbb{C}$ is irreducible.*

Proof. It was shown classically that there is a space $H_{k,b}$ parametrizing degree k covers of $\mathbf{P}^1$ simply branched over b points defined over the complex numbers. In [Fu], Fulton showed that the functor $F_{H_{k,b}}$ whose sections over a base B are families of smooth curves $C \to B$ together with a degree k map $C \to \mathbf{P}^1_B$ expressing each geometric fiber as a cover of $\mathbf{P}^1$ simply branched over b points is represented by a scheme which we also call $H_{k,b}$. In characteristic greater than k it is a finite étale cover of $P_b = (\mathbf{P}^1)^b - \Delta$, where Δ is the union of all diagonals. (This fact was known classically over $\mathbb{C}$.) Since P_b is obviously irreducible, it can be proved that $H_{k,b}$ is irreducible in high characteristic by showing that the monodromy of the covering $H_{k,b} \to P_b$ acts transitively on the fiber over a base point in P_b for all k, b. Since there is a universal family of branched covers $C_{k,b} \to H_{k,b}$ there is a map $H_{k,b} \to F_{\mathcal{M}_g}$ (where $g = b/2 - k + 1$). By the Riemann–Roch theorem for smooth curves, every curve of genus g can be expressed as a degree k cover of $\mathbf{P}^1$ with b simple branch points, as long as $k > g+1$. Thus for k (and thus b) sufficiently large, the map is surjective. Therefore $F_{\mathcal{M}_g}$ is irreducible in characteristic greater than k, and thus $F_{\mathcal{M}_g} \times_{\mathbb{Z}} \mathbb{C}$ is irreducible. □

Proof. (Outline of the proof of Theorem 3.5) Since $F_{\overline{\mathcal{M}}_g} - F_{\mathcal{M}_g}$ is a divisor, $F_{\overline{\mathcal{M}}_g}$ is irreducible if and only if $F_{\mathcal{M}_g}$ is as well. The stack $F_{\overline{\mathcal{M}}_g}$ is smooth, so it suffices to show that it is connected. The morphism $F_{\overline{\mathcal{M}}_g} \to \operatorname{Spec}\mathbb{Z}$ is proper, flat and has smooth geometric fibers, so by Theorem 2.4 the number of connected components of the geometric fibers is constant. By the proposition the geometric fiber $F_{\overline{\mathcal{M}}_g} \times_{\operatorname{Spec} \mathbb{Z}} \operatorname{Spec} \mathbb{C}$ is connected, so every geometric fiber is connected. □

Remark 3.5. In [HM] Harris and Mumford constructed a compactification of $H_{k,b}$ where the boundary represents stable curves expressed as branched covers of chains of $\mathbf{P}^1$'s. The existence of this compactification implies that every smooth curve admits degenerations to singular stable curves. Fulton [Fu82] used this fact to resurrect an argument of Severi giving a purely algebraic proof of the irreducibility of $F_{\mathcal{M}_g}$ in characteristic 0. This combined with the results of [DM] gives a purely algebraic proof that $F_{\mathcal{M}_g}$ is irreducible in arbitrary characteristic.

4 Construction of the moduli scheme

As we have previously seen, the moduli stack is a quotient stack of a smooth scheme $\overline{H}_g$ by $\mathbf{PGL}\,(N+1)$. In this, the final section, we discuss the construction of a quotient scheme $\overline{H}_g/\mathbf{PGL}\,(N+1)$ over an algebraically closed field k. We first prove that such a scheme is unique and is the coarse moduli space for the quotient stack. We then briefly discuss Gieseker's GIT construction of a quotient scheme.

4.1 Moduli schemes and geometric quotients

This definition is completely analogous to Mumford's definition ([GIT, p. 99]) of a coarse moduli scheme mentioned above.

Definition 4.1. The moduli space of a Deligne–Mumford stack F is a scheme M together with a proper morphism $\pi : F \to M$, such that

(*) for any algebraically closed field k there is a bijection between the set of isomorphism classes of objects in the groupoid $F(\Omega)$ and $M(\Omega)$, where $\Omega = \operatorname{Spec} k$.

Furthermore, M is universal in the sense that if N is a scheme then any morphism $F \to N$ factors through a morphism $M \to N$.

Remark 4.1. The universal property guarantees that the moduli scheme is unique if it exists. Though it is sufficient for our purpose, there are two drawbacks to this definition.

(1) In characteristic p, the property of being a moduli space is not invariant under base change; i.e. if $M' \to M$ is a morphism then M' need not be the moduli space of $M' \times_M F$. As a result, Gieseker [Gi] gave an alternative definition: Namely, M is a moduli space for F if the morphism $F \to M$ is proper and a bijection on geometric points. This notion is clearly preserved by base change but such a scheme is not unique. However, two moduli spaces are universally homeomorphic as schemes.

(2) If F is a Deligne–Mumford stack, then F may not have a moduli scheme (in either sense). If one is willing to look in the category of algebraic spaces a theorem of Keel and Mori [KM] states that, under very mild assumptions, if F is a Deligne–Mumford stack[3] there is an algebraic space M such that M is a moduli space for F; i.e. there is a morphism $F \to M$ which is a bijection on geometric points and which is universal for maps of *algebraic spaces*. However, these two notions may differ. For example, $\mathbf{A}^1$ is the moduli scheme (in our sense) of the non-separated algebraic space $[X/\mathbf{SL}(2, \mathbb{C})]$ discussed in Example 2.13.

Definition 4.2. [GIT, Definitions 0.5, 0.6] Let X/S be a scheme and let G/S be a smooth affine group scheme acting on X. An S-scheme Y is a *geometric quotient* of X by G if there is a morphism $X \to Y$ such that

(1) f is G invariant.

(2) The geometric fibers of f are orbits. (In particular f is surjective.)

(3) f is universally submersive, i.e. $U \subset Y$ is open iff $f^{-1}(U)$ is open, and this property is preserved by base change.

(4) $f_*(\mathcal{O}_X)^G = \mathcal{O}_Y$.

Remark 4.2. The purpose of the geometric invariant theory developed by Mumford is to construct geometric quotients for the action of a geometrically reductive group The definition of a geometrically reductive group is given in [GIT, Appendix A]. In characteristic 0 this notion is the same as the notion of *linear* reductivity; i.e. every representation decomposes as a direct sum of irreducibles. However, in characteristic p the only linear reductive groups are extensions of tori by finite groups of order prime to p. However, for the purpose of these notes, it suffices to know that $\mathbf{SL}(N+1, k)$ is reductive for a field k.

The following is a restatement of [GIT, Prop 0.1].

Proposition 4.1. *A geometric quotient is a categorical quotient. That is, if* $X \xrightarrow{f} Y$ *is a geometric quotient and if* $X \xrightarrow{g} Z$ *is a* G *invariant morphism, then there is a unique morphism* $\phi : Y \to Z$ *such that* $g = \phi \circ f$.

Note that the proposition implies that if a geometric quotient exists then it is unique.

Now let X be a scheme with a G action such that the stabilizers of geometric points are finite and reduced. We have seen that the groupoid $[X/G]$ is a Deligne–Mumford stack.

Proposition 4.2. *If* $f : X \to Y$ *is a geometric quotient of* X *by* G *then* Y *is the moduli space of the stack* $[X/G]$.

If in addition the action of G *on* X *is proper then the morphism* $[X/G] \to Y$ *is proper.*

---

[3]The theorem of Keel and Mori also applies to a certain class of Artin stacks.

Proof. If $\Omega = \operatorname{Spec} K$ where K is algebraically closed, then $Hom(\Omega, Y)$ is, by Condition (2) of the definition, the set of orbits of K-valued points of X. This is exactly the set of isomorphism classes in $[X/G](\Omega)$. Therefore, condition (*) is satisfied.

Next suppose that $[X/G] \to N$ is a morphism to a scheme. It is easy to see that the induced morphism $X \to N$ is G-invariant. By the universal mapping property of the quotient, the morphism $X \to N$ factors through Y. Thus, the morphism $[X/G] \to N$ also factors through Y, so Y is the moduli scheme for $[X/G]$.

If the action is proper then $[X/G]$ is separated, so the morphism $[X/G] \to Y$ is also separated. Then the the universal submersiveness of $f : X \to Y$ implies that the morphism $[X/G] \to Y$ satisfies the valuative criterion of properness. The proof is given in [Vi, Proof of Prop. 2.11]. □

4.2 Construction of quotients by geometric invariant theory

From now on we will assume that all schemes are defined over an algebraically closed field k.

In this paragraph we discuss the geometric invariant theory necessary to construct $\mathcal{M}_g$ and and $\overline{\mathcal{M}}_g$ as quotients of Hilbert schemes of n-canonically embedded (stable) curves. Our source is [Gi, Chapter 0]. For a full treatment of geometric invariant the classic reference is Mumford's [GIT].

Let $X \subset \mathbf{P}^N$ be a projective scheme, and let G be a reductive group acting on X via a representation $G \to \mathbf{GL}\,(N+1)$.

Definition 4.3. (1) A closed point $x \in X$ is called semi-stable if there exists a non-constant G-invariant homogeneous polynomial F such that $F(x) \neq 0$.

(2) $x \in X$ is called stable if $\dim o(x) = \dim G$ (where $o(x)$ denotes the orbit of x) and there exists a non-constant G-invariant polynomial such that $F(x) \neq 0$ and for every y_0 in $X_F = \{y \in X | F(y) \neq 0\}$, $o(y_0)$ is closed in X_F.

Let X^{ss} denote the semi-stable points of X, and let X^s denote the stable points. Then $X^s \subset X^{ss}$ are both open in X. However, they may be empty.

The following is the first main theorem of geometric invariant theory.

Theorem 4.1. *There exists a projective scheme Y and an affine, universally submersive morphism $f_{ss} : X^{ss} \to Y$ such that Y is a categorical quotient (such a morphism is often called a good quotient in the literature). Furthermore, there exists $U \subset Y$ open such that $f^{-1}(U) = X^s$ and $f_s : X^s \to U$ is a geometric quotient of X^s by G.*

Remark 4.3. Proposition 4.2 implies that U is the moduli space of $[X^s/G]$. Moreover, geometric invariant theory also says that G acts properly on $[X^s$ so the morphism $[X^s/G] \to U$ is proper.

4.3 Criteria for stability

Let $X \subset \mathbf{P}^N$ be a projective scheme, and let $\tilde{X} \subset \mathbf{A}^{N+1}$ be the affine cone over X. Assume as above that a reductive group G acts on X via a representation $G \to \mathbf{GL}(N+1)$. Then G acts on $\tilde{X}$ as well. The stability of $x \in X$ can be rephrased in terms of the stability of the points $\tilde{x} \in \tilde{X}$ lying over x.

Proposition 4.3. [GIT, Chapter 1, Proposition 2.2 and Appendix B] *A geometric point $x \in X$ is semi-stable if for one (and thus for all) $\tilde{x} \in X$ lying over X, $0 \notin o(\tilde{x})$. The point x is stable if $o(\tilde{x})$ is closed in $\mathbf{A}^{N+1}$ and has dimension equal to the dimension of G.*

The second main theorem of geometric invariant theory is Mumford's numerical criterion for stability which we now discuss.

Definition 4.4. A 1-parameter subgroup of G is a homomorphism $\lambda : G_m \to G$. This will be abbreviated to "λ is a 1-PS of G".

Now if λ is a 1-PS of G, then since λ is 1-dimensional, there is a basis $\{e_0, \ldots, e_N\}$ of $\mathbf{A}^{N+1}$ such that the action of λ is diagonalizable with respect to this basis; i.e. $\lambda(t)e_i = t^{r_i}e_i$ where $t \in G_m$ and $r_i \in \mathbb{Z}$. If $\tilde{x} = \sum x_i e_i \in \tilde{X}$, then the set of r_i such that x_i is non-zero is called the λ-weights of $\tilde{x}$. Note that if $x \in \mathbf{P}^N$ then the λ-weights are the same for all points in $\mathbf{A}^{N+1} - 0$ lying over x.

Definition 4.5. $x \in X$ is λ-semi-stable if for one (and thus for all) $\tilde{x} \in \tilde{X}$ lying over x, $\tilde{x}$ has a non-positive λ weight. A point x is λ-stable if $\tilde{x}$ has a negative λ-weight.

Theorem 4.2. [GIT] *A point $x \in X$ is (semi-)stable if and only if x is λ-(semi-)stable for all 1-PS $\lambda : G_m \to G$.*

Remark 4.5. [Remark on the Proof]It is easy to see that if x is unstable (i.e. not semi-stable) with respect to $\lambda : G_m \to G$ then x is unstable. The reason is that if all the weights of λ are positive then 0 will be in the closure of the G-orbit of $\tilde{x}$ in $\mathbf{A}^{N+1} - 0$. The converse is more difficult.

Example 4.1. (cf. [GIT, Proposition 4.1]). The set of homogeneous forms of degree 4 in two variables forms a five-dimensional vector space V. We will view $\mathbf{P}(V)$ as the space parametrizing 4-tuples of (not necessarily) distinct points in $\mathbf{P}^1$. There is a natural action of $\mathbf{SL}(2)$ on V inducing an action on $\mathbf{P}(V)$. Let us use the numerical criterion to determine the stable and semi-stable locus in $\mathbf{P}(V)$.

If $v \in V$ is a form of degree 4 and λ is a 1-PS subgroup of $\mathbf{SL}(2)$, then we can write $v = a_4X_0^4 + a_3X_0^3X_1 + a_2X_0^2X_1^2 + a_1X_0X_1^3 + a_0X_1^4$, and λ acts by $\lambda(t)(X_0) = t^rX_0$, $\lambda(t)(X_1) = t^{-r}X_0$ and $r > 0$ (the weight on X_1 must be the negative of the weight on X_0, since λ maps to $\mathbf{SL}(2)$). The possible weights of v are $\{4r, 2r, 0, -2r, -4r\}$. In order for v to be λ-stable one of a_1 or a_0 must be non-zero. It is λ-semi-stable if one of a_2, a_1 or a_0

is non-zero. On the other hand, we can consider the 1-PS, τ which acts by $\tau(t)X_0 = t^{-r}X_0$ and $\tau(t)X_1 = t^r X_1$. In order for v to be τ-stable one of a_4 and a_3 must be non-zero, while it is τ-semi-stable if a_2 is non-zero. Combining the conditions imposed by λ and τ we see that if v is stable, then one of a_0 or a_1 is non-zero and one of a_3 or a_4 is non-zero. This condition is equivalent to the condition that $(1:0)$ and $(0:1)$ are not multiple points of the subscheme of $\mathbf{P}^1$ cut out by the form v. Likewise, v is semi-stable if $(1:0)$ or $(0:1)$ is cut out with multiplicity no more than 2. Finally v is unstable if $(1:0)$ or $(0:1)$ is cut out with multiplicity more than 2.

From this analysis it is clear that if $v \in V$ cuts out four distinct points then it will be stable for every 1-PS. Likewise if v cuts out a subscheme of $\mathbf{P}^1$ with each point having multiplicity 2 or less then it is semi-stable for every 1-PS. Conversely, if v cuts a point of multiplicity 3 or more then $v = X_0^3(a_0X_0+a_1X_1)$ for some choice of coordinates on $\mathbf{P}^1$. Then v will have strictly positive weights for a 1-PS λ acting diagonally by $\lambda(t)X_0 = t^r X_0$ for $r > 0$.

4.4 Gieseker's construction of $\overline{\mathcal{M}}_g$

Let $Hilb^{N+1}_{P(t)}$ be the Hilbert scheme of curves in $\mathbf{P}^N$ with Hilbert polynomial $P(t)$. Grothendieck's uniform m-lemma says that if $X \subset \mathbf{P}^N$ is a curve with Hilbert polynomial $P(t)$, then there exists $m >> 0$ (independent of X) such that the restriction map $H^0(\mathbf{P}^N, \mathcal{O}_{\mathbf{P}^N}(m)) \to H^0(X, \mathcal{O}_X(m))$ is surjective and $\dim H^0(X, \mathcal{O}_X(m)) = P(m)$. Taking the $P(m)$-th exterior power of ϕ_m we obtain a linear map $V^m = \bigwedge^{P(m)} H^0(\mathbf{P}^N, \mathcal{O}_{\mathbf{P}^N}(m)) \to \bigwedge^{P(m)} H^0(X, \mathcal{O}_X(m)) = k$ unique up to scalars; i.e. an element of $\mathbf{P}(V^m)$. The corresponding point in $\mathbf{P}(V^m)$ is called the m-th Hilbert point of X and is denote $H_m(X)$. In this way we obtain a map $Hilb^{N+1}_{P(t)} \to \mathbf{P}(V^m)$. For m sufficiently large this map is an embedding.

Both $\mathbf{SL}\,(N+1)$ and $\mathbf{PGL}\,(N+1)$ act on $\mathbf{P}(V^m)$ via the m-th exterior power representation of $\mathbf{SL}\,(N+1) \to \mathbf{GL}\,(V^m)$. Now the action of $\mathbf{SL}\,(N+1)$ factors through the action of $\mathbf{PGL}\,(N+1)$ (the stabilizer of $\mathbf{SL}\,(N+1)$ at a geometric point is the group of $N+1$ roots of unity) so we have the following proposition.

Proposition 4.4. *If $X \subset \mathbf{P}(V^m)$ then $X \to Y$ is a geometric quotient by $\mathbf{SL}\,(N+1)$ if and only if it is a geometric quotient by $\mathbf{PGL}\,(N+1)$.*

Proof. If $X \to Y$ is a geometric quotient by $\mathbf{SL}\,(N+1)$ then the geometric fibers are $\mathbf{SL}\,(N+1)$ orbits. These orbits are the same as the $\mathbf{PGL}\,(N+1)$ orbits. Likewise, $\mathcal{O}_X^{\mathbf{SL}\,(N+1)} = \mathcal{O}_X^{\mathbf{PGL}\,(N+1)}$. Thus, $\mathcal{O}_Y \simeq f_*\mathcal{O}_X^{\mathbf{PGL}\,(N+1)}$. Finally if W and V are $\mathbf{PGL}\,(N+1)$ invariant, they are also $\mathbf{SL}\,(N+1)$ invariant. Thus if they are disjoint, then since $X \to Y$ is an $\mathbf{SL}\,(N+1)$ quotient, their images will be disjoint as well. Hence $X \to Y$ is a $\mathbf{PGL}\,(N+1)$ quotient. The converse is similar. $\square$

Let $g \geq 3$ and $d \geq 20(g-1)$ be integers. Consider the Hilbert scheme $H^{N+1}_{P(t)}$ of curves in $\mathbf{P}^{N=d-g}$ with Hilbert polynomial $P(t) = dt - g + 1$ (the curves parametrized necessarily have arithmetic genus g). The first step in Gieseker's construction is to prove the following theorem. The proof is ten pages long and uses the numerical criterion.

Theorem 4.3. [Gi, Theorem 1.0.0] *There exists an integer $m_0 >> 0$ such that if X is smooth then $H_{m_0}(X)$ is $\mathbf{SL}\,(N+1)$ stable.*

Remark 4.6. The theorem is not necessarily true for arbitrary $m_0 >> 0$. However there are infinitely many m_0 for which the theorem is true ([Gi, Remark after Theorem 1.0.0]).

The next, and technically most difficult step is to prove the following theorem. The proof takes 50 pages!

Theorem 4.4. [Gi, Theorem 1.0.1] *For the same integer m_0, every point in $Hilb^{N+1}_{P(t)} \cap \mathbf{P}(V^{m_0})^{ss}$ parametrizes a Deligne–Mumford semi-stable curve.*

Let $U \subset Hilb^{N+1}_{P(t)}$ be the subscheme of semi-stable curves with respect to the m_0-th Hilbert embedding. Let $Z_U \subset \mathbf{P}^N_U$ be the restriction of the universal family of projective curves. As before, view a point $h \in U$ as parametrizing a curve X_h and a very ample line bundle L_h of degree d on X_h. Set $U_c = \{h \in U | L_h \simeq \omega^n_{X_h}\}$. This is a locally closed subscheme of U which is empty unless $2g-2$ divides d. Gieseker then proves that U_c is in fact closed in U. He also proves that U_c is smooth ([Gi, Theorem 2.0.1]) and parametrizes only all Deligne–Mumford stable curves; thus, $U_c \simeq \overline{H}_{g,n}$. Since U_c is closed in U there is a projective quotient $U_c/\mathbf{SL}\,(N+1)$. Finally note that $\mathbf{PGL}\,(N+1)$ (and thus $\mathbf{SL}\,(N+1)$) acts with finite stabilizers on points of U_c because the curves parametrized have finite automorphism groups. Hence the points of U_c are in fact $\mathbf{SL}\,(N+1)$ stable. Thus a geometric quotient $U_c/\mathbf{SL}\,(N+1)$ exists. Since this is isomorphic to a geometric quotient $U_c/\mathbf{PGL}\,(N+1) \simeq \overline{H}_{g,n}/\mathbf{PGL}\,(N+1)$ we have succeeded in constructing a coarse moduli scheme for the stack of stable curves. □

References

[AW] M. Artin, G. Winters, Degenerate fibres and stable reduction of curves, *Topology* **10** (1971), 373–383

[DM] P. Deligne, D. Mumford, Irreducibility of the space of curves of given genus, *Publ. Math. IHES* **36** (1969), 75–110

[Di] S. Diaz, A bound on the dimensions of complete subvarieties of $\mathcal{M}_g$, *Duke Math. J.* **51** (1984), 405–408

[Do] I. Dolgachev, *Rationality of fields of invariants*, in *Algebraic geometry, Bowdoin, 1985 (Brunswick, Maine, 1985)*, Proc. Sympos. Pure Math. **46** Part 2, 3–15 (1987)

[EGA4] A. Grothendieck, Etudes locales des schémas et des morphismes des schemas, *Publ. Math. IHES*, **32** (1967)

[Fu] W. Fulton, Hurwitz schemes and irreducibility of moduli of of algebraic curves, *Ann. Math.* **90** (1969), 542–575

[Fu82] W. Fulton, On the irreducibility of the moduli space of curves, appendix to a paper of Harris and Mumford, *Invent. Math.* **67** (1982), 87–88

[Gi] D. Gieseker, *Tata lectures on moduli of curves*, Springer-Verlag, NY, (1982)

[Gil] H. Gillet, *Intersection theory on algebraic stacks and Q-varieties*, J. Pure Appl. Alg. **34** (1984), 193–240

[HM] J. Harris, D. Mumford, On the Kodaira dimension of the moduli space of curves, *Invent. Math.* **67** 23–86 (1982)

[KM] S. Keel, S. Mori, Quotients by groupoids, *Ann. Math.* **145** (1997), 193–213

[L-MB] G. Laumon, L. Moret-Bailly, *Champs algébriques*, preprint

[GIT] D. Mumford, J. Fogarty, F. Kirwan, *Geometric Invariant Theory, Third Enlarged Edition*, Springer-Verlag, NY, (1994)

[Vi] A. Vistoli, Intersection theory on algebraic stacks and their moduli spaces, *Invent. Math.* **97** (1989), 613–670

Department of Mathematics
University of Missouri
Columbia, MO 65211
E-mail: `edidin@math.missouri.edu`

The Excess Intersection Formula and Gravitational Correlators

Lars Ernström

ABSTRACT The genus zero Gromov–Witten invariants of a homogeneous variety X generally solve an enumerative problem of rational curves with conditions of incidence to cycles in general position in X. In this paper we compute the same Gromov–Witten invariants with cycles in special position. The invariants can then be reinterpreted, using the excess intersection formula, as the sum of a contribution from a component of excess dimension and a contribution of the expected dimension. The excess contribution is identified with a sum of gravitational correlators, which can be computed using well-known recursive equations from the Gromov–Witten invariant. As a result, the contribution of the right dimension, which solves the enumerative problem requiring rational curves in X to have a node on a given cycle, can be computed in terms of Gromov–Witten invariants and gravitational correlators.

1 Introduction

The theory of *Gromov–Witten invariants* on a smooth algebraic complex variety X has intuitively a classical interpretation as counting maps from (at worst) nodal curves to X subject to incidence conditions to cycles in X. This intuition is realized perfectly by the Gromov–Witten theory in genus zero on a homogenous variety whereas in higher genera, or for a non-homogenous smooth varieties, the theories display everything you might expect from a general intersection theory such as excess and self-intersection. There are however a set of fundamental properties (divisor, gluing, dimension, genus reduction), reflecting the classical and physical intuition which always are satisfied (see [KM94], [Beh97], [BF97]).

There is an extended set of invariants called *gravitational correlators* which contains the Gromov–Witten invariants as a subset. The name stems from the fact that these invariants are ingredients of a two-dimensional topological field theory coupled to gravity—a physical fact that will not be touched upon in this paper. The gravitational correlators are natural geometric invariants but lack the intuition of Gromov–Witten invariants as counts of enumerative problems.

In this paper, we use gravitational correlators to solve enumerative problems of counting rational curves on a homogenous variety, subject to the condition that the curve should have a node on a given cycle in addition to standard incidence conditions. The strategy is to compute a Gromov–Witten invariant with cycles in a specific position in contrast to having cycles of the same classes in a general position, giving the answer to a counting problem. With the specific position, we get two contributions, one the answer to a counting problem involving conditions on a node (referred to as *node invariant*), and the other is a contribution from excess intersection. The contribution from excess intersection is identified with a sum of gravitational correlators, thus we get an expression for a Gromov–Witten as the sum of a node invariant and gravitational correlators. Since gravitational correlators in genus zero can be reduced to Gromov–Witten invariants, in terms of explicit Witten integrals the node invariant can be expressed solely in terms of Gromov–Witten invariants and Witten integrals.

For applications of gravitational correlators to counting problems involving tangency conditions see [Gat98], [Koch] and [Pan].

The outline of the paper is as follows. In §2 we recall the formalism of gravitational correlators and give an argument showing how the gravitational correlators can be expressed in terms of Gromov–Witten invariants and Witten integrals. In §3 we recall the structure of tangent spaces to the space of stable maps of genus zero on a smooth convex variety. In §4 the main result Theorem 4.2 is proven following the outline above, and in §5 we discuss possible generalizations.

The author thanks Andreas Gathmann for interesting and helpful discussions.

2 Gromov–Witten invariants and Gravitational descendents

Let X be a homogenous variety. Denote by $\overline{M}_{0,n}(X,\beta)$ the moduli stack of stable maps $\mu\colon (C, x_1, \dots, x_n) \to X$ from an n-pointed arithmetic genus zero curve with class $\mu_*[C] = \beta \in A_1(X)$. Denote by

$$e_1, \dots, e_n \colon \overline{M}_{0,n}(X,\beta) \to X$$

the evaluation maps; the map e_i is defined by associating to the map $\mu\colon (C, x_1, \dots, x_n) \to X$, the point $\mu(x_i)$ in X.

The universal curve of $\overline{M}_{0,n}(X,\beta)$ is canonically isomorphic to $\overline{M}_{0,n+1}(X,\beta)$ and is equipped with n sections

$$s_1, \dots, s_n \colon \overline{M}_{0,n}(X,\beta) \to \overline{M}_{0,n+1}(X,\beta),$$

and a projection map π. The section s_i is defined by sending the map $\mu\colon (C, x_1, \ldots, x_n) \to X$) to the ith marking x_i on the fiber (identified with C) of the universal curve. However, since we view the universal curve as stable maps with $n+1$ markings, the image is rather the map with domain taken to be the marked curve constructed from C by gluing a $\mathbf{P}^1$ component at x_i and placing the ith and $n+1$st markings at distinct points of the $\mathbf{P}^1$. The map to X is given by μ extended constantly along the new $\mathbf{P}^1$. We denote by ψ_i, for $i = 1, \ldots, n$, the first Chern class of the the (invertible) sheaf $s_i^*\omega_\pi$, where ω_π is the relative dualizing sheaf of π. The fiber of the sheaf $s_i^*\omega_\pi$ over $\mu\colon (C, x_1, \ldots, x_n) \to X$ is the cotangent line of C at x_i. It is easy to see that the class ψ_i is equal to the negative of the first Chern class of the conormal (invertible) sheaf of $s_i\colon \overline{M}_{0,n}(X,\beta) \to \overline{M}_{0,n+1}(X,\beta)$. We can now define the *gravitational correlators*. Given homogenous classes $\gamma_1, \ldots, \gamma_n$ in $A^*(X)$ and nonnegative integers $a_1, \ldots, a_n$, we denote

$$\langle \tau_{a_1}(\gamma_1) \ldots \tau_{a_n}(\gamma_n) \rangle_\beta := \int e_1^*(\gamma_1)\psi_1^{a_1} \ldots e_n^*(\gamma_n)\psi_n^{a_n}[\overline{M}_{0,n}(X,\beta)].$$

The invariants are zero unless

$$\sum_{i=1}^{n}(\operatorname{codim}(\gamma_i) + a_i) = \dim(\overline{M}_{0,n}(X,\beta)) = \dim(X) + \int c_n(T_X)\beta + n - 3.$$

The *Gromov–Witten invariants* are those invariants which have all the a_i's equal to zero. When X is a point we recover some integrals on $\overline{M}_{0,n}$. The behavior of these integrals (for all genera) were formulated in the so-called Witten conjecture [Wit91], later established by Kontsevich [Kon92]. The integrals for genus zero are particularly simple, given by the formula

$$\int \prod_{i=1}^{n} \psi_i^{a_i}[\overline{M}_{0,n}] = \frac{(n-3)!}{a_1!a_2!\cdots a_n!}.$$

Proposition 2.1. *There is an algorithm for computing the gravitational correlators from the Gromov–Witten invariants and the integrals on $\overline{M}_{0,n}$ (the case when X is a point.)*

Proof. This proof was explained to the author by Pandharipande (see [Pan97]). It is contained here since we will be using the expressions occuring in the proof for the computation of some gravitational correlators. For other proofs of the proposition see [Get98] or [KM98].

The proof is based on the cross-ratio argument and the following properties of descendants. In the formulas, the invariants with a negative a_i in some τ_i are conventionally set to be zero.

String equation:

$$\langle \tau_{a_1}(\gamma_1) \ldots \tau_{a_n}(\gamma_n) 1 \rangle_\beta$$
$$= \sum_{i=1}^{n} \langle \tau_{a_1}(\gamma_1) \ldots \tau_{a_{i-1}}(\gamma_{i-1}) \tau_{a_i-1}(\gamma_i) \tau_{a_{i+1}}(\gamma_{i+1}) \ldots \tau_{a_n}(\gamma_n) \rangle_\beta.$$

Dilaton equation:

$$\langle \tau_{a_1}(\gamma_1)\dots\tau_{a_n}(\gamma_n)\tau_1(1)\rangle_\beta = (-2+n)\langle \tau_{a_1}(\gamma_1)\dots\tau_{a_n}(\gamma_n)\rangle_\beta.$$

Divisor equation:

$$\langle \tau_{a_1}(\gamma_1)\dots\tau_{a_n}(\gamma_n)D\rangle_\beta = \Big(\int D\beta\Big)\langle \tau_{a_1}(\gamma_1)\dots\tau_{a_n}(\gamma_n)\rangle_\beta$$
$$+\sum_{i=1}^{n}\langle \tau_{a_1}(\gamma_1)\dots\tau_{a_{i-1}}(\gamma_{i-1})\tau_{a_i-1}(\gamma_i D)\tau_{a_{i+1}}(\gamma_{i+1})\dots\tau_{a_n}(\gamma_n)\rangle_\beta.$$

We consider invariants $\langle \tau_{a_1}(\gamma_1)\dots\tau_{a_n}(\gamma_n)\rangle_\beta$ and prove by induction over β and $\sum a_i$ that such invariants can be computed in terms of Gromov–Witten invariants (the $\sum a_i = 0$ case). Assume $a_n > 0$ and $\beta > 0$. The case $\beta = 0$ is covered by the known Witten integrals over $\overline{M}_{0,n}$. Consider the class

$$\omega = e_1^*(\gamma_i)\psi_1^{a_1}\dots e_n^*(\gamma_n)\psi_n^{a_n}e_{n+1}^*(H)e_{n+2}^*(H)e_{n+3}^*(1)$$

on $\overline{M}_{0,n+3}(X,\beta)$, where H is an ample divisor class on X and $\sum_{i+1}^{n}(\mathrm{codim}(\gamma_i)+a_i)+2 = \dim(\overline{M}_{0,n+3}(X,\beta)) - 1$ (so evaluation on a divisor class is possibly non-zero). Next, we use the cross-ratio argument. Consider the morphism

$$\overline{M}_{0,n+3}(X,\beta) \to \overline{M}_{0,n,n+1,n+2,n+3},$$

which sends a map $\mu\colon (C,x_1,\dots,x_n) \to X$ to the stabilization of $(C,x_n,x_{n+1},x_{n+2},x_{n+3})$. The cross-ratio of the four marked points on $\overline{M}_{0,n,n+1,n+2,n+3}$ gives an isomorphism to $\mathbf{P}^1$. Pulling back the linearly equivalent divisors $[1]$ and $[\infty]$ on $\mathbf{P}^1$ all the way to $\overline{M}_{0,n+3}(X,\beta)$ yields two linearly equivalent divisors

$$D(n,n+1 \mid n+2,n+3) \text{ and } D(n,n+3 \mid n+1,n+2).$$

We have the following explicit expressions for divisors of the above kind (see [FP97, Prop. 8])

$$D(i,j \mid k,l) = \sum D(A_1,A_2;\beta_1,\beta_2) \tag{2.2}$$

where the sum is over disjoint unions $A_1 \coprod A_2 = A$ with $i,j \in A_1$ and $k,l \in A_2$ and $\beta = \beta_1+\beta_2$.

Next, we intersect the two linearly equivalent divisors with the class ω to get an equality

$$\int D(n,n+1 \mid n+2,n+3)w = \int D(n,n+3 \mid n+1,n+2)w,$$

which we, with the use of 2.2, express as a relation between gravitational correlators

$$\sum_{\substack{\beta_1+\beta_2=\beta \\ A_1 \coprod A_2=\{1,\dots,n-1\}}} \left\langle \left(\prod_{i\in A_1} \tau_{a_i}(\gamma_i) \right) \tau_{a_n}(\gamma_n) H T_e \right\rangle_{\beta_1} g^{ef} \left\langle \left(\prod_{i\in A_2} \tau_{a_i}(\gamma_i) \right) H 1 T_f \right\rangle_{\beta_2}$$
$$= \sum_{\substack{\beta_1+\beta_2=\beta \\ A_1 \coprod A_2=\{1,\dots,n-1\}}} \left\langle \left(\prod_{i\in A_1} \tau_{a_i}(\gamma_i) \right) \tau_{a_n}(\gamma_n) 1 T_e \right\rangle_{\beta_1} g^{ef} \left\langle \left(\prod_{i\in A_2} \tau_{a_i}(\gamma_i) \right) H H T_f \right\rangle_{\beta_2} \tag{2.3}$$

The sums are also taken over all $e, f = 0, \dots, m$. By induction, we may assume that the invariants with class less than β or sum of τ-numbers less than $\sum a_i$ is known. By use of the string and divisor equations above, the only term which is not known comes from the left correlator in the first sum with $\beta_1 = \beta$ and $A_1 = \{1, \dots, n-1\}$. A dimension count shows that the correlator

$$\left\langle \left(\prod_{i=1,\dots,n} \tau_{a_i}(\gamma_i) \right) H T_e \right\rangle_{\beta}$$

is zero unless T_e is a divisor class. Applying the divisor formula twice to the above correlator, we get a nonzero multiple of the correlator we were trying to compute, plus correlators known by induction, thereby finishing the proof. □

As an example we use the equations in the proof to write down some particular recursive equations for gravitational correlators of $\mathbf{P}^2$. We will later use these invariants for computations of the number of rational nodal plane curves, subject to certain conditions.

$$\left\langle \underbrace{h^2 \cdots h^2}_{3d-3 \text{ times}} \cdot (h^2, 1) \right\rangle_d^{\mathbf{P}^2} = \frac{1}{d^2} \left(N_d + \sum_{d_1+d_2=d} \binom{3d-3}{3d_1-2} d_1 d_2^3 N_{d_1} N_{d_2} \right) \tag{2.4}$$

$$\left\langle \underbrace{h^2 \cdots h^2}_{3d-2 \text{ times}} \cdot (h, 1) \right\rangle_d^{\mathbf{P}^2} = \frac{1}{d^2} \left(\sum_{d_1+d_2=d} \binom{3d-2}{3d_1-1} d_1^2 d_2^3 N_{d_1} N_{d_2} - d N_d \right) \tag{2.5}$$

3 The tangent space of the moduli space of stable curves

In this section we describe the tangent space at a point of $\overline{M}_{0,n}(X, \beta)$ or, which amounts to the same thing, the first-order deformations of pointed maps of genus zero $\mu\colon (C, x_1, \dots, x_n) \to X$ with fixed convex target X (recall that a smooth variety X is convex if $H^1(\mathbf{P}^1, \mu * T_X) = 0$ for all maps $\mu\colon \mathbf{P}^1 \to X$). Such deformations are classified by a vector space (we denote it by T_μ) which fits into an exact sequence to be described next. Let Θ_C denote the sheaf of derivatives on C. If C is nonsingular it is the tangent sheaf of C, and when C is nodal, it can be identified with the sheaf $\nu_*(\Theta_{\tilde{C}}(-\sum(P_i - Q_i))$ where $\nu\colon \tilde{C} \to C$ is the normalization of C and P_i, Q_i the pair of points lying over the ith node of C. Let $\mathcal{N} = \operatorname{coker}(d\mu\colon \Theta_C(-x_1 - \dots - x_n) \to \mu^* T_X)$. Then there is an exact sequence (compare [Kon95]),

$$0 \to H^0(C, \mathcal{N}) \to T_\mu \to \oplus_{\text{nodes of } C} T_{P_i}\tilde{C} \otimes T_{Q_i}\tilde{C} \to 0. \tag{3.1}$$

The vector space $H^0(C, \mathcal{N})$ parametrizes deformations of μ, preserving the combinatorial type of the arrangement of the components of C and $\oplus_{\text{nodes of } C} T_{P_i}\tilde{C} \otimes T_{Q_i}\tilde{C}$ the deformations of the nodes of C. Using Riemann–Roch we can now easily find the dimension of $\overline{M}_{0,n}(X, \beta)$ and of loci of maps from curves of fixed combinatorial type. If C is of arithmetic genus zero with q nodes ($q+1$ components each isomorphic to $\mathbf{P}^1$), then the dimension of $H^0(C, \mathcal{N})$, is given by

$$\begin{aligned} h^0(C, \mathcal{N}) &= h^0(C, \mu^* T_X) - \chi(C, \Theta_C(-x_1 - \dots - x_n)) \\ &= \chi(C, \mu^* T_X) - \chi(\tilde{C}, T_{\tilde{C}}(-\tilde{x}_1 - \dots - \tilde{x}_n - \textstyle\sum(p_i + q_i))) \\ &= \int_\beta c_1(T_X) + \dim(X)(1-g) - (3 - n + q) \\ &= \int_\beta c_1(T_X) + \dim(X) + n - 3 - q. \end{aligned}$$

The following simple lemma will be used in next section.

Lemma 3.2. *Assume that x and x' are distinct points on $\mathbf{P}^1$ and $\mu\colon \mathbf{P}^1 \to X$ is a stable map such that $h^1(\mathbf{P}^1, \mu^* T_X(-x-x') = 0$. Then the natural map*

$$H^0(C, \mu^* TX(-x')) \to T_{\mu(x)}X$$

is surjective.

Proof. The result is obtained from the long exact sequence associated to

$$0 \to \mu^* TX(-x-x') \to \mu^* TX(-x') \to T_{\mu(x)}X \to 0$$

and the vanishing of the first cohomology group. □

The vanishing of the first cohomology groups of Lemma 3.2 holds for projective space when the degree of $\mu_*[\mathbf{P}^1]$ is non-zero.

4 Gromov–Witten invariants computed using excess intersection

In this section we show the main result of the paper, which is an expression for the enumerative number of rational curves in a homogenous variety with the condition that the curve should have a node on a given cycle, in additon to usual Gromov–Witten incidence conditions. The formula is derived by means of excess intersection, expressing the usual Gromov–Witten invariants as a linear combination of gravitational descendants and a *node invariant.*

We start by giving a precise definition of the node invariants. Given a non-zero class β in $A_1(X)$, consider the closure in $\overline{M}_{0,n+2}(X,\beta)$ of maps $\mu\colon \mathbf{P}^1 \to X$ of class β satisfying $\mu(x_{n+1}) = \mu(x_{n+2})$. The image of μ is then necessarily a curve with a node or worse singularity. Denote the closure Z with the induced reduced structure. Given homogenous classes $\gamma_1, \dots, \gamma_n$ and λ in $A^*(X)$ we define

$$\Delta_\beta(\gamma_1 \dots \gamma_n; \lambda) := \int e_1^*(\gamma_1) \dots e_n^*(\gamma_n) e_{n+1}^*(\lambda)[Z].$$

Proposition 4.1. *Let $\Gamma_1, \dots, \Gamma_n$ and Λ be cycles in general position representing the classes $\gamma_1, \dots, \gamma_n$ and λ. Then $\Delta_\beta(\gamma_1 \dots \gamma_n; \lambda)$ is equal to the number of maps*

$$\mu\colon (\mathbf{P}^1, x_1, \dots, x_{n+2}) \to X$$

satisfying $\mu(x_1) \in \Gamma_1, \dots, \mu(x_n) \in \Gamma_n$ and $\mu(x_{n+1}) = \mu(x_{n+2}) \in \Lambda$.

Proof. The proof is a standard application of the Kleiman–Bertini theorem [Kle74]. Compare with [FP97, Lemma 14]. □

We can now formulate the relation between a given node invariant, a corresponding Gromov–Witten invariant and a sum of gravitational correlators. We need to fix β and X such that the condition of Lemma 3.2 is satisfied, i.e., that $h^1(\mathbf{P}^1, \mu^*TX(-x-x')) = 0$ for any stable map $\mu\colon \mathbf{P}^1 \to X$, with $\mu_*[\mathbf{P}^1] = \beta$.

Theorem 4.2. *With notation and conditions as above, we have that*

$$\begin{aligned} &\langle \gamma_1 \dots \gamma_n, \lambda, \lambda \rangle_\beta \\ &\quad = \Delta_\beta(\gamma_1 \dots \gamma_n; \lambda) + \sum_{i=0}^{\dim(X)-1} \langle \gamma_1 \dots \gamma_n \tau_{\dim(X)-1-i}(c_i(T_X)\lambda) \rangle_\beta. \end{aligned}$$

In the proof we will use the excess intersection formula to compute the Gromov–Witten invariant to the left in 4.2. Since the computation of gravitational correlators can be reduced to that of Gromov–Witten invariants, Theorem 4.2 makes it possible to compute node invariants.

We start by using the formalism of [Ful84] for intersections; given that the cycles $\Gamma_1, \ldots, \Gamma_n$ are regularly embedded in X, we can express Gromov–Witten invariants, using the Cartesian square

$$\begin{array}{ccc} W & \longrightarrow & \overline{M}_{0,n+2}(X,\beta) \\ \downarrow & & \downarrow{\scriptstyle e_1 \times \cdots \times e_{n+2}} \\ \Gamma_1 \times \cdots \times \Gamma_n \times [\Lambda] \times [\Lambda] & \xrightarrow{i} & \underbrace{X \times \cdots \times X}_{n+2 \text{ times}}, \end{array}$$

as

$$\langle [\Gamma_1] \cdots [\Gamma_n][\Lambda][\Lambda] \rangle_\beta = \int i^! [\overline{M}_{0,n}(X,\beta)].$$

Since X is homogenous there is no loss of generality in requiring the cycles Γ_i to be regularly embedded; the (Chow) cohomology ring is generated by effective divisors. The Gromov–Witten invariants depend only on the rational equivalence class of the cycles and thus we have the freedom to choose cycles, within the equivalence class, as one pleases. Here, we have chosen two cycles to be equal. This way, the cycles are of course not in general position.

To understand the intersection W, we divide the cartesian diagram above, defining W, into two—first intersecting with the diagonal $\Delta: X \to X \times X$ and then with $\Gamma_1 \times \cdots \times \Gamma_n \times \Lambda \to X^{n+1}$

$$\begin{array}{ccccc} W & \longrightarrow & W' & \longrightarrow & \overline{M}_{0,n+2}(X,\beta) \\ \downarrow & & \downarrow & & \downarrow{\scriptstyle e_1 \times \cdots \times e_n} \\ \Gamma_1 \times \cdots \times \Gamma_n \times \Lambda & \xrightarrow{j} & X^{n+1} & \xrightarrow{\mathrm{id}_{X^{n+1}} \times \Delta} & X^{n+2}. \end{array}$$

We claim that W' consists of irreducible components W_0' and $\{W_A'\}$ where the index A is a subset of $\{1, \ldots, n+2\}$ containing $n+1$ and $n+2$. The component W_0' is the closure of maps $\mu\colon (\mathbf{P}^1, x_i) \to X$ satisfying $\mu(x_{n+1}) = \mu(x_{n+2})$, with the reduced structure and expected $\operatorname{codim}(W_0', \overline{M}_{0,n+2}(X,\beta)) = \operatorname{codim}(\Delta) = \dim(X)$. W_A' is scheme-theoretically identical with the divisor $D(\{1,2,\ldots,n+2\} \setminus A, A, \beta, 0)$ (see [FP97]) consisting of maps μ from a curve with two components, such that generically μ is constant on one component containing the markings in A. A component W_A' thus has excess dimension equal to $\dim(X) - 1$. We first prove the set theoretic claim. We need to show that a stable map $\mu\colon (C, x_i) \to X$ satisfying the condition $\mu(x_{n+1}) = \mu(x_{n+2})$ is either in W_0' or in one of the W_A''s. Recall that a curve C of arithmetic genus zero has component structure according to the combinatorial configuration of a tree. Let C' be the minimal subtree of components containing x_{n+1} and x_{n+2}. Now, either μ restricted to C' is constant or not constant. In the non-constant case we can smooth all the nodes of C within W', then the source curve is $\mathbf{P}^1$ and μ is in W_0'. In the constant case we can smooth all nodes except one within W' so that the source curve has two components and the map is constant on one of them. The map μ is then in W_A where A is the set of markings which are contained in C'.

Next, we will prove, using tangent space considerations, that all the components of W' are reduced. The tangent space of the diagonal inside $X \times X$ at a point (x,x) is of course a pair of identical tangent vectors in $T_{(x,x)}(X \times X) = T_xX \oplus T_xX$. The restriction imposed on tangent vectors of $\overline{M}_{0,n+2}(X,\beta)$ at points μ (denoted by T_μ) of W' is thus that the images of tangent maps induced by the evaluation maps are equal. It remains to show that this restricts the tangent vectors sufficiently for the different components. In the case of W_0' this follows from Lemma 3.2 which assures that there are enough global sections in $H^0(\mathbf{P}^1, \mu^*TX(-x_{n+1}))$ to span $T_{\mu(x_{n+2})}$ and conversely (with x_{n+1} and x_{n+2} interchanged). In all there are $2\dim(X)$ independent sections of $H^0(\mathbf{P}^1, \mu^*TX)$ representing tangent vectors in T_μ corresponding to deforming the image of the markings x_{n+1} and x_{n+2} in arbitrary directions in X. Thus, restricting the directions to be equal cuts the dimension of T_μ down by $\dim(X)$ and thus down to the tangent space of W_0'.

Since the components W_A' are divisors it is enough to show that the dimension of the tangent space T_μ at a point μ of W_A is cut down by one. This is easily seen since in deformations of a map $\mu\colon C \to X$ in W_A the deformed images of the markings x_{n+1} and x_{n+2} can clearly move independenly along a limit tangent vector of $\mu(C)$. Thus the dimension of T_μ is restricted by one.

In the next step, we compute the intersection with the diagonal (Gysin homomorphism), using a residual intersection formula [Ful84]. The component $W_{\{n+1,n+2\}}$ is identical with the image of the $(n+1)$st canonical section $s_i\colon \overline{M}_{0,n+1}(X,\beta) \to \overline{M}_{0,n+2}(X,\beta)$. Let R denote the union of the other components of W' (the residual to $W_{\{n+1,n+2\}}$ in W'), and consider the residual intersection diagram

$$\begin{array}{ccccc} & & R & & \\ & & \downarrow & & \\ W_{\{n+1,n+2\}} & \longrightarrow & W' & \longrightarrow & \overline{M}_{0,n+2}(X,\beta) \\ & & \downarrow & & \downarrow{\scriptstyle e_1\times\cdots\times e_{n+2}} \\ & & X^{n+1} & \xrightarrow{\mathrm{id}_{X^{n+1}}\times\Delta} & X^{n+2}. \end{array}$$

We have that [Ful84, Thm. 9.2]

$$
\begin{aligned}
&(\mathrm{id}_{X^{n+1}} \times \Delta)^!([\overline{M}_{0,n+2}(X,\beta)]) \\
&\quad = \{c(N_{\Delta(X)}X \times X)s(W_{\{n+1,n+2\}}, \overline{M}_{0,n+2}(X,\beta))\}^{\dim(X)} \\
&\qquad + \{c(N_{\Delta(X)}X \times X) \otimes \mathcal{O}(-W_{\{n+1,n+2\}})s(R, \overline{M}_{0,n+2}(X,\beta))\}^{\dim(X)} \\
&\quad = \{c(TX)c(N_{s_{n+1}(\overline{M}_{0,n+1}(X,\beta))}\overline{M}_{0,n+2}(X,\beta))\}^{\dim(X)} + [W_0'] + R',
\end{aligned}
$$

where R' is a class supported on the union of the components W_A' for $A \neq \{n+1, n+2\}$. We simplify the first term further, identifying $W_{\{n+1,n+2\}}$ with $\overline{M}_{0,n+1}(X,\beta)$

$$
\begin{aligned}
&\{c(TX)c(N_{s_{n+1}(\overline{M}_{0,n+1}(X,\beta))}\overline{M}_{0,n+2}(X,\beta))\}^{\dim(X)} \\
&\quad = \{c(TX)\frac{1}{1+\psi_{n+1}}\}^{\dim(X)} \\
&\quad = \sum_{i=0}^{\dim(X)-1} c_i(TX)\psi_{n+1}^{\dim(X)-1-i}.
\end{aligned}
$$

In the second part of the intersection, we have a standard situation with cycles $\Gamma_1, \ldots, \Gamma_n$ and Λ in general position. It follows first of all, from a dimension count, that there is no contribution from R' (there are too many conditions on the node). It is easily verified that the contribution from W_0' in the second intersection is the node invariant in the statement of the theorem, and also that the sum of contributions from $W_{\{n+1,n+2\}}$ is identical to the sum of gravitational correlators of the statement.

Example 4.3. We finish this section by deriving some formulas for node invariants of the projective plane. We take $X = \mathbf{P}^2$, h to be the class of a line and $\beta = dh$. Set

$$N_d = \langle \underbrace{h^2 \cdots h^2}_{3d-1 \text{ times}} \rangle.$$

The node invariant

$$\Delta_d(\underbrace{h^2 \cdots h^2}_{3d-3 \text{ times}}; h^2)$$

is then equal to twice (because of the symmetry between the two markings mapping to the node) the number of rational plane degree d points with a node at a point, through $3d-3$ other points. We denote this number (half the node invariant) by P_d. Theorem 4.2 gives us the relation

$$N_d = 2P_d + \langle \underbrace{h^2 \cdots h^2}_{3d-3 \text{ times}} \tau_1(h^2) \rangle,$$

and using 2.4 we can express P_d solely interms of Gromov–Witten invariants:

$$P_d = \frac{1}{2}\Big[N_d - \sum_{d_1+d_2=d} \binom{3d-3}{3d_1-2} d_1 d_2^3 N_{d_1} N_{d_2}\Big].$$

The numbers $P_3 = 1$, $P_4 = 96$ etc. were calculated before by Aluffi [Alu91] for $d = 3$ using complete cubics and by Pandharipande–Göttsche [GP98] and by [Gat98] for all d using quantum cohomology of the blow-up of $\mathbf{P}^2$ at a point.

Similarly, we can express, using Theorem 4.2 and 2.5, the number Q_d of rational plane curves of degree d with a node on a line and through $3d-2$ pts. Q_d is half of the node invariant $\Delta_d(\underbrace{h^2 \cdots h^2}_{3d-2 \text{ times}}; h)$. The expression for Q_d is

$$Q_d = \frac{1}{2}\Big[(2d - 3 - \frac{1}{d})N_d - \frac{1}{d^2} \sum_{d_1+d_2=d} \binom{3d-2}{3d_1-1} d_1^2 d_2^3 N_{d_1} N_{d_2}\Big],$$

and the number $Q_3 = 6$ was computed by Aluffi [Alu91]. The next number is $Q_4 = 768$.

5 Generalizations

A natural generalization is to allow for more than one node condition. The author considered this but realized that there will be more excess components contributing to the intersection and that it is necessary to consider cuspidal invariants analogous to nodal invariants.

Another possible generalization is to define invariants, analogous to the node invariants used in this paper, on the stack of stable lifts (see [EK98]) which were used to define Gromov–Witten invariants of rational curves with tangency conditions. It seems feasible that one could define invariants with bitangent or tac-node conditions combined with standard tangency conditions. A necessity for this development is, however, that it is possible to develop a theory of gravitational descendents for stable lifts.

References

[Alu91] P. Aluffi, The enumerative geometry of plane cubics. II. Nodal and cuspidal cubics, *Math. Annalen.* **289** (1991), 543–572

[Beh97] K. Behrend, Gromov–Witten invariants in algebraic geometry, *Invent. Math.* (1997), 601–617

[BF97] K. Behrend and B. Fantechi, The intrinsic normal cone, *Invent. Math.* **127** (1997), 45–88

[EK98] L. Ernström and G. Kennedy, Recursive formulas for the characteristic numbers of rational plane curves, *J. Algebraic Geom.* **7** (1998), 141–181

[FP97] W. Fulton and R. Pandharipande, *Notes on stable maps and quantum cohomology*, Algebraic geometry—Santa Cruz 1995, Proc. Sympos. Pure Math., vol. 62, Amer. Math. Soc., Providence, RI, 1997, pp. 45–96

[Ful84] W. Fulton, *Intersection Theory*, Springer-Verlag, Berlin, 1984

[Gat98] A. Gathmann, *Gromov–Witten and degeneration invariants: computation and enumerative significance*, Ph.D. thesis, Hannover, 1998

[Get98] E. Getzler, *Topological recursion relations in genus 2*, alg-geom/ 9801003, 1998

[GP98] L. Göttsche and R. Pandharipande, The quantum cohomology of blow-ups of $\mathbf{P}^2$ and enumerative geometry, *J. Differential Geom.* **48**:1 (1998), 61–90

[Kle74] S. L. Kleiman, The transversality of a general translate, *Comp. Math.* **28**:3 (1974), 287–297

[KM94] M. Kontsevich and Yu. Manin, Gromov–Witten classes, quantum cohomology, and enumerative geometry, *Comm. Math. Phys.* **164**:3 (1994), 525–562

[KM98] M. Kontsevich and Yu. Manin, Relations between the correlators of the topological sigma-model coupled to gravity, *Comm. Math. Phys.* **196**:2 (1998), 385–398

[Koch] J. Koch, Recursion for twisted descendents and characteristic numbers of rational plane curves, Math. AG/9902021

[Kon92] M. Kontsevich, Intersection theory on the moduli space of curves and the matrix Airy function, *Comm. Math. Phys.* **147**:1 (1992), 1–23

[Kon95] M. Kontsevich, *Enumeration of rational curves via torus actions*, The moduli space of curves (Texel Island, 1994), Birkhäuser Boston, Boston, MA, 1995, pp. 335–368

[Pan] R. Pandharipande, e-mail with recursive equation for characteristic numbers

[Pan97] R. Pandharipande, *Quantum differential equations, II*, Quantum Cohomology (Mittag Leffler Institute) (P. Aluffi, ed.), 1997, pp. 78–83

[Wit91] E. Witten, *Two-dimensional gravity and intersection theory on moduli space*, Surveys in differential geometry (Cambridge, MA, 1990), Lehigh Univ., Bethlehem, PA, 1991, pp. 243–310

Department of Mathematics
Royal Institute of Technology
S-100 44 Stockholm
Sweden
ernstrom@math.kth.se

Join Varieties and Intersection Theory

Hubert Flenner

0 Introduction

In classical intersection theory as represented by the work of Severi, van der Waerden, Weil, Samuel and many others, one starts with projective varieties X and Y of the projective space which meet properly. The famous Bezout theorem states that

$$\deg X \deg Y = \sum_C i_C \deg C,$$

where C runs through the irreducible components of $X \cap Y$. The intersection number i_C can be defined in the Weil–Samuel approach as a certain multiplicity of the diagonal ideal of the local ring $\mathcal{O}_{X\times Y,C}$ of $X \times Y$ at the generic point of C or, alternatively, in the classical approach of Severi and van der Waerden, by moving the local intersections and counting points. For improper intersections the situation is much more complicated. In the classical theory there was no satisfying treatment of arbitrary intersections.

Nowadays we have the important intersection theory of Fulton and MacPherson which developed during the 1970s and the beginning of the 1980s. By this it is possible to assign to an arbitrary intersection of subvarieties $X \cap Y$ a cycle on $X \cap Y$ which is unique up to linear equivalence and which represents the intersection product in the Chow groups. This approach works very generally for arbitrary algebraic manifolds.

In 1981 Vogel and Stückrad gave another treatment of improper intersections. Although their approach is restricted to intersections of subvarieties in projective space their method is very interesting and produces a cycle in a quite natural and simple way. In this paper we will give an overview of their method and some of the main applications.

Slightly more generally their method can be described as follows. Let V be an equidimensional algebraic scheme, $\mathcal{L}$ a line bundle on V and $\sigma_0, \dots, \sigma_d$ sections in $\mathcal{L}$ with common zero set $W \subseteq V$. Consider generic linear combinations $s_i = \sum u_{ij}\sigma_j$ and the divisors $D_i := \{s_i = 0\}$ on V (to make the approach more canonical we will use indeterminates as coefficients u_{ij}). The following simple algorithm gives the Stückrad–Vogel cycle.

Step 1: Let B^0 be the closure of $V \backslash W$ in V. Moreover, let V^0 be the closure of $V \backslash B^0$ in V.

Step i: For $i \geq 1$, let B^i be the closure of $D_{i-1} \cap B^{i-1} \backslash W$ and V^i the closure of $D_{i-1} \cap B^{i-1} \backslash B^i$.

By the genericity of the hyperplanes, the intersection $D_{i-1} \cap B^{i-1}$ is proper at each step. Thus V^i is the part of this intersection that is set-theoretically contained in W, and B^i is the union of the remaining components. Let $v^i = \sum j_C[C]$ and β^i be the cycles on W associated to V^i, B^i respectively. By construction these are cycles of dimension $\dim V - i$. In the first section of this paper we will study these cycles carefully. We will describe the multiplicities j_C for the irreducible components C of W. The main result due to Stückrad, Vogel and, in this generalized form, due to van Gastel, will be that for a proper variety we have

$$\deg_{\mathcal{L}} V = \sum \deg_{\mathcal{L}} v_i + \deg(V/\sigma(V)) \deg \sigma(V),$$

where $\sigma : V \dashrightarrow \mathbb{P}^d$ is the rational map $[\sigma_0 : \cdots : \sigma_d]$ and where $\deg V/\sigma(V)$ is the usual mapping degree.

In Section 2 we will use this construction to deduce a refined Bezout theorem for arbitrary intersections. The key observation for this is the join construction. Let X, Y be subvarieties of $\mathbb{P}^n$ and let J be their join which is the subvariety of $\mathbb{P}^{2n+1}$ given by all points $[x : y]$ such that $[x] \in X$ or $x = 0$, and $[y] \in Y$ or $y = 0$. There is a natural diagonal embedding $\mathbb{P}^n \cong \Delta \subseteq \mathbb{P}^{2n+1}$ given by $[x] \mapsto [x : x]$. Then Δ is a linear subspace of $\mathbb{P}^{2n+1}$ that intersects J in $X \cap Y \hookrightarrow \Delta$. If we choose homogeneous coordinates $[x_0 : \cdots : x_n]$ on $\mathbb{P}^n$ and take $[x_0 : \cdots : x_n : y_0 : \cdots : y_n]$ as homogeneous coordinates on $\mathbb{P}^{2n+1}$ then Δ is given by the equations $\sigma_i := x_i - y_i = 0$, $i = 0, \dots, n$. Applying the construction above to $V := J$ and these σ_i we get cycles v_i defined on $\Delta \cap J \cong X \cap Y$. Using the fact that the degree of J is just the product of the degrees of X and Y we deduce from the above formula that the Bezout theorem holds in the following form:

$$\deg X \deg Y = \sum \deg v^i + \deg J/XY \deg XY.$$

Here XY is the embedded join variety which is by definition the image of J under the rational map $\pi : J \dashrightarrow \mathbb{P}^n$ with $[x : y] \mapsto [x - y]$. We will study in Section 2 the cycle $v = \sum j_C[C]$ more carefully. For irreducible components C of the intersection we will characterize as to when $j_C = 1$ or when the j-multiplicity is equal to the product of the multiplicities of the local rings $\mathcal{O}_{X,C}$, $\mathcal{O}_{Y,C}$.

In Sections 3 and 4 we will describe the connection of the cycle $v = \sum v^i$ with Segre classes. In Section 3 we will focus our attention to the case of cones $V = C$, i.e., W will be the zero-section of C. The main result due to van Gastel is that v is equal to the Segre class $s(C)$. We will give some applications to the positivity of Segre classes of negative vector bundles.

This will be generalized further in Section 4 to the general case. The main observation for this is the result of van Gastel that the Stückrad–Vogel cycle is invariant under passage to the normal cone. In an appendix to this section we will give a self-contained algebraic proof of this result. We use this to study the K-rational part of v and will show that the K-rational components are closely related to the irreducible components of the normal cone $C_W V$.

Following [FVo2] we will introduce in Section 5 the limit of join variety which is a generalization of the join variety to the non-homogeneous case. By this we assign to every pair of subschemes X, Y of, say, $\mathbb{P}^n$ and a subvariety $C \subseteq X \cap Y$ a cone $\mathrm{LJoin}_C(X,Y)$. In the case that the ground field are the complex numbers and $C = \{p\}$ is a point this is the union of all lines $\ell = \lim \overline{x_\nu y_\nu}$, with $x_\nu \in X$, $y_\nu \in Y$ converging to 0. We will show that this cone has maximal dimension if and only if C belongs to the v-cycle, see 5.11. Thus we obtain a geometric characterization of the K-rational components of v.

In Section 6 we consider the question when the join variety has minimal dimension $\dim X + \dim Y - \dim X \cap Y$. Following [FGV] we will show that this holds if and only if

$$\deg X \deg Y = \sum_{\mathrm{irr}} j_C \deg C,$$

where the sum is taken over all the irreducible components of $X \cap Y$. We also discuss a converse to Bezout's in the Cohen–Macaulay case (cf. [FVo1]) and show that one has always

$$\deg X \deg Y \geq \sum_{\mathrm{irr}} \lg(\mathcal{O}_{X\cap Y,C}) \deg C + e,$$

where e is the excess dimension of this intersection. Thus the intersection $X \cap Y$ is proper if and only if $\deg X \deg Y = \deg X \cap Y$. We will give a largely simplified proof of these results in the case that X is irreducible.

In Section 7 we will treat some applications to linear projections. Partly this material is taken from [BHSS] with some simplified proofs. We will give estimates for the dimension of the image of a linear projection $\pi : V \dashrightarrow \mathbb{P}^d$ of a subvariety $V \subseteq \mathbb{P}^n$. Moreover some of the extremal cases are characterized.

An important application of the Stückrad–Vogel cycle is given in Sections 7 and 8. Following [vGa3, FMa1] we will study generic linear projections $\pi : X \cup Y \to \mathbb{P}^{n+m-k}$, where n and m are the dimensions of X, $Y \subseteq \mathbb{P}^N$ respectively. We will show that the scheme of double points (which we will place into the join) can be identified with the cycle β_k, see 8.5. Using this and the formulas for v_i in terms of Segre classes of Sections 3 and 4, we recover classical formulas for the number of double points of generic linear projections due to [Adl, PSi]. Moreover, in 9.1 we will show that in the case $X = Y$ the ramification locus of π is just the cycle v_{k-1} on the regular part of X. We will give a simplified treatment for these facts. Applying again the formulas of Sections 3 and 4 we will get a very natural approach to the classical embedding obstructions of Johnson and Holme [Joh, Hol1, Hol2].

In the final section we present some results of [SUV, FUV] on when the limit of join variety at a given subvariety C of the intersection has maximal dimension so that it belongs to the intersection cycle. We present some of the main results of [FUV] which rely on the theory of residual intersections of Artin–Nagata, Huneke and many others.

I have tried to make this text as self-contained as possible in the hopes that it will serve as a readable overview of this theory. As general references for basic facts and notations in commutative algebra, algebraic geometry and intersection theory I refer the reader to the books [Mat, Har2, Ful1]. A largely extended presentation of this material will appear in the forthcoming monograph [FOV].

1 The intersection algorithm

The central topic of this lecture is to describe the basic intersection algorithm which is due to Stückrad and Vogel.

1.1. Let V be an equidimensional algebraic scheme over the base field K and $\mathcal{L}$ a line bundle on V. Let $\sigma_0, \dots, \sigma_d$ be fixed elements of $H^0(V, \mathcal{L})$ and set $W := \{\sigma_0 = \cdots = \sigma_d = 0\}$.

We will construct from these data cycles v^i, $i \geq 0$, with v^i (if non-zero) an effective cycle of dimension $\dim V - i$ defined on $W_L := W \otimes_K L$, where L is a pure transcendental field extension of K. We introduce indeterminates U_{ij} and set

$$L := K(U_{ij} : 0 \leq i, j \leq d),$$
$$D_i := V(s_i) \text{ with } s_i := \textstyle\sum_{j=0}^{d} U_{ij}\sigma_j, \quad 0 \leq i \leq d.$$

By Cramer's rule, the set of zeros of $s_0, \dots, s_d$ again is W. To construct v^i, we define for $i \geq 0$ closed subschemes V^i, B^i of $V_L := V \otimes_K L$ by the following simple algorithm:

Step 0: Let B^0 be the closure of $V_L \backslash W_L$ in V_L. We will always equip the closure with the structure given by the ideal sheaf $\ker(\mathcal{O}_{V_L} \to j_*(\mathcal{O}_{V_L \backslash W_L}))$, where $j : V_L \backslash W_L \to V_L$ is the inclusion. (We sometimes refer to this as the *scheme-theoretic closure.*) Moreover, we define V^0 to be the closure of $V_L \backslash B^0$ in V_L.

Step i: For $i \geq 1$, let B^i be the closure of $D_{i-1} \cap B^{i-1} \backslash W_L$ and V^i the closure of $D_{i-1} \cap B^{i-1} \backslash B^i$.

Observe that $B^{d+1} = \emptyset$. For $i > d+1$ we set $B^i = V^i = \emptyset$. Note that at each step of the construction the following properties are satisfied:

1. $D_i \cap B^i$ is a Cartier divisor on B^i.
2. $V^i \subseteq W_L$ set-theoretically, whereas no component (irreducible or embedded) of B^i is set-theoretically contained in W_L.
3. $D_0, \ldots, D_{i-1}$ and V^i, B^i are defined over $L_{i-1} := K(U_{\mu\nu}, 0 \leq \mu \leq i-1, 0 \leq \nu \leq d)$.

Clearly (2) and (3) are immediate from the construction. Let us show (1). Because of (2) and (3) this is a consequence of the following proposition, which is a kind of generic Bertini theorem.

Proposition 1.2. *Let X be an algebraic K-scheme, $\mathcal{L}$ a line bundle on X and $\sigma_0, \ldots, \sigma_d$ sections of $\mathcal{L}$ without common zero. For indeterminates $U_0, \ldots, U_d$ consider $s := \sum U_i \sigma_i$ as a section of $\mathcal{L} \otimes_K L$ where $L := K(U_0, \ldots, U_d)$. Then $s = 0$ defines a Cartier divisor D on $X_L := X \otimes_K L$. Moreover, if X is irreducible (resp. reduced, normal, regular, satisfies Serre's condition S_k) then D has the same property.*

Proof. We may assume that $X = \operatorname{Spec} A$ is affine and that $\mathcal{L}$ is generated by, say, σ_0 so that $\sigma_i = f_i \sigma_0$ with suitable elements $f_i \in A$. Then D is the spectrum of the ring

$$A \otimes_K L/(U_0 + \sum_{i \geq 1} U_i f_i).$$

Obviously this is a localization of the ring

$$A[U_0, \ldots, U_d]/(U_0 + \sum_{i \geq 1} U_i f_i) \cong A[U_1, \ldots, U_d].$$

This immediately gives that D is a Cartier divisor. As the listed properties are stable under polynomial ring extensions and localization, the second part also follows. □

It is clear from the construction that B^{i+1} is the closure of $D_0 \cap \cdots \cap D_i \backslash W_L$ in V_L (in the scheme-theoretic sense). By construction, $V^i = B^i = \emptyset$ for $i > \dim V$.

Definition 1.3. The cycles associated to V^i, B^i will always be denoted by

$$v^i = v(\underline{\sigma}, V), \quad \beta^i := \beta(\underline{\sigma}, V)$$

respectively. Moreover v_i and β_i denote the part of dimension i of $v := \sum v^i$ and $\beta := \sum \beta^i$, so that the upper index denotes the codimension in V and the lower one the dimension of the cycle. We will refer to v, β as the *v-cycle*, resp. *β-cycle*; note that v is a cycle on W_L. Assume that X is a locally closed subscheme of V. Then there is also a well-defined v-cycle $v(\underline{\sigma}|X, X)$ which we will denote in brief by $v(\underline{\sigma}, X)$, and similarly for the β-cycle.

The following proposition shows that v depends only on the cycle class $[V]$ of V:

Proposition 1.4. *Assume that $[V] = \sum n_i[V_i]$, where the V_i are the irreducible components of V equipped with their reduced structure. Then*

$$v(\underline{\sigma}, V) = \sum n_i v(\underline{\sigma}, V_i), \qquad \beta(\underline{\sigma}, V) = \sum n_i \beta(\underline{\sigma}, V_i).$$

Proof. Let $V^j(V), B^j(V), v^j(V), \beta^j(V)$ be the subschemes, resp. cycles, given by the above construction. By induction on j we will show that

$$v^j(V) = \sum_i n_i v^j(V_i), \quad \beta^j(V) = \sum_i n_i \beta^j(V_i).$$

For $j = 0$ this is obvious. Assume that the assertion has been shown for the index $j < d+1$. By property (1) above, D_j gives a Cartier divisor, say, $\tilde{D}_j$ on $B^j(V)$. Then

$$\begin{aligned} \beta^{j+1}(V) + v^{j+1}(V) &= \beta^j(V) \cdot_{B^j(V)} \tilde{D}_j \quad \text{by 1.5 below} \\ &= \sum_i n_i \beta^j(V_i) \cdot_{B^j(V)} \tilde{D}_j \\ &= \sum_i n_i(\beta^{j+1}(V_i) + v^{j+1}(V_i)) \quad \text{by 1.5,} \end{aligned}$$

where for the second equality we have used the induction hypothesis. □

In the proof we have used the following simple lemma.

Lemma 1.5. *Let V be an equidimensional algebraic K-scheme and D an effective Cartier divisor on V. Then $D \cdot [V] = [D]$ in $Z_*(V)$.*

Proof. Let $C \subseteq D \cap V$ be an irreducible component of $D \cap V$, equipped with its reduced structure. Let $A := \mathcal{O}_{V,C}$ be the local ring at the generic point of C and $f \in A$ the local equation of D. By our assumptions, f is not a zerodivisor in A, and since A is of dimension 1 we have

$$\lg(A/fA) = e(fA, A).$$

Observe that this integer is the coefficient of C in $[D]$. Let $\mathfrak{p}_1, \ldots, \mathfrak{p}_r \subseteq A$ be the minimal primes of A. Then the coefficient of C in $D \cdot [V]$ is

$$\sum_{\rho=1}^{r} \lg(A_{\mathfrak{p}_\rho}) \cdot \lg(A/(\mathfrak{p}_\rho + fA)) = \sum_{\rho=1}^{r} \lg(A_{\mathfrak{p}_\rho}) \cdot e(f \cdot A/\mathfrak{p}_\rho, A/\mathfrak{p}_\rho).$$

Now the well-known associativity formula for multiplicities gives that this is just $e(fA, A)$, proving 1.5. □

For a k-cycle $\alpha = \sum n_i[X_i]$ on V we set

$$v(\underline{\sigma}, \alpha) := \sum n_i v(\underline{\sigma}, X_i) \quad \text{and} \quad \beta(\underline{\sigma}, \alpha) := \sum n_i \beta(\underline{\sigma}, X_i).$$

Note that $v(\underline{\sigma}, \alpha)$ and $\beta(\underline{\sigma}, \alpha)$ are linear in α. By Proposition 1.4 we have $v(\underline{\sigma}, X) = v(\underline{\sigma}, [X])$ if X is a closed subscheme of V, and similarly for the β-cycle.

In the special case $|\alpha| \subseteq W$, one gets $v^0 = \alpha$ and $v^i = 0$ otherwise, since $\beta^0 = 0$ by construction.

Lemma 1.6. *In $A_*(D_0 \cap \cdots \cap D_{i-1})$ we have*

$$c_1(\mathcal{L} \otimes_K L)^i \cap [V_L] = \sum_{j=0}^{i} c_1(\mathcal{L} \otimes_K L)^{i-j} \cap v^j + \beta^i, \quad i \leq d+1.$$

Proof. From the construction we obtain $D_i \cdot \beta^i = v^{i+1} + \beta^{i+1}$. Using the fact that $D_i \cdot \beta^i = c_1(\mathcal{L} \otimes_K L) \cap \beta^i$, the assertion follows by induction on i. □

By construction, $v(\underline{\sigma}, V)$ is a cycle on W_L. The next result, due to Fulton [Ful2], shows that modulo rational equivalence $v(\underline{\sigma}, V)$ can be considered as a cycle on W, justifying our construction with indeterminates.

Proposition 1.7. *Let X be an algebraic K-scheme and $L : K$ a pure transcendental field extension. Then*

$$A_k(X) \xrightarrow{\sim} A_k(X_L) \quad \text{for all } k.$$

Proof. By a simple direct limit argument it is sufficient to treat the case of a finitely generated extension $K \to L$, i.e., $L = K(T_1, \ldots, T_n)$. It is easily seen that

$$A_k(X_L) = \varinjlim A_{k+n}(X \times_K U),$$

where the limit is over all nonempty open subsets U of $\mathbb{A}^n_K$. Thus it is sufficient to show that the canonical map

$$A_k(X) \xrightarrow{\pi^*} A_{k+n}(X \times U)$$

is bijective for all open dense subsets U of $\mathbb{A}^n_K$. This map factors as follows: $A_k(X) \to A_{k+n}(X \times \mathbb{A}^n_K) \to A_{k+n}(X \times U)$ and both of the latter maps are surjective by [Ful1, Th. 3.3 (a), resp. 1.8]. If there is a K-rational point $x \in U$ and $i_x : X \hookrightarrow X \times U$ is the embedding at x then $i_x^* : A_{k+n}(X \times U) \to A_k(X)$ satisfies $i_x^* \pi^* = id$ and thus π^* is also injective. This proves the lemma unless K is a finite field. But in this case one can find finite extensions K_a and K_b of K of relatively prime degrees a and b such that U has rational points over K_a and K_b, and π^* becomes an isomorphism after base change $K \to K_a$, resp. $K \to K_b$. Now the composed map

$$A_*(X) \xrightarrow{j^*} A_*(X \otimes_K K_a) \xrightarrow{j_*} A_*(X)$$

where $j : X \otimes_K K_a \to X$ is the canonical morphism, is multiplication by a (cf. [Ful1, Ex. 13.12]). Using the diagram

$$\begin{array}{ccc} A_*(X) & \longrightarrow & A_*(X \otimes_K K_a) \\ \downarrow\pi^* & & \downarrow (\pi\otimes K_a)^* \\ A_*(X \times U) & \longrightarrow & A_*((X \times U) \otimes_K K_a) \end{array}$$

it follows that $\ker \pi^*$ is annihilated by a. Similarly, $b \cdot \ker \pi^* = 0$. Since $(a, b) = 1$, this implies $\ker \pi^* = 0$, as required. □

We now continue our study of the intersection algorithm. We note the following simple properties.

Proposition 1.8. (0) *Let $L' : K$ be a field extension and $U'_{ij} \in L'$ $(0 \le i, j \le d)$ be algebraically independent elements. Set $s'_i := \sum U'_{ij}\sigma_j$. Performing the intersection algorithm with s'_i instead of s_i yields a cycle v' on $V \otimes L'$. Then $v' = v \otimes_L L'$, under the embedding $L \hookrightarrow L'$ given by $U_{ij} \mapsto U'_{ij}$.*

(1) *If σ_d is identically zero on V then $v(\sigma_0, \dots, \sigma_{d-1}, V)$ and $v(\sigma_0, \dots, \sigma_d, V)$ are equal up to a pure transcendental field extension.*

(2) *Assume that $\tau_i = \sum_{j=0}^{d} \alpha_{ij}\sigma_j$, $0 \le i \le d$, with (α_{ij}) a nonsingular $(d+1) \times (d+1)$-matrix over K. Then $v(\underline{\sigma}, V)$ and $v(\underline{\tau}, V)$ are equal up to a K-automorphism of L.*

(3) *If $\pi : V' \to V$ is a flat morphism of algebraic K-schemes then $\pi_L^*(v(\underline{\sigma}, V)) = v(\pi^*(\underline{\sigma}), V')$, where $\pi_L := \pi \otimes_K L$. Similarly, the formation of the subschemes V^i, B^i and cycles β^i is compatible with π^*.*

Proof. (1) and (2) follow immediately from the construction. For the proof of (3), it is sufficient to show that taking the scheme-theoretic closure of a closed subscheme Z of $V \backslash W$ in V is compatible with pull-back under flat morphisms. But this follows from the fact that the ideal sheaf $\mathcal{I}_{\bar{Z}}$ of the closure $\bar{Z}$ is given by the kernel of the map $\mathcal{O}_V \to j_*(\mathcal{O}_Z)$, where j denotes the inclusion $Z \hookrightarrow V$, and from the exactness of the functor π^*. □

Example 1.9. Let $V \subseteq \mathbb{P}^n$ be a linear subspace and let $\sigma_0, \dots, \sigma_d$ be linear forms on $\mathbb{P}^n$ considered as sections of $\mathcal{L} := \mathcal{O}_{\mathbb{P}^n}(1)$. Then up to a pure transcendental field extension $v(\underline{\sigma}, V)$ is given by $[E]$, where $E = V \cap \{\sigma_0 = \cdots = \sigma_d = 0\}$. In fact, we may assume that $V = \mathbb{P}^n$ and, by 1.8 (1), (2), that $\sigma_0, \dots, \sigma_d$ are part of a homogeneous coordinate system of $\mathbb{P}^n$. But then the assertion is obvious.

Proposition 1.10. *Assume that V is equidimensional and let $C \subseteq W$ be an irreducible component. Then C_L appears in the v-cycle with coefficient $e(I, \mathcal{O}_{V,C})$ where $I \subseteq \mathcal{O}_{V,C}$ is the ideal of W at C and $e(I, \mathcal{O}_{V,C})$ denotes the multiplicity. Moreover, if $e \in \mathcal{L}_C$ is a local generator of $\mathcal{L}$ and $s_i = f_i e$ with $f_i \in \mathcal{O}_{V_L, C_L}$ then*

$$e(I, \mathcal{O}_{V,C}) = e((f_0, \ldots, f_{k-1}), \mathcal{O}_{V_L, C_L}). \tag{$*$}$$

Proof. As the statement is local we may assume that $e = 1$ and $\mathcal{L} \cong \mathcal{O}_V$, so that $s_i = f_i$ and $I = (\sigma_0, \ldots, \sigma_d)$. Since C is an irreducible component of W, the ideal I is primary to the maximal ideal, and the ring $\mathcal{O}_{V,C}$ has dimension $k := \dim V - \dim C$. Moreover, the elements $s_0, \ldots, s_{k-1}$ form a minimal reduction of $I\mathcal{O}_{V_L, C_L}$ and so $(*)$ holds (cf. [FOV, 1.2.12 and 1.5.16 (1)]). Comparing the intersection algorithm with the usual algorithm to compute multiplicities (see e.g., [FOV, 1.2.16]), $e((s_0, \ldots, s_{k-1}), \mathcal{O}_{V_L, C_L})$ is just the coefficient of C_L in the v-cycle. □

Corollary 1.11. (Criterion for multiplicity one) *Assume that V is strictly equidimensional and let C be an irreducible component of W. Then C_L appears with coefficient 1 in the v-cycle iff $\mathcal{O}_{V,C}$ and $\mathcal{O}_{W,C}$ are regular.*

The proof follows immediately from Nagata's criterion for multiplicity 1 (see e.g. [Nag, FOV, 1.2.21]) and 1.10.

Exercise 1.12. (1) Consider the projective plane $V = \mathbb{P}^2$ with homogeneous coordinate functions X_0, X_1, X_2 and compute $v((\sigma_0, \sigma_1), V)$ with $\sigma_0 := X_1^2 X_2$ and $\sigma_1 := X_1 X_2^2$. Here $\mathcal{L} = \mathcal{O}_{\mathbb{P}^2}(3)$. Show that

$$v(\underline{\sigma}, V) = 3P + l_1 + l_2,$$

where $P := [1 : 0 : 0]$ and where l_i is the line $\{X_i = 0\}$.

(2) In contrast, compute the cycle $v(\sigma_0, V)$ with σ_0 as in (1) and $V := \{X_1 X_2^2 = 0\}$.

For the formulation of the next result we remind the reader of the notion of $\mathcal{L}$-degree (cf. [Ful1]). Let V be a proper K-scheme of dimension n and $\mathcal{L}$ a line bundle on V. Then

$$\deg_{\mathcal{L}} V := \deg c_1(\mathcal{L})^n \cap [V]$$

is called the *$\mathcal{L}$-degree* of V. Similarly, for a cycle $\alpha = \sum_i \alpha_i [C_i]$ of dimension k we set

$$\deg_{\mathcal{L}} \alpha := \sum_i \alpha_i \deg_{\mathcal{L}|C_i} C_i.$$

It is clear from this definition that $\deg_{\mathcal{L}^{\otimes r}} V = r^n \deg_{\mathcal{L}} V$. In the case that $\mathcal{L}$ is ample on V, this gives the usual degree of V. Moreover, if $\mathcal{L} \cong \mathcal{O}_X$, this is the degree $\int_X \alpha$ of a cycle.

The following is the main result of this section and will in the next section immediately yield Bezout's theorem for improper intersections.

Theorem 1.13. *Let V be an irreducible proper K-scheme and $\mathcal{L}, \underline{\sigma} = (\sigma_0, \cdots, \sigma_d)$, W be as before. Let $\sigma : V \dashrightarrow \mathbb{P}^d$ denote the rational map given by $\underline{\sigma}$. Then*

$$\deg_{\mathcal{L}} V = \sum_i \deg_{\mathcal{L}} v^i(\underline{\sigma}, V) + \deg(V/\sigma(V)) \deg \sigma(V),$$

where $\sigma(V)$ denotes in brief the closure of $\sigma(V \backslash W)$ in $\mathbb{P}^d$ equipped with its reduced structure.

Proof. Applying 1.6 to the case $i = n := \dim V$ and taking degrees we obtain

$$\deg_{\mathcal{L}} V = \sum_{j=0}^{n} \deg_{\mathcal{L}} v^j + \deg \beta^n.$$

Hence 1.13 is a consequence of part (3) of the following proposition. □

Proposition 1.14. *Let V be as in 1.13. Then the following hold.*

1. *$v^i = 0$ unless i is in the range* $\dim V - \dim W \leq i \leq \dim \sigma(V) + 1$.
2. *$\beta^i \neq 0$ iff* $0 \leq i \leq \dim \sigma(V)$.
3. $\deg \beta^{\dim V} = \deg(V/\sigma(V)) \deg \sigma(V)$.

Proof. Clearly (1) is a consequence of (2) and the intersection algorithm. For the proof of (2) let $\xi_0, \ldots, \xi_d$ be homogeneous coordinates on $\mathbb{P}^d$ such that $\sigma^*(\xi_i) = \sigma_i$. Then the elements $X_i := \sum_{j=0}^{d} U_{ij} \xi_j$ are coordinates of $\mathbb{P}^d_L$ with $\sigma^*(X_i) = s_i$. Thus, if H_i denotes the hyperplane of $\mathbb{P}^d_L$ given by $X_i = 0$, then D_i is the proper transform of H_i under σ. Clearly each hyperplane H_i meets

$$Z^i := \sigma(V_L) \cap H_0 \cap \cdots \cap H_{i-1}$$

properly provided that the intersection is nonempty. Now (2) follows from the fact that by the intersection algorithm $B^i \backslash W_L = \sigma^{-1}(Z^i)$ and $Z^i = \emptyset$ iff $i > \dim \sigma(V)$.

In order to show (3), set $k := \dim V$. If $k > \dim \sigma(V)$ then β^k as well as $\deg V/\sigma(V)$ vanish, and (3) is clear. If $k = \dim \sigma(V)$ then the scheme Z^k has dimension 0 and is of degree $\deg \sigma(V)$. The preimage of Z^k in $V_L \backslash W_L$ is just B^k and again of dimension 0. Using the fact that the degree of a map is compatible with taking generic hyperplane sections (see e.g. [FOV, 1.5.13]) yields that $\deg(V/\sigma(V)) = \deg(B^k/Z^k)$. Counting points gives that $\deg(B^k/Z^k) \deg Z^k = \deg B^k$, proving (3). □

Example 1.15. A typical application of 1.13 is to compute the degree of a linear projection of a complete variety V. Using 1.13 we can compute $\deg(V/\sigma(V)) \deg \sigma(V)$ from the knowledge of the degrees $\deg_{\mathcal{L}} V$ and $\deg_{\mathcal{L}} v^i$. In particular, we obtain the following results:

(a) If $W = \emptyset$ then $\deg(V/\sigma(V)) \deg \sigma(V) = \deg_{\mathcal{L}} V$.
(b) If $W = \{P_1, \ldots, P_s\}$ is a finite set then

$$\deg(V/\sigma(V)) \deg \sigma(V) = \deg_{\mathcal{L}} V - \sum_j e(\mathcal{I}_{W,P_j}, \mathcal{O}_{V,P_j}),$$

where $\mathcal{I}_W$ denotes the ideal sheaf of W.

(c) For a projective variety $V \subseteq \mathbb{P}^N$ and a point P of V consider the linear projection $\sigma : V \dashrightarrow \mathbb{P}^{N-1}$ from P. Then (b) implies the well-known formula

$$\deg(V/\sigma(V)) \deg \sigma(V) = \deg V - e(\mathcal{O}_{V,P}).$$

Exercise 1.16. In this exercise we will describe the behaviour of the v-cycle under the cone construction. Assume that $V \subseteq \mathbb{P}^n$ is a subvariety and $\mathcal{L} = \mathcal{O}_V(1)$. Let $c(V) \subseteq \mathbb{P}^{n+1}$ and $c(W)$ be the projective cone over V, resp. W with vertex, say, $P \in \mathbb{P}^{n+1}$. If $\alpha = \sum \alpha_i[C_i]$ is a cycle on W the cone $c(\alpha) := \sum \alpha_i[c(C_i)]$ is a cycle on $c(W)$. Let $\sigma_0, \ldots, \sigma_d \in H^0(V, \mathcal{L})$ be sections. Using the natural inclusion

$$H^0(V, \mathcal{O}_V(1)) \hookrightarrow H^0(c(V), \mathcal{O}_{c(V)}(1))$$

we can consider $\sigma_0, \ldots, \sigma_d$ also as sections of $\mathcal{O}_{c(V)}(1)$. Show the following facts:

(1) $v^i(\underline{\sigma}, c(V)) = c(v^i(\underline{\sigma}, V))$ for $i = 0, \ldots, \dim V$, and

$$v^{\dim c(V)}(\underline{\sigma}, c(V)) = j_P \cdot [P] \quad \text{with} \quad j_P = \deg(V/\sigma(V)) \deg \sigma(V).$$

Similarly, $\beta^i(\underline{\sigma}, c(V)) = c(\beta^i(\underline{\sigma}, V))$ for all i.

(2) Let $[x_0 : \cdots : x_{n+1}]$ be homogeneous coordinates on $\mathbb{P}^{n+1}$ so that V is the hyperplane $\{x_{n+1} = 0\}$ in $c(V)$. If we set $\sigma_{d+1} := x_{n+1}$ then $W \subseteq c(V)$ is the zero set of $\sigma_0, \ldots, \sigma_{d+1}$. Show that up to a pure transcendental field extension

$$v^{i+1}(\sigma_0, \ldots, \sigma_{d+1}, c(V)) = v^i(\sigma_0, \ldots, \sigma_d, V)$$

for $i = 0, \ldots, \dim V$, and $v^0(\sigma_0, \ldots, \sigma_{d+1}, c(V)) = 0$.

2 The Refined Bezout Theorem

In this section we will apply the results of the preceding section to obtain a refined version of Bezout's theorem. The principal idea for this is to replace the intersection of two subschemes by the intersection of the ruled join with the diagonal.

Recall that the ruled join of two subschemes X, Y of $\mathbb{P}^n$ is the subscheme of $\mathbb{P}^{2n+1}$ given by

$$J(X, Y) := \{[x : y] \in \mathbb{P}^{2n+1} \mid x \in X \text{ and } y \in Y\}.$$

More precisely, let $A = K[X_0, \dots, X_n]/\mathfrak{a}$ and $B = K[X_0, \dots, X_n]/\mathfrak{b}$ be the homogeneous coordinate rings of X, Y respectively. (Note that we always take the maximal homogeneous ideals $\mathfrak{a}$, $\mathfrak{b}$ defining X, Y, respectively.) Then $J(X,Y) := \operatorname{Proj}(A \otimes_K B)$, where $A \otimes_K B$ is graded by $(A \otimes_K B)_k = \sum_{i+j=k} A_i \otimes_K B_j$. In other words,

$$J = J(X,Y) \subseteq \mathbb{P}^{2n+1} = \operatorname{Proj}(K[X_0, \dots, X_n, Y_0, \dots, Y_n])$$

is the subscheme given by the sum of the extension ideals of $\mathfrak{a}$ and $\mathfrak{b}(\underline{Y})$, where $\mathfrak{b}(\underline{Y})$ is the set of polynomials $g(\underline{Y})$ with $g = g(\underline{X}) \in \mathfrak{b}$.

Let $\hat{X} := \operatorname{Spec} A \subseteq \mathbb{A}^{n+1}$ and $\hat{Y} := \operatorname{Spec} B \subseteq \mathbb{A}^{m+1}$ be the affine cones, i.e., geometrically speaking, $\hat{X} = \{x \in \mathbb{A}^{n+1} \mid x = 0 \text{ or } [x] \in X\}$, and similarly for Y. Then $\hat{J} = \hat{X} \times \hat{Y}$. In this definition, we consider the empty set as a projective scheme via $\emptyset = \operatorname{Proj} K$. Thus $J(X, \emptyset) = X$.

Proposition 2.1. 1. $\dim J = \dim X + \dim Y + 1$.
2. $\deg J = \deg X \deg Y$.

Proof. (1) is an immediate consequence of the fact that $\hat{J} = \hat{X} \times \hat{Y}$. In order to show (2), we may suppose that K is algebraically closed. First we reduce (2) to the case that X and Y have dimension 0. In fact, if $H \subseteq X$ is a hyperplane section of X then the reader may easily verify that $J(H,Y)$ is a hyperplane section of J. Cutting X repeatedly with hyperplanes we are reduced to the case that $\dim X = 0$. In the same way we can reduce to the case that $\dim Y = 0$.

So assume now that X and Y have dimension 0. Because of $\hat{J} = \hat{X} \times \hat{Y}$ the join construction is compatible with taking cycles, i.e., if $[X] = \sum n_i[X_i]$ then $[J] = \sum n_i[J(X_i, Y)]$, and similarly for Y. Using this remark we can also suppose that X and Y are reduced points. But in this case the ruled join is a line and so the assertion follows. □

We will use the construction in the preceding section to consider the intersection of the join $J \subseteq \mathbb{P}^{2n+1}$ of subschemes $X, Y \subseteq \mathbb{P}^n$ with the linear subspace which is defined by the "diagonal" embedding $\mathbb{P}^n \xrightarrow{\sim} \Delta \subseteq \mathbb{P}^{2n+1}$ given by $[x] \mapsto [x : x]$. Note that $X \cap Y \cong \Delta \cap J(X,Y)$ and that Δ is given by the equations $\sigma_0 = \cdots = \sigma_n = 0$, where $\sigma_i := X_i - Y_i$.

2.2. The intersection algorithm in $\mathbb{P}^n$. Fix closed equidimensional subschemes $X, Y \subseteq \mathbb{P}^n := \mathbb{P}^n_K$ and consider the join variety

$$J := J(X,Y) \subseteq \mathbb{P}^{2n+1}.$$

Applying the construction in 1.1 to J and the linear forms

$$\sigma_i = X_i - Y_i \in H^0(\mathbb{P}^{2n+1}, \mathcal{O}_{\mathbb{P}^{2n+1}}(1)),\ 0 \le i \le n,$$

we obtain a cycle

$$v_{\mathbb{P}^n}(X,Y) := v(X,Y) := v(\underline{X} - \underline{Y}, J) = v^0 + \cdots + v^n.$$

Again, we define v_i to be the part of dimension i of v. Similarly, we obtain a cycle

$$\beta_{\mathbb{P}^n}(X,Y) := \beta(X,Y) := \beta(\underline{X} - \underline{Y}, J) = \beta^0 + \cdots + \beta^n,$$

and we let β_i be its part of dimension i. $v(X,Y)$ is called the *v-cycle* of the intersection of X and Y. The coefficient of a subvariety C in $v(X,Y)$ is denoted by $j(X,Y;C)$. Thus

$$v(X,Y) = \sum_C j(X,Y;C)[C].$$

Remark 2.3. (1) By 1.8 (2), the construction is independent of the choice of coordinates on $\mathbb{P}^n_K$, up to a K-automorphism of L. Moreover it is symmetric in X and Y.

(2) Assume that X, Y are contained in a linear subspace Λ of $\mathbb{P}^n_K$. Then, by 1.8 (1), the v-cycles $v_{\mathbb{P}^n}(X,Y)$ on $\mathbb{P}^n$ and $v_\Lambda(X,Y)$ are equal, up to a pure transcendental field extension.

(3) The formation of $v(X,Y)$ is compatible with taking cycles. Thus, if $[X] = \sum n_i[X_i]$ then $v(X,Y) = \sum n_i v(X_i,Y)$, and similarly for the other component. This follows from the fact that $[J(X,Y)] = \sum n_i[J(X_i,Y)]$, and from 1.4.

Example 2.4. (1) If $X, Y \subseteq \mathbb{P}^n$ are linear subspaces, then by 1.9 we have $v(X,Y) = [X \cap Y]$ up to the field extension $K \to L$.

(2) Consider the intersection of $X := V_+(X_1^2 X_2)$ and $Y := V_+(X_1 X_2^2)$ in $\mathbb{P}^2$. By linearity and (1) we obtain $v(X,Y) = 5P + 2l_1 + 2l_2$. Taking degrees, we have

$$9 = \deg X \cdot \deg Y = 5 \deg P + 2 \deg l_1 + 2 \deg l_2$$

(compare with 1.12 (1), (2)).

For the next result, we remind the reader of the *embedded join* XY of subschemes X, Y of $\mathbb{P}^n$. Geometrically speaking, XY is the closure of the union of all lines $\ell = \overline{xy}$ with $x \in X$ and $y \in Y$. Equivalently, if $\sigma : J := J(X,Y) \dashrightarrow \mathbb{P}^n$ is the map $[\sigma_0 : \cdots : \sigma_n]$ then $XY = \sigma(J(X,Y))$. In the special case $X = Y$ this gives the well-known secant variety of X. For further facts on joins and secants we recommend the book of J. Harris [Harr].

We are now able to give a version of Bezout's theorem for improper intersections.

Theorem 2.5. (Refined Bezout theorem). *Let $X, Y \subseteq \mathbb{P}^n_K$ be subvarieties over an arbitrary field K. Then*

$$\deg X \deg Y = \sum_{C \in \mathcal{C}} j(X,Y;C) \deg C + \deg(J/XY) \deg XY.$$

Proof. By 2.1 (2) we have $\deg J = \deg X \deg Y$. Thus the theorem follows immediately from 1.13, applied to the irreducible components of J and $\sigma_i := X_i - Y_i$. □

Remark 2.6. (1) When $k := \dim J < n+1$ the proof of 1.13 shows that

$$\deg \beta^k = \deg \beta_0 = \deg(J/XY) \deg XY.$$

(2) Using 1.14 it follows that $\beta^i \neq 0$ iff $0 \leq i \leq \dim XY$ and so $v^i = 0$ for $i > \dim XY + 1$. Thus for each C with $j(X,Y;C) \neq 0$ we have

$$\dim X \cap Y \geq \dim C \geq \dim X + \dim Y - \dim XY.$$

(3) Recall that X and Y are said to meet properly if

$$\dim X \cap Y = \dim X + \dim Y - n.$$

If X and Y intersect properly in a non-empty set then $\dim J = \dim X + \dim Y + 1 \geq n+1$. Moreover, by (2) above only irreducible components of the intersection $X \cap Y$ appear in v and $XY = \mathbb{P}^n$. In particular, $\deg J/XY$ vanishes. Hence we recover Bezout's theorem for proper intersections.

(4) Let $\hat{}$ denote the affine cone, so that $\hat{J} = \hat{X} \times \hat{Y}$. Using 1.16, it follows that for $0 \leq i \leq k := \dim J$ the cone $v^i(X,Y)\hat{}$ is just $\hat{v}^i := v^i(\underline{\sigma}, \hat{J})$, where as above $\sigma_j = X_j - Y_j$. Moreover the vertex $P = 0$ appears in $\hat{v}^{k+1}$ with coefficient

$$j_P := \deg(J/XY) \deg XY.$$

In the original paper of Stückrad–Vogel [SVo] the refined Bezout theorem was stated with $j(\hat{X}, \hat{Y}, 0)$ instead of $\deg(J/XY) \deg XY$. The interpretation of this number in terms of joins was observed by van Gastel.

(5) Consider now the projective cones $c(X)$ and $c(Y)$ in $\mathbb{P}^{n+1}$ with vertex, say P. Then, up to a pure transcendental field extension, for $1 \leq i \leq \dim X + \dim Y + 2$ we have $v^i(c(X), c(Y)) = c(v^{i-1}(X,Y))$, and $v^0(c(X), c(Y)) = 0$. Moreover $v_0(c(X), c(Y)) = j_P[P]$ where j_P is given by the formula in (5). This follows from 1.16 (1) and (2).

As in the case of proper intersections, the j-multiplicity for irreducible components of the intersection can be described in terms of multiplicities of local rings.

Proposition 2.7. *Let C be an irreducible component of the intersection $X \cap Y$. Then $j(X,Y;C)$ is equal to Weil's i-symbol, i.e.,*

$$j(X,Y;C) = e(\mathfrak{d}, \mathcal{O}_{J,C}) = e(\hat{\mathfrak{d}}, \mathcal{O}_{\hat{J},\hat{C}}) = e(\mathfrak{c}, \mathcal{O}_{X \times Y, C})$$

where $\mathfrak{c}$, $\mathfrak{d}$ and $\hat{\mathfrak{d}}$ are the ideals of the diagonal in the respective rings.

Proof. The first equality follows from 1.10. The canonical map $\mathcal{O}_{J,C} \to \mathcal{O}_{\hat{J},\hat{C}}$ is smooth. Moreover the map $[x:y] \mapsto ([x],[y])$ is easily seen to induce a smooth morphism $\mathcal{O}_{Y\times Y,C} \to \mathcal{O}_{J,C}$. Therefore the remaining equalities are consequences of the well-known fact that multiplicities are stable under regular extensions of local rings. □

We obtain the following criterion for when the j-multiplicity is 1 (cf. [AHV]).

Proposition 2.8. *Assume that K is a perfect field. Let $C \subseteq X \cap Y$ be an irreducible component such that $\mathcal{O}_{X,C}$, $\mathcal{O}_{Y,C}$ are strictly equidimensional. Then the following are equivalent.*

1. *$\mathcal{O}_{X,C}$, $\mathcal{O}_{Y,C}$ and $\mathcal{O}_{X\cap Y,C}$ are smooth.*
2. *$j(X,Y;C) = 1$.*

Proof. It is well known that $\mathcal{O}_{X,C}$, $\mathcal{O}_{Y,C}$ are smooth resp. strictly equidimensional iff $\mathcal{O}_{X\times Y,C}$ has the same property. In view of 2.7 we need to show that $e(\mathfrak{d}, \mathcal{O}_{J,C}) = 1$ if and only if $\mathcal{O}_{J,C}$ and $\mathcal{O}_{X\cap Y,C}$ are regular. But this is a consequence of 1.11. □

In the case of a proper intersection the condition that $X \cap Y$ is smooth at C just means that X and Y meet transversally.

A local analysis at intersection points was considered by L. Berzolari [Ber] in 1896. His work was extended by O. Zariski [Zar] in 1937 where the following result was proved: if the origin is an m_i-fold point of n hypersurfaces F_i of $\mathbb{P}^n$, $i = 1, \ldots, n$, having only a finite number of common points and if further the origin is an isolated point of intersection of the F_i, then the intersection multiplicity at the origin is at least $m_1 m_2 \ldots m_n$. In the result 2.10 below we get necessary and sufficient conditions for when the intersection multiplicity at a point, which is an isolated component of an intersection, say, $X \cap Y$, is the product of the multiplicities of the local rings of X, Y at this point. This generalizes a result of Fulton and Lazarsfeld (see [Ful1]) to the case of improper intersections. Our main tool in proving this will be the following theorem of Rees.

Theorem 2.9. *Let $(A, \mathfrak{m})$ be an equidimensional local complete ring and $\mathfrak{q}' \subseteq \mathfrak{q}$ be $\mathfrak{m}$-primary ideals such that $e(\mathfrak{q}, A) = e(\mathfrak{q}', A)$. Then $\mathfrak{q}'$ is a reduction of $\mathfrak{q}$.*

Recall that $\mathfrak{q}'$ is called a *reduction* of $\mathfrak{q}$ if $G_{\mathfrak{q}}(A)/\mathfrak{q}'TG_{\mathfrak{q}}(A)$ is of dimension 0, where $G_{\mathfrak{q}}(A) = \bigoplus_{i\geq 0} \mathfrak{q}^i/\mathfrak{q}^{i+1}T^i$ denotes the associated graded ring.

For a proof we refer the reader to [FOV, 1.2.17]. In the next result we will denote by $C_P V$ the tangent cone of an algebraic variety V at a point $P \in V$ which is given by the spectrum of the associated graded ring $G_{\mathfrak{m}_A} A$, where $A := \mathcal{O}_{V,P}$.

Theorem 2.10. *Assume that K is algebraically closed and that $P \in X \cap Y$ is an isolated point of the intersection. Then $j(X, Y; P) \geq e(\mathcal{O}_{X,P})e(\mathcal{O}_{Y,P})$ with equality if and only if $C_P X \cap C_P Y = \{P\}$ in $T_{\mathbb{P}^n,P}$.*

Proof. It is easy to see that

$$G_{\mathfrak{m}_A}(A) \cong G(\mathcal{O}_{X,P}) \otimes_K G(\mathcal{O}_{Y,P}),$$

where $A := \mathcal{O}_{X \times Y,P}$ and the associated graded rings are taken with respect to the maximal ideals. Comparing e.g. the Hilbert functions of these rings it follows that

$$e(A) = e(\mathcal{O}_{X,P})e(\mathcal{O}_{Y,P}). \tag{1}$$

Let $x_1, \ldots, x_n$ be elements of A defining the diagonal. By 2.7

$$j(X, Y; P) = e((x_1, \ldots, x_n), A). \tag{2}$$

Since $(x_1, \ldots, x_n) \subseteq \mathfrak{m}_A$ we obtain from (1), (2) that $j(X, Y; C) \geq e(\mathcal{O}_{X,P})e(\mathcal{O}_{Y,P})$. Moreover, by the result of Rees above equality holds if and only if $(x_1, \ldots, x_n)$ is a reduction of $\mathfrak{m}$, i.e.,

$$G_{\mathfrak{m}_A}(A)/(\bar{x}_1 T, \ldots, \bar{x}_n T) \text{ has dimension } 0. \tag{3}$$

The initial forms of $x_1, \ldots, x_n$ in $G_{\mathfrak{m}_A}(A)$ define the intersection of

$$\operatorname{Spec}(G_{\mathfrak{m}_A} A) = C_P(X \times Y) \cong C_P X \times C_P Y \subseteq T_{\mathbb{P}^n,P} \times T_{\mathbb{P}^n,P}$$

with the diagonal $T_{\mathbb{P}^n,P} \hookrightarrow T_{\mathbb{P}^n,P} \times T_{\mathbb{P}^n,P}$. Hence (3) holds iff $C_P X \cap C_P Y = \{P\}$. □

Now assume that $X \subseteq \mathbb{P}^n$ is an equidimensional closed subscheme and that $Y = \Lambda$ is linear and given by equations $\sigma_0, \ldots, \sigma_d = 0$. Intersecting X with Λ as in the intersection algorithm 1.1, we obtain a cycle $v(\underline{\sigma}, X)$. In the next proposition we will compare this cycle with the cycle $v(X, Y)$ defined in 2.2.

Proposition 2.11. *(see [AVo]) Let $X \subseteq \mathbb{P}^n = \mathbb{P}^n_K$ be an equidimensional closed subscheme and $\Lambda \subseteq \mathbb{P}^n$ a linear subspace of codimension $d+1$ given by the linear equations $\sigma_0 = \cdots = \sigma_d = 0$. Then, up to a pure transcendental field extension, $v(\underline{\sigma}, X)$ and $v(X, \Lambda)$ are equal.*

Proof. By 2.3 (5), when calculating $v(X, \Lambda)$, we may assume that $\sigma_i = X_i$ is the i-th homogeneous coordinate function on $\mathbb{P}^n$ for $0 \leq i \leq d$. As usual we equip $\mathbb{P}^{2n+1}$ with homogeneous coordinates X_i, Y_i, $0 \leq i \leq n$. By definition,

$$\begin{aligned} v(X, \Lambda) &= v(X_0 - Y_0, \ldots, X_n - Y_n, J(X, \Lambda)) \\ &= v(X_0, \ldots, X_d, X_{d+1} - Y_{d+1}, \ldots, X_n - Y_n, J) \end{aligned}$$

since $Y_\lambda = 0$ on $J := J(X, \Lambda)$ for $\lambda \le d$. Consider J as an iterated cone where we embed $X \hookrightarrow J$ via

$$[x_0 : \cdots : x_n] \mapsto [x_0 \cdots : x_n : 0 : \cdots : 0 : x_{d+1} : \cdots : x_n],$$

so that X is the set of zeros of $X_{d+1} - Y_{d+1}, \ldots, X_n - Y_n$. Applying 1.16 repeatedly we get that, up to a pure transcendental field extension,

$$v(X_0, \ldots, X_d, X_{d+1} - Y_{d+1}, \ldots, X_n - Y_n, J) = v(X_0, \ldots, X_d, X).$$

□

Remark 2.12. If X is as in 2.11 and Y is a complete intersection given by equations $\sigma_0, \ldots, \sigma_d$ of the same degree $k \ge 2$, then the cycles $v(\underline{\sigma}, X)$ and $v(X, Y)$ are not equal in general, cf. 1.12 (1), (2) and 2.4 (2).

At the end of this section let us consider the following instructive example.

Example 2.13. (van Gastel) We consider the self-intersection of a hypersurface $X = \{f = 0\}$ in $\mathbb{P}^n = \mathbb{P}^n_K$ which is generically smooth. As usual take homogeneous coordinates $X_0, \ldots, X_n$ (resp. $X_0, \ldots, X_n, Y_0, \ldots, Y_n$) on $\mathbb{P}^n$ (resp. $\mathbb{P}^{2n+1}$). Then $J := J(X, X)$ is given by $\{f(\underline{X}) = f(\underline{Y}) = 0\}$ in $\mathbb{P}^{2n+1}$. We have to intersect $J \otimes_K L$ with the generic hyperplanes $D_i \subseteq \mathbb{P}^{2n+1}_L$ given by

$$l_i = \sum_{j=0}^{n} U_{ij}(X_j - Y_j) = 0, \ 0 \le i \le n,$$

where as usual $L = K(U_{ij})$.

Since J is a complete intersection and $\dim(J \cap D_0 \cap \cdots \cap D_{n-1}) = \dim J - n$, we have $B^n = J \cap D_0 \cap \cdots \cap D_{n-1}$, where B^n is the subscheme introduced in the intersection algorithm, i.e., $[B^n] = \beta^n$. Resolving the equations $l_i = 0$, $0 \le i \le n-1$ (e.g. using Cramer's rule) we obtain

$$X_i - Y_i = U_i(X_0 - Y_0), \ 1 \le i \le n \tag{$*$}$$

where the U_i are algebraically independent functions in the U_{ij}. Hence, if we set $U_0 := 1$, we have (modulo $l_0, \ldots, l_{n-1}$)

$$\begin{aligned} f(\underline{X}) - f(\underline{Y}) &\equiv f(\underline{Y} + \underline{U}(X_0 - Y_0)) - f(\underline{Y}) \\ &= \sum_{\nu=0}^{n} U_\nu \tfrac{\partial f}{\partial Y_\nu}(\underline{Y})(X_0 - Y_0) + (X_0 - Y_0)^2 G(X_0, \underline{Y}), \end{aligned}$$

by the Taylor expansion. We note that after eliminating the X_i, $i = 1, \ldots, n$, with the help of the equations $(*)$, l_n becomes a multiple of $X_0 - Y_0$.

The above equation shows that

$$\begin{aligned} J \cap D_0 \cap \cdots \cap D_{n-1} \\ &= \{f(\underline{Y}) = \sum_{\nu=0}^{n} U_\nu \tfrac{\partial f}{\partial Y_\nu} \cdot (X_0 - Y_0) + (X_0 - Y_0)^2 G(X_0, \underline{Y}) = 0\} \\ &= \{f(\underline{Y}) = X_0 - Y_0 = 0\} \\ \cup \{f(\underline{Y}) &= \sum_{\nu=0}^{n} U_\nu \tfrac{\partial f}{\partial Y_\nu} + (X_0 - Y_0) G(X_0, \underline{Y}) = 0\}, \end{aligned}$$

if we consider this variety in $\mathbb{P}^{n+1}$ with coordinates $X_0, Y_0, \ldots, Y_n$. Thus in the intersection algorithm $V^n = \{f(\underline{Y}) = X_0 - Y_0 = 0\} = X$ and

$$B^n = \{f(\underline{Y}) = \sum_{\nu=0}^{n} U_\nu \frac{\partial f}{\partial Y_\nu} + (X_0 - Y_0) G(X_0, \underline{Y}) = 0\}.$$

Intersecting B^n with $X_0 - Y_0 = 0$ gives

$$V^{n+1} = \{f(Y) = \sum_{\nu=0}^{n} U_\nu \frac{\partial f}{\partial Y_\nu} = 0\}$$

in $\mathbb{P}^n_L$. By the algorithm, $B^{n+1} = \emptyset$. Hence

$$v(X, X) = [X] + [V^{n+1}].$$

Thus v contains all 1-codimensional components, say, C of the singular set counted with multiplicity

$$j(X, X; C) = \lg(\mathcal{O}_{X_L, C} / (\sum_{\nu=0}^{n} U_\nu \frac{\partial f}{\partial Y_\nu})) = e((\frac{\partial f}{\partial Y_0}, \ldots, \frac{\partial f}{\partial Y_n}), \mathcal{O}_{X_L, C}).$$

Moreover, if X is smooth then V^{n+1} has no K-rational components and has degree $d(d-1)$ by 2.5, where $d = \deg f$.

3 Segre Classes, v-Cycles and Positivity

The aim of this section is to interpret the Segre classes of a cone as a v-cycle. As an application we will derive positivity results for Segre classes. In the next section, this will be generalized further to Segre classes of arbitrary embeddings. We first fix some notation.

3.1. Let X be an algebraic K-scheme and $\mathcal{S} = \bigoplus_{\nu=0}^{\infty} \mathcal{S}_\nu$ a graded quasi-coherent $\mathcal{O}_X$-algebra such that $\mathcal{S} = \mathcal{S}_0[\mathcal{S}_1]$ and $\mathcal{S}$ is finitely generated over $\mathcal{S}_0$. For simplicity we assume that $\mathcal{S}_0 = \mathcal{O}_X$. Let $C = \operatorname{Spec} \mathcal{S}$ be the associated cone and

$$\mathbb{P}(C \oplus 1) := \operatorname{Proj} \mathcal{S}[T] \xrightarrow{q} X$$

be its projective completion. Obviously C is canonically embedded into $\mathbb{P}(C\oplus 1)$ as the set $D_+(T) = \operatorname{Spec} \mathcal{S}[T]/(T-1)$. The complement $\mathbb{P}(C\oplus 1)\setminus C$ is isomorphic to $\mathbb{P}(C)$ and may be considered as the hyperplane at infinity. Because $\mathcal{S}_0 = \mathcal{O}_X$, the scheme X is canonically contained in C and thus also in $\mathbb{P}(C \oplus 1)$.

We recall the notion of the Segre class of a cone, see [Ful1, Section 4.1]. We set

$$s^i(C) := q_*(c_1(\mathcal{O}(1))^i \cap [\mathbb{P}(C \oplus 1)]),$$

where $\mathcal{O}(1)$ is the tautological line bundle on $\mathbb{P}(C \oplus 1)$. Thus, if C_j are the irreducible components of C with geometric multiplicities m_j then $s^i(C) = \sum_j m_j s^i(C_j)$. Note that in general this cycle can have a mixed dimension. However, in most of our applications we will have that C is equidimensional, and then $s^i(C)$ has dimension $\dim C - i$.

Definition 3.2. The Segre class of C, denoted by $s(C)$, is the class in $A_*(X)$ defined by $s(C) := \sum_{i\geq 0} s^i(C)$. By $s_i(C) := s^{\dim V - i}(C)$ we denote the part of $s(C)$ of dimension i.

In the case that $\mathbb{P}(C_i) \neq \emptyset$ for all irreducible components C_i of C one can also express the Segre classes as

$$s^i(C) := p_*(c_1(\mathcal{O}(1))^{i-1} \cap [\mathbb{P}(C)]),$$

where $p : \mathbb{P}(C) \to X$ is the projection (see [Ful1, 4.1.2]).

If $\mathcal{E}$ is a locally free sheaf on X and

$$E := \mathbb{V}(\mathcal{E}) := \operatorname{Spec}(S(\mathcal{E}))$$

is the associated vector bundle then it is convenient to set

$$s(\mathcal{E}) := s(E^\vee) \quad \text{and} \quad c(\mathcal{E}) := c(E^\vee).$$

Note that with this convention the Segre classes and Chern classes of a vector bundle are the same as the corresponding classes of its sheaf of sections. Moreover, in this case the Segre classes can be expressed by the Chern classes of E. Namely, $s(E) = c^{-1}(E) \cap [X]$ where $c^{-1}(E)$ is the inverse of the Chern class of E, see [Ful1]. For instance, if X has pure dimension n and $c(E) = 1 + c_1 + c_2 + c_3 + \dots$ then $s_n = [X]$ and

$$\begin{aligned} s_{n-1} &= -c_1 \cap [X] \\ s_{n-2} &= (-c_2 + c_1^2) \cap [X] \\ s_{n-3} &= (-c_3 + 2c_2c_1 - c_1^3) \cap [X]. \end{aligned}$$

In the main result of this section due to van Gastel we consider the following situation. Let $\mathcal{S}, X$ be as in 3.1. Assume that $\mathcal{S}_1$ is globally generated by sections $\sigma_0, \dots, \sigma_d$. With these σ_i considered as sections of $\mathcal{O}(1)$

on $\mathbb{P}(C \oplus 1)$ we will perform the intersection algorithm of 1.1. Note that the set of zeros $\{\sigma_0 = \cdots = \sigma_d = 0\}$ is obviously just X_L under its canonical embedding into $\mathbb{P}(C \oplus 1)_L$. From the intersection algorithm we obtain cycles $v^i = v^i(\underline{\sigma}, \mathbb{P}(C \oplus 1))$, β^i on $\mathbb{P}(C \oplus 1)_L$, where $\operatorname{supp} v^i \subseteq X_L$ and no irreducible component of β^i lies in X_L.

Theorem 3.3. *In $A_*(X_L) \cong A_*(X)$ we have $v^i = s^i(C)$. Thus if we set $v := \sum v^i$, then $v = s(C)$.*

Proof. By 1.6 we know

$$c_1(\mathcal{O}(1))^i \cap [\mathbb{P}(C \oplus 1)_L] = \sum_{\nu=0}^{i} c_1(\mathcal{O}(1))^{i-\nu} \cap v^\nu + \beta^i.$$

Thus it suffices to show that

(1) $q_*(\beta^i) = 0$

(2) $q_*(c_1(\mathcal{O}(1))^{i-\nu} \cap v^\nu) = 0$ if $i > \nu$.

For the proof of (1) observe that every component, say B, of β^i is itself a cone not contained in X_L, and thus the fibre dimension of $q : B \to q(B)$ is positive. Hence $q_*(\beta^i) = 0$.

In order to show (2) it is sufficient to prove that for every subvariety, say Z, of X_L we have $c_1(\mathcal{O}(1))^i \cap j_*(Z) = j_*(c_1(\mathcal{O}_X)^i \cap Z)$, where j denotes the canonical embedding $X_L \hookrightarrow \mathbb{P}(C \oplus 1)_L$. But this is just the projection formula (cf. [Ful1, Prop. 2.5 (c)]) since $j^*(\mathcal{O}(1)) = \mathcal{O}_X$. □

Example 3.4. Let $\mathcal{E}$ be a locally free sheaf of rank $r \geq 1$ on X and consider $C := \mathbb{V}(\mathcal{E})$. Let $\sigma_0, \ldots, \sigma_d$ be sections of $\mathcal{E}$ which generate $\mathcal{E}$. Then it follows in particular that $v(\underline{\sigma}, C) = s(C) = c(\mathcal{E}^\vee)^{-1} \cap [X]$.

Assume now that X is a proper variety over K. Let us compute the degree $\deg_{\mathcal{O}_{\mathbb{P}(\mathcal{E})}(1)} \mathbb{P}(\mathcal{E})$. By definition (see Section 1), it is given by the degree of the the zero-cycle

$$c_1(\mathcal{O}_{\mathbb{P}(\mathcal{E})}(1))^{n-r+1} \cap [\mathbb{P}(\mathcal{E})]$$

and so is just the degree of the Segre class $v_0 = s_0(C)$. Hence we get

$$\deg \sigma(\mathbb{P}(\mathcal{E})) \deg \mathbb{P}(\mathcal{E})/\sigma(\mathbb{P}(\mathcal{E})) = \deg_{\mathcal{O}_{\mathbb{P}(\mathcal{E})}(1)} \mathbb{P}(\mathcal{E}) = \int_X c(\mathcal{E}^\vee)^{-1} \cap [X],$$

where the first equality follows from 1.13 and where $\sigma : \mathbb{P}(\mathcal{E}) \to \mathbb{P}^d$ is the map given by $[\sigma_0 : \cdots : \sigma_d]$.

Example 3.5. Assume that $C = E = \mathcal{O}_X^{\oplus r}$ is a trivial vector bundle and $\beta \in A_k(E)$ is a positive k-cycle. With the same argument as in the proof of 3.3,

$$v^i(\underline{\sigma}, \beta) = q_*(c_1(\mathcal{O}_Z(1))^i \cap \bar{\beta}),$$

where $\bar{\beta}$ denotes the projective closure of β in $Z := \mathbb{P}(E \oplus 1)$. Let F be the universal rank r quotient of $q^*(E \oplus 1)$ so that there is an exact sequence

$$0 \to \mathcal{O}_Z(-1) \to q^*(E \oplus 1) \to F \to 0.$$

Then $c(F) = (1 - c_1(\mathcal{O}_Z(1)))^{-1}$ and therefore $c_r(F) = c_1(\mathcal{O}_Z(1))^r$. In particular we obtain

$$v^r(\underline{\sigma}, \beta) = q_*(c_r(F) \cap \bar{\beta}) = s_E^*(\beta),$$

where s_E denotes the Gysin homomorphism (see [Ful1, Prop. 3.3]). Therefore this cycle is semi-positive.

(2) More generally, assume that E is a vector bundle that is generated by its global sections. Thus there is a surjective map $\pi : G := \mathcal{O}_X^{\oplus N} \to E$ for some N. If β is a positive cycle then $s_G^*(\pi^*(\beta)) = s_E^*(\beta)$. It follows that s_E^* also transforms positive cycles into semi-positive cycles.

In general, of course, $\mathcal{S}_1$ will not be globally generated. But if X is a quasi-projective scheme and $\mathcal{L} \in \operatorname{Pic}(X)$ is sufficiently ample then $\mathcal{S}_1 \otimes \mathcal{L}$ becomes globally generated. Thus it is important to know how the Segre class $s(C)$ can be expressed in terms of $s(C \otimes L)$ where $L := \mathbb{V}(\mathcal{L})$ is the associated line bundle. Clearly $C \otimes L = \operatorname{Spec}(\bigoplus_{\nu \geq 0} \mathcal{S}_\nu \otimes \mathcal{L}^{\otimes \nu})$. We will also write $C \otimes \mathcal{L}$ instead of $C \otimes L$.

Proposition 3.6. *Assume that C is a cone over X and $\mathcal{L}$ is a locally free $\mathcal{O}_X$-module of rank 1. Then*

$$s^k(C \otimes \mathcal{L}) = \sum_{i \geq 0} \binom{k-1}{i-1} c_1(\mathcal{L})^{k-i} \cap s^i(C).$$

Here we use the convention that $\binom{m}{-1} := 0$ for $m \geq 0$ and $\binom{-1}{-1} := 1$.

Proof. Let C_ν run through the irreducible components of C with geometric multiplicity m_ν. Then $[\mathbb{P}(C \oplus 1)] = \sum m_\nu [\mathbb{P}(C_\nu \oplus 1)]$ and

$$s^k(C) = \sum m_\nu \cdot s^k(C_\nu).$$

Thus it is sufficient to treat the case that C is irreducible and reduced. If $\mathbb{P}(C) = \emptyset$ then $\mathbb{P}(C \oplus 1) = C$ and $\mathcal{O}(1) = \mathcal{O}_C$, so the assertion is obvious in this case.

Now assume that $\mathbb{P}(C) \neq \emptyset$. Then

$$\begin{aligned} s^k(C) &= q_*(c_1(\mathcal{O}(1))^{k-1} \cap [\mathbb{P}(C)]) \\ s^k(C \otimes \mathcal{L}) &= q_*(c_1(\tilde{\mathcal{O}}(1))^{k-1} \cap [\mathbb{P}(C \otimes \mathcal{L})]) \end{aligned} \tag{1}$$

where $\tilde{\mathcal{O}}(1)$ is the tautological sheaf on $\mathbb{P}(C \otimes \mathcal{L})$. But it is well known that $\mathbb{P}(C) \cong \mathbb{P}(C \otimes \mathcal{L})$ and $\mathcal{L} \otimes \mathcal{O}(1) = \tilde{\mathcal{O}}(1)$, see e.g. [Har2, II (7.9)]. Thus using $c_1(\mathcal{L} \otimes \mathcal{O}(1)) = c_1(q^*\mathcal{L}) + c_1(\mathcal{O}(1))$ we get

$$c_1(\tilde{\mathcal{O}}(1))^{k-1} = c_1(\mathcal{L} \otimes \mathcal{O}(1))^{k-1} = \sum_{i=0}^{k} \binom{k-1}{i-1} c_1(q^*\mathcal{L})^{k-i} c_1(\mathcal{O}(1))^{i-1}$$

for $k \geq 1$. Taking the cap product with $[\mathbb{P}(C)]$ on both sides and using (1), the result follows. □

Let $X, C = \operatorname{Proj} \mathcal{S}$ be as in 3.1 and $\mathcal{L} \in \operatorname{Pic} X$. Assume that $\mathcal{S}_1 \otimes \mathcal{L}$ is globally generated by sections, say $\sigma_0, \ldots, \sigma_d$. Then we can apply the intersection algorithm to $\mathcal{S} \otimes \mathcal{L}$ and obtain an intersection cycle $v = \sum_{i \geq 0} v^i$ such that $v^i = s^i(C \otimes \mathcal{L})$, see 3.3. Applying the preceding proposition we obtain

Corollary 3.7. (1) $v^k = \sum_{i=0}^{k} \binom{k-1}{i-1} c_1(\mathcal{L})^{k-i} \cap s^i(C)$.

(2) $s^k(C) = \sum_{i=0}^{k} \binom{k-1}{i-1} (-1)^{k-i} c_1(\mathcal{L})^{k-i} \cap v^i$.

Proof. Since $v = s(C \otimes \mathcal{L})$, (1) follows from 3.6. Moreover $s^k(C) = s^k((C \otimes \mathcal{L}) \otimes \mathcal{L}^{-1})$. Applying 3.6 again to $C \otimes \mathcal{L}$, $\mathcal{L}^{-1}$ instead of C, $\mathcal{L}$ gives (2). □

Remark 3.8. It is often useful to rewrite the above formula (1) in the form

$$v_k = \sum_{j=k}^{\dim X} \binom{m-k-1}{m-j-1} c_1(\mathcal{L})^{j-k} \cap s_j(C).$$

Example 3.9. Let us consider the special case that X is of pure dimension n, that $C = E$ is a vector bundle over X, and that $\mathcal{L}$, σ_i, $i = 0, \ldots, d$, are as above. Write $h := c_1(\mathcal{L})$. As remarked after 3.2, we can express the Segre classes by the Chern classes c_1, c_2, $c_3 \ldots$ of E. Combining this with 3.8 gives for instance that $v_n = [X]$ and

$$\begin{aligned} v_{n-1} &= (-c_1 + (n+1)h) \cap [X] \\ v_{n-2} &= \left(-c_2 + c_1^2 - (n+2)c_1 h + \binom{n+2}{2} h^2\right) \cap [X] \\ v_{n-3} &= \left(-c_3 + 2c_2 c_1 - c_1^3 + (n+3)(c_1^2 h - c_2 h) - \binom{n+3}{2} c_1 h^2 + \binom{n+3}{3} h^3\right) \\ &\quad \cap [X]. \end{aligned}$$

In the final part of this section we will apply these results in order to get positivity statements for Segre classes. Let $X, \mathcal{S}, C$ be as in 3.1 and assume that $\mathcal{S}_1$ is globally generated. Then the Segre classes $s^k(C)$ are equal to the cycle v^k and so are semipositive cycles, i.e., $s^k(C)$ can be written as a linear combination of subvarieties of X with non-negative coefficients. In the next result we want to extend this to the case that only $\mathcal{S}_d$ is globally generated for some $d \geq 1$.

Theorem 3.10. *Let $X, \mathcal{S}, C$ be as in 3.1 and assume that $\mathcal{S}_d$ is globally generated for some $d \geq 0$. Then $s(C)$ is a positive cycle.*

Proof. As in the proof of 3.6 we can reduce the assertion to the case that C is irreducible and reduced. If $\mathbb{P}(C) = \emptyset$ then $s(C) = [C]$ obviously is positive. Now assume that $\mathbb{P}(C) \neq \emptyset$ so that

$$s^k(C) = q_*(c_1(\mathcal{O}(1))^{k-1} \cap [\mathbb{P}(C)]),$$

as in the proof of 3.6. Let $\mathcal{S}^{(d)} := \oplus_{i\geq 0}\mathcal{S}_{id}$ be the d-th Veronese transform of $\mathcal{S}$. Then $\mathcal{S}_1^{(d)} = \mathcal{S}_d$ is globally generated and so the cone $C^{(d)} := \operatorname{Spec}\mathcal{S}^{(d)}$ has positive Segre class $s(C^{(d)})$. Thus it remains to show that

$$s^k(C^{(d)}) = d^{k-1}s^k(C).$$

But $\mathbb{P}(C^{(d)}) = \mathbb{P}(C)$ and $\mathcal{O}_{\mathbb{P}(C^{(d)})}(1) = \mathcal{O}_{\mathbb{P}(C)}(d)$, which shows that

$$\begin{aligned} s^k(C^{(d)}) &= q_*(c_1(\mathcal{O}_{\mathbb{P}(C^{(d)})}(1))^{k-1} \cap [\mathbb{P}(C)]) \\ &= d^{k-1}q_*(c_1(\mathcal{O}_{\mathbb{P}(C)}(1))^{k-1} \cap [\mathbb{P}(C)]). \end{aligned}$$

□

Corollary 3.11. *Assume that with the notations and assumptions as in 3.10 $\mathcal{L}$ is an ample line bundle on X. Then $\deg_{\mathcal{L}} s(C)$ is a positive integer.*

Applying this to the special case of a vector bundle gives the following result.

Corollary 3.12. *Let $\mathcal{E}$ be a locally free sheaf on X such that the d-th symmetric power $S^d(\mathcal{E})$ is globally generated for some $d \geq 0$. Then $s(\mathcal{E}^\vee)$ is a positive cycle.*

4 Segre Classes: The General Case

In this section we will again consider the v-cycle $v(\underline{\sigma}, V)$ on an arbitrary algebraic K-scheme V which is not necessarily a cone. We will show that in this case the Segre class of the normal cone $C_W V$ can be also expressed in terms of the v-cycle and vice versa. The central observation for this is a result of van Gastel stating that the intersection algorithm is compatible with passing to the normal cone. We will use the following notation.

4.1. Let V be an equidimensional algebraic K-scheme, $\mathcal{L}$ a line bundle on V and $\sigma_0, \ldots, \sigma_d \in H^0(V, \mathcal{L})$. As usual W denotes the zero locus of $\sigma_0, \ldots, \sigma_d$ and $\mathcal{I}$ the ideal sheaf of W. Applying the intersection algorithm, we obtain a cycle $v(\underline{\sigma}, V)$.

On the other hand, consider the normal cone $C := C_W V$ of W in V. For a coherent $\mathcal{O}_V$-module $\mathcal{M}$ let $G(\mathcal{M})$ be the sheaf of $\mathcal{O}_C$-modules associated to $G_{\mathcal{I}}(\mathcal{M}) = \bigoplus_{n\geq 0}\mathcal{I}^n\mathcal{M}/\mathcal{I}^{n+1}\mathcal{M}$. The initial forms $\sigma_i^* \in H^0(C, \mathcal{L}\otimes\mathcal{I}/\mathcal{I}^2)$ of σ_i, $0 \leq i \leq d$, define sections in $G(\mathcal{L})$ whose set of zeros is just $W \hookrightarrow C$. Applying the intersection algorithm to C and $\underline{\sigma}^*$ we obtain a v-cycle $v^* := v(\underline{\sigma}^*, C_W V)$.

The central result of this section is:

Theorem 4.2. $v(\underline{\sigma}, V) = v(\underline{\sigma}^*, C_W V)$ *as cycles on W.*

The original proof of van Gastel [vGa3] uses deformation to the normal cone. An algebraic proof was given by [AMa] and also [FOV]. We will give a simplified version of the latter proof in an appendix to this section. It is based on an algebraic version of the deformation to the normal cone using Rees algebras.

In the remaining part of the section we will give several applications of 4.2. As an almost immediate corollary we obtain more precise information about the range in which v^i vanishes.

Corollary 4.3. *Let $\sigma^* : \mathbb{P}(C_W V) \to \mathbb{P}^d$ be the map $[\sigma_0^* : \cdots : \sigma_d^*]$. Then*

$$\max\{i \mid v^i \neq 0\} = \dim \sigma^*(\mathbb{P}(C_W V)) + 1.$$

Proof. As $v_*^i = v^i$ it is sufficient to show the above formula for v_*^i instead of v^i. But it is easily seen and follows from homogeneity that $\beta_*^i \neq 0$ implies $\beta_*^i \cap D_i^* \neq 0$. Thus using 1.14 we obtain

$$\max\{i \mid v_*^i \neq 0\} = \max\{i \mid \beta_*^{i-1} \neq 0\} = \dim \sigma^*(\mathbb{P}(C_W V)) + 1. \qquad \square$$

Combining 4.2 with the results of the last section (see 3.7) we are able to express the v-cycle in terms of Segre classes.

Corollary 4.4. *In $A_*(W)$ we have*

1. $v^k(\underline{\sigma}, V) = s^k(C_W V \otimes \mathcal{L}) = \sum_{i=0}^{k} \binom{k-1}{i-1} c_1(\mathcal{L})^{k-i} \cap s^i(C_W V)$
2. $s^k(C_W V) = \sum_{i=0}^{k} \binom{k-1}{i-1} (-1)^{k-i} c_1(\mathcal{L})^{k-i} \cap v^i(\underline{\sigma}, V)$.

Example 4.5. As a typical application, let us study an equidimensional subscheme $V \subseteq \mathbb{P}^N$ and the projection $\sigma : V \dashrightarrow \mathbb{P}^d$ from a linear subspace $L \subseteq \mathbb{P}^N$ of codimension $d+1$. Set $n := \dim V$. Using the refined Bezout theorem we get that $\deg V/\sigma(V) \deg \sigma(V)$ is just $\deg V - \deg v$. Thus, in order to compute $\deg V/\sigma(V) \deg \sigma(V)$ we need a handy formula for $\deg v$. Taking degrees in 4.4 (1) and summing up gives

$$\deg v = \sum_{k=0}^{\dim W} \sum_{i=k}^{\dim W} \binom{n-k-1}{n-i-1} \deg s_i(C_W V) = \sum_{i \geq 0} \binom{n}{i} \deg s_i(C_W V),$$

where for the second equality we have used 4.6 below.

Assume now that $W := L \cap V$ is locally a complete intersection in V (not necessarily of codimension $d+1$). Then $N := C_W V$ is a vector bundle on W, so that the Segre class $s(C_W V)$ is just the inverse Chern class $c^{-1}(N) \cap [W]$. For instance, in the case $1 \leq \dim W \leq 3$ we have the following formulas (see p. 147), where c_1, c_2, c_3 are the Chern classes of N and where h denotes the first Chern class of $\mathcal{L}$:

$$\begin{aligned}
\deg v &= 3\deg W - c_1 && \text{if } \dim W = 1,\\
\deg v &= 10\deg W - 5c_1h + (c_1^2 - c_2) && \text{if } \dim W = 2,\\
\deg v &= 35\deg W - 21c_1h^2 + 7(c_1^2h - c_2) + (2c_1c_2 - c_1^3 - c_3) && \text{if } \dim W = 3.
\end{aligned}$$

The proof of the following simple identity is left to the reader.

Lemma 4.6. $\sum_{k=0}^{m} \binom{n-k-1}{n-m-1} = \binom{n}{n-m} = \binom{n}{m}$.

As an application of 4.2, we will deduce below a characterization of distinguished components in the sense of Fulton [Ful1, Def. 6.1.2]. Slightly more generally, we will introduce distinguished varieties as follows.

Definition 4.7. Let $\pi : C := C_W V \to W$ be the normal cone of W in V.

(a) A subvariety $Z \subseteq W$ is called a *distinguished component* for $W \subseteq V$ if there is an irreducible component C' of C with $\pi(C') = Z$.

(b) For a subvariety Z of W with generic point η_Z we denote by $(C_W V)_Z$ the fibre $\pi^{-1}(\eta_Z) \subseteq C_W V$. We will call $(C_W V)_Z$ the *fibre of the normal cone at* Z.

Assume that V is a subscheme of an algebraic K-scheme X and that $Y \subseteq X$ is locally a complete intersection in X. Then the distinguished components of $W := Y \cap V$ in V as introduced above are just the distinguished components of the intersection $Y \cap V$ in the sense of Fulton [Ful1].

In algebraic terms, let I in $A := \mathcal{O}_{V,Z}$ be the ideal of W and $G := \bigoplus_{n\geq 0} I^n/I^{n+1}$ its associated graded ring. Then $(C_W V)_Z$ is the spectrum of the ring $G \otimes_A A/\mathfrak{m}_A$. The dimension of $G \otimes_A A/\mathfrak{m}_A$ or, equivalently, of $(C_W V)_Z$ is often called the *analytic spread* of the ideal I in A.

The following result gives a simple estimate for the dimension of $(C_W V)_Z$ in terms of the dimensions of the subschemes involved and the minimal number of generators $\mu(I)$ of the ideal I.

Proposition 4.8. *With the above notation, the following hold.*

1. $\dim \mathcal{O}_{V,Z} - \dim \mathcal{O}_{W,Z} \leq \dim(C_W V)_Z \leq \min\{\mu(I), \dim \mathcal{O}_{V,Z}\}$.
2. *If* $\dim(C_W V)_Z = \dim \mathcal{O}_{V,Z}$ *then* Z *is distinguished.*
3. *If* $\dim(C_W V)_Z = \mu(I)$ *then* $(C_W V)_Z \cong \mathbb{A}^{\mu(I)}_{k(\eta_Z)}$.

Proof. The graded $A/\mathfrak{m}_A$-algebra $B := G \otimes_A A/\mathfrak{m}_A$ is generated by $\mu(I)$ elements and so $\dim B \leq \mu(I)$ with equality if and only if B is a polynomial ring, proving (3). Moreover $\dim B \leq \dim G = \dim A$, which gives the second inequality in (1). For the first one we apply the dimension formula to the morphism of rings $A/I \to G$ and obtain

$$\dim G \leq \dim A/I + \dim B,$$

see [Mat, 15.1 (i)].

Assume that $\dim B = \dim A$. Then every irreducible component of Spec B of maximal dimension is an irreducible component of Spec G and so corresponds to an irreducible component, say C_i, of $C_W V$ with $\pi(C_i) = Z$, proving (2). □

In the next result we will describe a local version of 4.3. For a cycle α on W_L and a subvariety $Z \subseteq W$ the germ of α at the generic point η_{Z_L} of Z_L will be denoted by α_{Z_L}. Thus $\alpha_{Z_L} = 0$ iff $\alpha|U = 0$ for some neighbourhood U of η_{Z_L} in Z_L.

Proposition 4.9. *Let $Z \subseteq W$ be a subvariety (defined over K). Then*

$$\max\{i \mid v^i_{Z_L} \neq 0\} = \dim(C_W V)_Z,$$

and $\beta^i_{Z_L} \neq 0$ iff $0 \leq i \leq \dim \mathbb{P}((C_W V)_Z)$.

Proof. With the same argument as in the proof of 4.3 it is sufficient to prove the second assertion. But this follows easily from the fact that

$$\dim D^*_0 \cap \cdots \cap D^*_{i-1} \cap (C_{W_L} V_L)_{Z_L} \backslash W_L) = \dim(C_W V)_Z - i$$

for $i \leq \dim(C_W V)_Z$. □

The following geometric criterion due to van Gastel [vGa3] gives a description of the K-rational components of v; for another proof see also [AMa].

Theorem 4.10. *Assume that V is strictly equidimensional and $Z \subseteq W_L$ is a subvariety. The following conditions are equivalent.*

1. *Z is a K-rational component of v.*
2. *Z is, up to the field extension $K \to L$, a distinguished subvariety.*
3. *$\dim(C_{W_L} V_L)_Z = \dim \mathcal{O}_{V_L, Z}$.*

Proof. Write $[C_W V] = \sum m_i [C_i]$, where $C_i \subseteq C := C_W V$ are the irreducible components. It is clear that every distinguished subvariety is K-rational and that it appears in the v-cycle because of 1.10, 4.2 and the fact that

$$v(\underline{\sigma}^*, C) = \sum m_i v(\underline{\sigma}^*, C_i). \qquad (*)$$

Thus (1) follows from (2). That (3) implies (2) is a consequence of 4.8 (2). To prove (1)⇒(3), observe first that $Z \in \mathcal{C}$ implies $v^{\dim \mathcal{O}_{V,Z}} \neq 0$. Thus by 4.9 $\dim \mathcal{O}_{V,Z} \leq \dim(C_W V)_Z$. As the converse inequality is always satisfied, this proves (3). □

Remark 4.11. Let C be a component of the intersection cycle and consider trdeg C which is the transcendence degree of the smallest subfield $L' \subseteq L$ such that C is defined over L'. It is an interesting question to characterize geometrically the transcendence degree of non-rational components of the intersection cycle. For some results in this direction see [FMa3].

4.1 Appendix: An algebraic proof of 4.2

4.12. We need a few preparations. We may assume that $\mathcal{L} \cong \mathcal{O}_V$ is trivial and that $V = \operatorname{Spec} A$ is affine. So the $\sigma_i \in A$ generate the ideal $I := \Gamma(V, \mathcal{I})$. For a finite A-module M consider the associated cycle on W

$$[M]_W = \sum_{\Gamma \subseteq W} \lg(M_\Gamma) \cdot [\Gamma]$$

where Γ runs through the irreducible components of $\operatorname{supp} M$ which are contained in W, i.e., if Γ corresponds to the prime ideal $\mathfrak{p} \in W$ then $\mathfrak{p}$ is a minimal prime of M and $\lg(M_\Gamma) = \lg M_{\mathfrak{p}}$. Let $G := G_I(A)$ be the associated graded ring and $G(M)$ the associated graded module. Then $W = \operatorname{Spec} A$ is naturally embedded in the cone $C = \operatorname{Spec} G$. Similarly as above, for a finite G-module N let $[N]_W$ be the part of the cycle associated to N that is supported on W. Observe that then

$$[M]_W = [G(M)]_W,$$

since for every irreducible component Γ of $\operatorname{supp} M$ which is contained in W we have $\sum_\nu \lg(I^\nu M / I^{\nu+1} M)_\Gamma = \lg M_\Gamma$ by the additivity of the length function.

For $s \in I$ let $s^* \in G$ denote its initial form of degree 1. Multiplication by s^* gives a natural map $G(M) \to G(M)$. Moreover, we set $\bar{M} := M/sM$.

Proposition 4.13. *Let M be an equidimensional finite A-module such that $M \xrightarrow{s} M$ is injective. Assume that the map $G(M) \xrightarrow{s^*} G(M)$ is also injective on $C \backslash W$. Then*

(1) the natural map $G(M)/s^*G(M) \to G(\bar{M})$ is bijective on $C \backslash W$.

(2) Let $K \subseteq G(M)$ denote the submodules of elements supported on $W \hookrightarrow C$. Then

$$[G(M)/(K + s^*G(M))]_W = [\bar{M}]_W.$$

Proof. Consider the so-called extended Rees ring and the Rees module

$$R := \sum_{i \in \mathbb{Z}} I^n T^n \subseteq A[T, T^{-1}] \quad \text{and} \quad N := R(M) := \sum_{i \in \mathbb{Z}} I^n T^n M,$$

where as usual $I^n = A$ for $i \leq 0$. Let $\bar{N} := R(\bar{M})$ be the Rees module associated to $\bar{M}$. Note that $N/T^{-1}N \cong G(M)$ and $\bar{N}/T^{-1}\bar{N} \cong G(\bar{M})$. Set $N' := \ker(N \to \bar{N})$, i.e., $(N')_\nu = s^*M \cap I^\nu M$. The diagram

$$\begin{array}{ccccccccc} 0 & \to & N' & \to & N & \to & \bar{N} & \to & 0 \\ & & \downarrow T^{-1} & & \downarrow T^{-1} & & \downarrow T^{-1} & & \\ 0 & \to & N' & \to & N & \to & \bar{N} & \to & 0 \end{array}$$

induces an exact sequence of cokernels

$$0 \to G' := N'/T^{-1}N' \to G(M) \to G(\bar{M}) \to 0. \qquad (*)$$

Denote the cokernel of the natural injection $N \overset{s^*}{\hookrightarrow} N'$ by L. The diagram

$$\begin{array}{ccccccccc} 0 & \to & N & \to & N' & \to & L & \to & 0 \\ & & \downarrow T^{-1} & & \downarrow T^{-1} & & \downarrow T^{-1} & & \\ 0 & \to & N & \to & N' & \to & L & \to & 0 \end{array}$$

and the snake-lemma gives an exact sequence

$$0 \to U \to G(M) \to G' \to V \to 0$$

where U, V are the kernel and cokernel of $T^{-1} : L \to L$ respectively. Together with $(*)$ it follows that

$$\begin{aligned} U &= \ker(s^* : G(M) \to G(M)) \\ V &= \ker(G(M)/s^*G(M) \to G(\bar{M})). \end{aligned} \tag{$**$}$$

By the lemma of Artin–Rees the module $L = \bigoplus_\nu (s^*M \cap I^\nu M)/I^\nu M$ is annihilated by T^{-k} for $k \gg 0$. Thus the modules U, V and L have the same support. Since by assumption U is supported on W the same holds for V. In particular (1) follows.

For (2) it is sufficient to prove that

$$[G(M)/K + s^*G(M)]_W = [G(\bar{M})]_W\,,$$

since $[G(\bar{M})]_W = [\bar{M}]_W$ by the remark in 4.12. Let $\Gamma \subseteq W$ be a subvariety of dimension $\dim M - 1$. We need to show that

$$\lg(G(M)/K + s^*G(M))_\Gamma = \lg G(\bar{M})_\Gamma\,,$$

where the length is taken as an $A := O_{C,\Gamma}$-module. Because of $(**)$ and $[U] = [V]$

$$\lg(G(M)/s^*G(M))_\Gamma = \lg U_\Gamma + \lg G(\bar{M})_\Gamma\,.$$

Applying the lemma below to $M := G(M)_\Gamma$, (2) follows. □

Lemma 4.14. *Let $(B, \mathfrak{m})$ be a 1-dimensional local Noetherian ring, M a finitely generated B-module and $s \in \mathfrak{m}$. Assume that $U := \ker(M \xrightarrow{s} M)$ has finite length and set $K := H^0_{\mathfrak{m}}(M)$. Then*

$$\lg M/sM = \lg M/(K + sM) + \lg U.$$

Proof. This is a standard equation for multiplicities. For instance, applying [FOV, 1.2.13] to the $\mathfrak{m}$-primary ideal sB gives that $\lg M/sM = e(sB, M) + \lg U$. Since

$$e(sB, M) = e(sB, M/K) = \lg M/(K + sM),$$

this proves the lemma. □

Proof of 4.2. As above we may assume that $\mathcal{L} = \mathcal{O}_V$ and $V = \operatorname{Spec} A$ is affine. Let V^i, $B^i = \operatorname{Spec} A_i$ be the subschemes of $A_L := A \otimes_K L$ defined by the intersection algorithm for $\sigma_0, \ldots, \sigma_d$, V, and denote by V_*^i, $B_*^i = \operatorname{Spec} G_i$ the subschemes of C_L given by the corresponding algorithm for $\sigma_0^*, \ldots, \sigma_d^*$, C. Let v^i, v_*^i be the corresponding cycles. Then

$$v^0 = [A_L]_W \quad \text{and} \quad v_*^0 = [G(A_L)]_W$$

with the notation as in 4.12, so these cycles are equal, see 4.12. The kernel of $A_L \to A_0$ has support in W. Using the lemma of Artin–Rees the reader may easily verify that $G(A) \to G(A_0)$ is bijective on $C \backslash W$. This proves that

$$G(A_0)|(C\backslash W) = G_0|(C\backslash W), \tag{$*$}$$

i.e., the localizations of $G(A_0)$ and G_0 are equal for all primes in $C\backslash W$. Assume that we have already shown that $G(A_i) = G_i$ on $C\backslash W$. We will prove that then

$$v^{i+1} = v_*^{i+1} \quad \text{and} \quad G(A_{i+1})|(C\backslash W) = G_{i+1}|(C\backslash W).$$

The multiplication by $s := s_i = \sum_{j=0}^{d} U_{ij}\sigma_j$ on A_i is injective, and the multiplication by s^* on $G(A_i)$ is injective on $C\backslash W$ as follows from 1.2. Applying 4.13(1) we obtain

$$(G(A_i)/s^* G(A_i))|(C\backslash W) = G(A_i/sA_i)|(C\backslash W).$$

The left-hand side is just $G_{i+1}|(C\backslash W)$, and the right-hand side is isomorphic to $G(A_{i+1})|(C\backslash W)$ by the same argument which establishes $(*)$. Moreover, if $K \subseteq G(A_i)$ is as in 4.13(2), then because of $G(A_i)|(C\backslash W) = G_i|(C\backslash W)$ we have $G(A_i)/K = G_i$, and so we get from 4.13(2) that

$$[G_i/s^* G_i]_W = [A_i/sA_i]_W.$$

Since the left-hand side is just v_*^{i+1} and the right-hand side is v^{i+1}, this proves 4.2.

5 Limits of Joins and Intersections

The aim of this section is to characterize distinguished subvarieties of the intersection of closed subschemes $X, Y \subseteq \mathbb{P}^n$ in terms of the so-called limit of join varieties. These varieties are introduced as follows.

Definition 5.1. Let X, Y be closed subschemes of an algebraic K-scheme Z and $C \subseteq X \cap Y$ a subvariety. Consider the diagonal embedding $X \cap Y \hookrightarrow X \times Y$. Then the scheme

$$\operatorname{LJoin}_C(X, Y) := C_{X\cap Y}(X \times Y)_C$$

is called the *limit of join variety* of X, Y at C. Here $C_{X\cap Y}(X\times Y)_C$ denotes the fibre of the normal cone as in 4.7 (b).

In the special case $X = Y$ we denote $\mathrm{LJoin}_C(X,Y)$ by $\mathrm{LSec}_C X$ and call this the *limit of secant variety* of X at C.

We note that the limit of join variety is not irreducible or reduced, in general, even in the case that X and Y are varieties. For closed points $C = \{P\}$ the limit of secant variety is a classical object, which is one of the cones studied by H. Whitney [Whi2]. The limit of join variety seems to appear first in [ATW] in the case that X and Y meet in one point. For the general case see [FVo2].

Remark 5.2. (1) $\mathrm{LJoin}_C(X,Y)$ depends only on the scheme-theoretic union of X and Y given by the intersection of the ideal sheaves of X and Y in Z, but not on the embedding in Z. In particular $\mathrm{LSec}_C X$ does not depend on embeddings.

(2) By definition, $\mathrm{LJoin}_C(X,Y)$ is a cone over $k(C)$, where $k(C)$ is the function field of C. In other words, we may write

$$\mathrm{LJoin}_C(X,Y) = \mathrm{Spec}\, A,$$

where A is a finitely generated graded $k(C)$-algebra with $A_0 = k(C)$. In particular $\mathbb{P}\,\mathrm{LJoin}_C(X,Y) = \mathrm{Proj}\, A$ is a projective $k(C)$-scheme.

(3) The definition of the limit of join variety can be extended to complex analytic spaces in the obvious way. Clearly, for algebraic varieties over $\mathbb{C}$ the formation of the limit of join variety is compatible with passing to the associated complex space.

(4) Let $X, Y \subseteq \mathbb{C}^n$ be locally closed complex subspaces with $0 \in X \cap Y$. Then $\mathrm{LJoin}_0(X,Y)$ has the following geometric meaning: *set-theoretically* $\mathrm{LJoin}_0(X,Y)$ *is the union of* 0 *and of all limits*

$$\ell = \lim_{\nu\to\infty} \overline{x_\nu y_\nu}, \qquad x_\nu \neq y_\nu$$

with sequences $x_\nu \in X$, $y_\nu \in Y$ *converging to zero.* This follows from the usual geometric interpretation of the normal cone: let $W \subseteq V$ is a subscheme of the algebraic $\mathbb{C}$-scheme V which is given by equations $f_1, \dots, f_d$, and let $f := (f_1, \dots, f_d) : V\backslash W \to \mathbb{C}^d$ be the induced map. Then $(C_W V)_P$ is the union of all lines $\lim \overline{0f(x_\nu)}$ where x_ν is a sequence in $V\backslash W$ converging to $P \in W$.

We will first show that $\mathrm{LJoin}_C(X,Y)$ admits a natural embedding into the tangent space $T_{X\cup Y,C}$. Let us recall the notion of tangent space at an arbitrary point of an algebraic K-scheme Z.

5.3. Let $T_Z := \mathbb{V}(\Omega^1_{Z/K})$ be the *tangent bundle* of Z, where $\mathbb{V}(\Omega^1_{Z/K})$ is the linear space associated to $\Omega^1_{Z/K}$. Let $\pi : T_Z \to Z$ be the canonical projection. For a point $\eta \in Z$ we denote by $T_{Z,\eta} := \pi^{-1}(\eta)$ the *tangent space* of Z at η. As usual, for a subvariety $C \subseteq Z$ with generic point η we will also write $T_{Z,C}$ instead of $T_{Z,\eta}$. We note the following well-known properties.

(1) $T_{Z,C} \cong \mathbb{A}^{\mu}_{k(C)}$ is an affine space where $\mu := \mu(\Omega^1_{Z/K,C})$.

(2) If K is a perfect field then $\dim T_{Z,C} = \operatorname{emdim} \mathcal{O}_{Z,C} + \dim C$. Furthermore, if $z \in Z$ is a closed point, then $\Omega^1_{Z,z}/\mathfrak{m}_z\Omega^1_{Z,z} \cong \mathfrak{m}_z/\mathfrak{m}_z^2$ (cf. [Eis1])and so $T_{Z,z}$ is just the classical Zariski tangent space $(\mathfrak{m}_z/\mathfrak{m}_z^2)^\vee$ of Z at z.

Proposition 5.4. *$C_{X\cap Y}(X\times Y)$ is canonically embedded into the restricted tangent bundle $T_{X\cup Y}|(X\cap Y)$. In particular, for a subvariety $C \subseteq X \cap Y$*

$$\mathrm{LJoin}_C(X,Y) \subseteq T_{X\cup Y,C}$$

is a cone in the tangent space.

This is an immediate consequence of the following lemma, which is stated in terms of commutative algebra. For a K-algebra R we briefly write Ω^1_R instead of $\Omega^1_{R/K}$.

Lemma 5.5. *Let R be a K-algebra, $\mathfrak{a}, \mathfrak{b} \subseteq R$ ideals, and denote by $I \subseteq R/\mathfrak{a} \otimes_K R/\mathfrak{b}$ the kernel of the multiplication map $R/\mathfrak{a} \otimes_K R/\mathfrak{b} \to R/\mathfrak{a}+\mathfrak{b}$. Then*

$$I/I^2 \cong \Omega^1_{R/\mathfrak{a}\cap\mathfrak{b}}/(\mathfrak{a}+\mathfrak{b})\Omega^1_{R/\mathfrak{a}\cap\mathfrak{b}}.$$

Proof. We may replace R by $R/\mathfrak{a}\cap\mathfrak{b}$ and thus are reduced to the case $\mathfrak{a}\cap\mathfrak{b} = 0$. Tensoring the exact sequence $0 \to \mathfrak{a} \to R \to R/\mathfrak{a} \to 0$ with $R/\mathfrak{b}$ gives

$$0 \to \mathrm{Tor}_1^R(R/\mathfrak{a}, R/\mathfrak{b}) \to \mathfrak{a}/\mathfrak{a}\mathfrak{b} \to R/\mathfrak{b} \to R/\mathfrak{a}+\mathfrak{b} \to 0,$$

and so

$$\mathrm{Tor}_1^R(R/\mathfrak{a}, R/\mathfrak{b}) = \mathfrak{a}\cap\mathfrak{b}/\mathfrak{a}\mathfrak{b} = 0. \qquad (*)$$

Set $J := \ker(R \otimes_K R/\mathfrak{a} \to R/\mathfrak{a})$. Then $J/J^2 \cong \Omega^1_R/\mathfrak{a}\Omega^1_R$. Considering $R\otimes_K R/\mathfrak{a}$ as an R-module with respect to the first component and tensoring

$$0 \to J \to R \otimes_K R/\mathfrak{a} \to R/\mathfrak{a} \to 0$$

with $R/\mathfrak{b}$ gives, because of $(*)$,

$$0 \to R/\mathfrak{b} \otimes_R J \to R/\mathfrak{b} \otimes_K R/\mathfrak{a} \to R/\mathfrak{a}+\mathfrak{b} \to 0,$$

i.e., $I = R/\mathfrak{b} \otimes_R J$. Thus $I/I^2 \cong J/J^2 \otimes_R R/\mathfrak{b} \cong \Omega^1_R/(\mathfrak{a}+\mathfrak{b})\Omega^1_R$. □

In our next result we will give elementary estimates for the dimension of the limit of join variety. For subschemes X, Y of a scheme Z we will denote by $X \cup Y$ the scheme-theoretic union of X and Y, i.e., the ideal sheaf of $X \cup Y$ in Z is the intersection of the ideal sheaves of X and Y.

Proposition 5.6. *Let Z be an algebraic K-scheme and $X, Y \subseteq Z$ closed subschemes. Then for a subvariety $C \subseteq X \cap Y$*

$$\dim \operatorname{LJoin}_C(X,Y) \le \min\{\mu, \dim \mathcal{O}_{X,C} + \dim \mathcal{O}_{Y,C} + \dim C\},$$

where $\mu := \dim_{k(C)} T_{X\cup Y/K,C}$; moreover

$$\dim \operatorname{LJoin}_C(X,Y) \ge \dim \mathcal{O}_{X,C} + \dim \mathcal{O}_{Y,C} - \dim \mathcal{O}_{X\cap Y,C} + \dim C.$$

If $\dim \operatorname{LJoin}_C(X,Y) = \mu$ *then* $\operatorname{LJoin}_C(X,Y) \cong \mathbb{A}^{\mu}_{k(C)}$.

Proof. The inequality $\dim \operatorname{LJoin}_C(X,Y) \le \dim_{k(C)} T_{X\cup Y/K,C}$ is an immediate consequence of 5.4. The remaining assertions follow from 4.8 using the following lemma. □

Lemma 5.7. *Let $X, Y \subseteq Z$ and C be as in 5.6. Then*

$$\dim \mathcal{O}_{X\times Y,C} = \dim \mathcal{O}_{X,C} + \dim \mathcal{O}_{Y,C} + \dim C.$$

We leave the straightforward proof to the reader.

Remark 5.8. The limit of join variety is compatible with morphisms: let $f: Z_1 \to Z_2$ be a morphism of algebraic K-schemes and X_i, $Y_i \subseteq Z_i$ closed subschemes for $i = 1,2$ such that f induces morphisms $X_1 \to X_2$ and $Y_1 \to Y_2$. There is an induced map of normal cones

$$C_{X_1\cap Y_1}(X_1 \times Y_1) \to C_{X_2\cap Y_2}(X_2 \times Y_2).$$

Therefore, for $\eta \in X_1 \cap Y_1$, the tangent map $T_\eta F : T_{X_1\cup Y_1,\eta} \to T_{X_2\cup Y_2,f(\eta)}$ maps $\operatorname{LJoin}_\eta(X_1,Y_1)$ into $\operatorname{LJoin}_{f(\eta)}(X_2,Y_2)$.

There is a close connection between the join variety of projective schemes and the limit of join variety of their affine cones.

Proposition 5.9. *Let $X, Y \subseteq \mathbb{P}^n$ be closed subschemes and $\hat{X}, \hat{Y} \subseteq \mathbb{A}^{n+1}$ the affine cones. Then $\operatorname{LJoin}_0(\hat{X},\hat{Y})$ is just the affine cone $(XY)\hat{}$ over the join variety XY.*

Proof. Here we consider XY with the following scheme structure: let $\mathfrak{a} \subseteq R := K[X_0,\dots,X_n]$ and $\mathfrak{b} \subseteq R$ be the saturated defining ideals of X resp. Y. Then XY is given by the ideal $\ker\psi$ where $\psi : R \to S := R/\mathfrak{a} \otimes_K R/\mathfrak{b}$ is the map $\psi(X_i) := X_i \otimes 1 - 1 \otimes X_i$. The homogeneous ideal $I = \bigoplus I_k$ of the diagonal is generated by the degree 1 elements $\psi(X_0),\dots,\psi(X_n)$. Therefore the canonical map $I^k \cap I_k \to I^k/\mathfrak{m}I^k$, $k \ge 0$, is bijective, where $\mathfrak{m}$ denotes the maximal homogeneous ideal of S. Moreover, since the elements $X_0,\dots,X_n$ generate R as a K-algebra, we also have

$$I^k \cap I_k = \psi(R_k), \quad k \ge 0.$$

Hence the canonical map $R/\ker\psi \to G_I(S)/\mathfrak{m}G_I(S)$ is bijective, proving the proposition. □

Before giving the main result of this section we need a description of the limit of join variety for projective schemes in terms of the normal cone of the ruled join. Consider closed subschemes $X, Y \subseteq \mathbb{P}^n$ and their ruled join $J := J(X,Y)$ in $\mathbb{P}^{2n+1}$. Let $X \cap Y \hookrightarrow J$ be the diagonal embedding and let $U \subseteq J$ be the subset of all $[x : y]$ such that $x \neq 0$ and $y \neq 0$. Clearly $X \cap Y \subseteq U$, and the morphism

$$\pi : U \longrightarrow X \times Y \quad \text{with} \quad [x : y] \longmapsto ([x], [y])$$

defines a regular map that restricts to the diagonal embedding on $X \cap Y$. The multiplicative group $\mathbb{G}_m$ of the ground field K acts on the fibres of π via $\lambda.[x : y] \mapsto [x : \lambda y]$, and U is a principal homogeneous space over $X \times Y$. This action maps $\{1\} \times (X \cap Y) \subseteq \mathbb{G}_m \times J$ onto $X \cap Y$ and therefore we get an induced map of cones

$$C_{\{1\}\times(X\cap Y)}(\mathbb{G}_m \times U) \cong \mathbb{A}^1 \times C_{X\cap Y}J \to C_{X\cap Y}J,$$

which is an action of the additive group $\mathbb{A}^1$ on $C_{X\cap Y}J$ so that $C_{X\cap Y}J$ becomes a principal homogeneous space over $C_{X\cap Y}(X \times Y)$. It is clear that $\mathbb{A}^1 \times (X \cap Y) \subseteq C_{X\cap Y}J$ is just the image of the relative tangent bundle $T_{U/X\times Y}|(X \cap Y)$ under the natural map. Rephrasing this in terms of exact sequences as in [Ful1, 4.1.6] we get the following result.

Lemma 5.10. *There is an exact sequence of cones*

$$0 \to \mathbb{A}^1 \times (X \cap Y) \to C_{X\cap Y}J \to C_{X\cap Y}(X \times Y) \to 0.$$

In particular, the Segre classes of $C_{X\cap Y}J$ and $C_{X\cap Y}(X \times Y)$ are the same. Moreover, for every subvariety $C \subseteq X \cap Y$,

$$(C_{X\cap Y}J)_C \cong \mathrm{LJoin}_C(X,Y) \times \mathbb{A}^1_{k(C)}.$$

Let us specialize to the case when $X = Y$ is a subscheme of $\mathbb{P}^n$. Then $C_X(X \times X)$ is just the tangent bundle, and $C_X J$ is isomorphic to the linear fibre space which is induced by the tangent bundle $T_{\hat{X}}$ of the cone $\hat{X}$ on X. Moreover, the action of $\mathbb{A}^1$ gives a trivial subbundle $X \times \mathbb{A}^1$ of $C_X J$ which is the kernel of the map $C_X J \to T_X$. For the sheaves this means that there is an exact sequence

$$0 \to \Omega^1_X \to \tilde{\Omega}^1_{\hat{X}} \to \mathcal{O}_X \to 0$$

which is just the Euler sequence. Thus the above lemma must be viewed as a generalization of the classical Euler sequence.

We are now able to characterize the distinguished components by limit of join varieties, see [FVo2].

Theorem 5.11. *Let $X, Y \subseteq \mathbb{P}^n$ be strictly equidimensional closed subschemes and $C \subseteq X \cap Y$ a K-rational subvariety. Then C appears in the v-cycle $v = v(X,Y)$ (up to a pure transcendental field extension) iff*

$$\dim \mathrm{LJoin}_C(X,Y) = \dim \mathcal{O}_{X,C} + \dim \mathcal{O}_{Y,C} + \dim C;$$

i.e., C is a distinguished component of the intersection $X \cap Y$.

Proof. Let $J := J(X,Y)$ be the join and $X \cap Y \hookrightarrow J$ the diagonal embedding. By the preceding lemma

$$\dim \mathrm{LJoin}_C(X,Y) = \dim\,(C_{X\cap Y}J)_C - 1.$$

Moreover $\dim \mathcal{O}_{X\times Y,C} = \dim \mathcal{O}_{J,C} - 1$, and so by 5.7

$$\dim \mathcal{O}_{X,C} + \dim \mathcal{O}_{Y,C} + \dim C = \dim \mathcal{O}_{J,C} - 1.$$

Now the theorem follows from 4.10. □

As an example consider the case of a hypersurface $H \subseteq \mathbb{A}^{n+1}$ given by an equation $f = 0$ where $f \in K[X_1, \ldots, X_{n+1}]$. Using the projective closure of H in $\mathbb{P}^{n+1}$ and 2.13 we obtain that for all irreducible components C of $\mathrm{Sing}\, H$ which are of codimension 1 in H, the limit of secant variety $\mathrm{LSec}_C\, H$ is just $\mathbb{A}^{n+1}_{k(C)}$, see 5.6.

6 Joins of Minimal Dimension and a Converse to Bezout's Theorem

This section focuses on the converse to Bezout's Theorem. More precisely, we examine the following question. Let X, Y be non-degenerate subvarieties of projective space and assume that

$$\deg X \deg Y = \sum j(X,Y;C) \deg C, \tag{$*$}$$

where the sum is taken over the *irreducible* components of $X \cap Y$. Is it true then that the intersection is proper? We will see that this is not the case in general. The main result of this section will be that *the above equality holds if and only if the join variety has minimal dimension.* Observe that we always have

$$\dim XY \geq \dim X + \dim Y - \dim X \cap Y \tag{$**$}$$

(cf. 2.6 (2)). Hence XY has minimal dimension if and only if equality holds in $(**)$. Note that this can occur without the intersection being proper as is seen by the following example.

Example 6.1. Let $Z \subseteq \mathbb{P}^a$ be a subvariety and consider $J := J(Z, \mathbb{P}^b) \subseteq \mathbb{P}^n$ with $n := a+b+1$, where $a, b \geq 2$. Let $H \subseteq \mathbb{P}^n$ resp. $H' \subseteq \mathbb{P}^b$ be generic hypersurfaces of degree ≥ 2 and regard $\mathbb{P}^b$ as a linear subspace of $\mathbb{P}^n$ via the canonical embedding $\mathbb{P}^b \hookrightarrow J$. Then the embedded join of

$$X := J \cap H \quad \text{and} \quad Y := H'$$

is contained in J, and since $\dim J = \dim X + \dim Y - \dim X \cap Y$, we have $J = XY$. In this example, if Z is not a cone then X is also not a cone. Of course Y is not a cone, too. Moreover, if Z is nondegenerate then the same holds for X.

The converse to Bezout's mentioned above will follow from a closer analysis of the v-cycle. Let $X, Y \subseteq \mathbb{P}^n_K$ be closed subvarieties, where the ground field K is algebraically closed. As was observed above, by 2.6 (2) we have

$$\tau := \dim X + \dim Y - \dim XY \leq \dim X \cap Y. \tag{6.1.1}$$

Moreover, $v_i \neq 0$ implies that i is in the range $\max(0, \tau) \leq i \leq \dim X \cap Y$. Note that $\tau \geq -1$ and that XY has minimal dimension iff $\tau = \dim X \cap Y$.

With this notation, the main result of this section is as follows.

Theorem 6.2. 1. *For every i in the range* $\max(0, \tau) \leq i < \dim X \cap Y$ *there is a component in the v-cycle which is not an irreducible component of $X \cap Y$.*

2. *Assume that $\tau \geq 0$. Then for every distinguished component C of v of dimension τ we have that $j(X, Y; C) \geq \deg XY$.*

3. *Assume that $0 \leq \tau < \dim X \cap Y$. Then $\deg XY \leq \sum j_C \deg C$, where the sum is taken over all components of v of dimension τ which are not irreducible components of $X \cap Y$.*

The proof will be given in 6.12. We first treat a few applications. In the following, let $\sum_{irr} j_C \deg C$ denote the sum taken over all *irreducible* components of $X \cap Y$.

Corollary 6.3. *Assume that the ground field K is algebraically closed. The following are equivalent.*

1. *XY has minimal dimension and $X \cap Y \neq \emptyset$.*

2. $\deg X \deg Y = \sum_{irr} j_C \deg C$.

Moreover, if the above conditions are satisfied then all irreducible components C of $X \cap Y$ have the same dimension and $j_C \geq \deg XY$.

Proof. If (1) is satisfied then by $(*)$ above $\tau = 0$ and all components of v have dimension τ. In particular $\dim XY < \dim X + \dim Y + 1$ and so $\deg J(X, Y)/XY$ vanishes. Hence (2) follows from the refined Bezout theorem 2.5. Conversely, assume that (2) holds. Comparing with the refined Bezout, we get that v contains only irreducible components and that $\deg J(X, Y)/XY = 0$. By 6.2, $\max(0, \tau) = \dim X \cap Y$. As $\deg J(X, Y)/XY = 0$, we must have $\dim J(X, Y) > \dim XY$ and so $\tau \geq 0$. Hence (1) follows. The remaining assertions now follow from 6.2 above. $\square$

Corollary 6.4. *Assume that the ground field K is algebraically closed and that XY is nondegenerate with* $\dim X \cap Y > \tau$, *i.e., the join has no minimal* *dimension. Then*

$$\deg X \deg Y \geq \sum\nolimits_{irr} j_C \deg C + e,$$

where $e := \dim X \cap Y - (\dim X + \dim Y - n)$ *is the excess dimension.*

Proof. For every i in the range $\max(0,\tau) \leq i < \dim X \cap Y$ there is a component, say C_i, of dimension i which is not an irreducible component of $X \cap Y$. Moreover, if $\tau \geq 0$ then $j(X,Y;C_\tau) \geq \deg XY$. Using the refined Bezout theorem we obtain, for $\tau \geq 0$,

$$\deg X \deg Y \geq \sum\nolimits_{irr} j_C \deg C + \deg XY + \dim X \cap Y - \tau - 1.$$

This inequality also holds if $\tau = -1$. Since XY is nondegenerate we have $\deg XY \geq \operatorname{codim}_{\mathbb{P}^n} XY + 1$. This easily implies the result. □

As a particular case we obtain the following result of R. Lazarsfeld.

Corollary 6.5. *Assume that the ground field K is algebraically closed. If X, Y are subvarieties of $\mathbb{P}^n$ and $X \cup Y$ is nondegenerate then*

$$\deg X \deg Y \geq \sum\nolimits_{irr} \deg C + e.$$

Proof. In the case that $\dim X \cap Y > \tau$ this follows from 6.4. Now assume that $\dim X \cap Y = \tau$, so that XY has minimal dimension. The refined Bezout theorem gives

$$\deg X \deg Y = \begin{cases} \sum_{irr} j_C \deg C & \text{if } \tau \geq 0 \\ \deg XY \deg(J(X,Y)/XY) & \text{if } \tau = -1. \end{cases}$$

By 6.2 $j_C \geq \deg XY$ for every irreducible component C. Since XY is nondegenerate, we have $\deg XY \geq \operatorname{codim}_{\mathbb{P}^n} XY + 1$. As $\operatorname{codim}_{\mathbb{P}^n} XY = e$, the result follows. □

In order to prove 6.2, we will reduce via the join construction to the following situation which was treated in the intersection algorithm in Section 1.

6.6. In the remaining part of this section, let V be a projective variety over an arbitrary field K and $W \subseteq V$ a closed non-empty subscheme. We fix an ample line bundle $\mathcal{L} \in \operatorname{Pic}(V)$ and consider sections $\sigma_0, \ldots, \sigma_d \in H^0(V,\mathcal{L})$ with common zero set $W = \{\sigma_0 = \cdots = \sigma_d = 0\}$. Let

$$v = v(\underline{\sigma}, V) = v_0 + \cdots + v_d$$

be the v-cycle constructed in Section 1. As before, it will turn out to be very useful to study the rational map

$$\sigma = [\sigma_0 : \cdots : \sigma_d] : V \dashrightarrow \mathbb{P}^d,$$

which is a morphism on $V \backslash W$. Let $\sigma(V)$ be the closure of $\sigma(V \backslash W)$ in $\mathbb{P}^d$.

In the next proposition we need the notion of connectedness in dimension d. Recall that a scheme X is said to be *connected in dimension* d if

(a) $\dim X > d$, and

(b) for every subscheme $Y \subseteq X$ of dimension $< d$ the set $X \backslash Y$ is connected.

By Grothendieck's connectedness theorem, if a closed subscheme of $\mathbb{P}^n$ is connected in dimension d then for arbitrary hyperplanes $H_1, \ldots, H_s$ with $s \leq d$ the intersection $X \cap H_1 \cap \cdots \cap H_s$ is connected in dimension $d - s$ (see e.g. [FOV, 3.1.1]).

In the next result we will obtain information about the range where the v_i are non-vanishing, in terms of the dimension of $\sigma(V)$. We set

$$\tau := \dim V - \dim \sigma(V) - 1.$$

Proposition 6.7. *In the dimension range*

$$\max(0, \tau) \leq i < \dim W$$

v_i *contains a component which is not an irreducible component of* W. *In particular,* $v_i \neq 0$ *iff* $\max(0, \tau) \leq i \leq \dim W$.

It follows that the v-cycle has no gaps, i.e., if $v_i \neq 0$ and $v_j \neq 0$ for some $i < j$ then also $v_k \neq 0$ for $i < k < j$.

Proof. Let D_i, L and B^j be as in the intersection algorithm 1.1. First we note that set-theoretically

$$V_L \cap D_0 \cap \cdots \cap D_{j-1} = B^j \cup W_L$$

and so, by the connectedness theorem of Grothendieck mentioned above, this scheme is connected in dimension $\dim V - j - 1$. This implies in particular that B^j intersects some irreducible component of W_L of dimension $\geq \dim V - j$ in a set of dimension $\dim V - j - 1$ if $\dim B^j \leq \dim W$ and $B^j \neq \emptyset$, i.e., if

$$\dim V - j \leq \dim W \quad \text{and} \quad j \leq \dim \sigma(V).$$

Thus v^{j+1} contains a component which is not an irreducible component for $j + 1$ in the range

$$1 + \dim V - \dim W \leq j + 1 \leq \dim \sigma(V) + 1.$$

This proves the first part. The second statement is a consequence of the first one and 1.14. □

6.8. Let $\mathcal{I}_W \subseteq \mathcal{O}_V$ be the ideal sheaf of W in V. What is important in the following is the equality of the v-cycle $v(\underline{\sigma}, V)$ with $v(\underline{\sigma}^*, C)$, where $C = C_W V$ is the normal cone of W in V and where $\sigma_0^*, \ldots, \sigma_d^* \in H^0(\mathcal{I}_W / \mathcal{I}_W^2 \otimes \mathcal{L})$

are the initial forms of $\sigma_0 \dots, \sigma_d$, see 4.2. For this it is convenient to consider the blowing up of V along W

$$B_W(V) = \operatorname{Proj}(\bigoplus_{\nu=0}^{\infty} \mathcal{I}_W^\nu) \xrightarrow{\pi} V.$$

Clearly, the rational map $\sigma : V \dashrightarrow \mathbb{P}^d$ lifts to a morphism $\tilde{\sigma} : B_W(V) \to \mathbb{P}^d$. Let

$$E = \operatorname{Proj}(\bigoplus_{\nu=0}^{\infty} \mathcal{I}_W^\nu / \mathcal{I}_W^{\nu+1}) \subseteq B_W(V)$$

be the exceptional set, which is just the projectivized normal cone $\mathbb{P}(C)$. Let $\sigma^* : C \backslash W \to \mathbb{P}^d$ be the map given by the initial forms $[\sigma_0^* : \cdots : \sigma_d^*]$. This map induces a morphism $\mathbb{P}(C) \to \mathbb{P}^d$, also denoted by σ^*, which is the restriction of $\tilde{\sigma}$ to $E \cong \mathbb{P}(C)$.

For a subcone C' of C we call the set $\sigma^*(\mathbb{P}(C'))$ the *span* of C' and denote it by $\operatorname{Span} C'$. Observe that by our construction $\operatorname{Span} C' \subseteq \operatorname{Span} C \subseteq \sigma(V)$ and $\dim \operatorname{Span} C \leq \dim E = \dim V - 1$. In 4.3 we characterized the range in which v^i vanishes, in terms of the dimension of $\operatorname{Span} C$. Comparing this with 6.7 gives the following result which was essentially shown in [FGV].

Corollary 6.9. *If $W \neq \emptyset$ then either $\sigma(V) = \operatorname{Span} C$ and $\dim \sigma(V) < \dim V$, or*

$$\dim \operatorname{Span} C + 1 = \dim \sigma(V) = \dim V.$$

Proof. By 4.3 and 6.7 we have

$$\max(0, \dim V - \dim \sigma(V) - 1) = \dim V - \dim \operatorname{Span} C - 1.$$

This easily implies the result. □

Next we prove an estimate of j_C for distinguished components of the v-cycle.

Proposition 6.10. *Assume that K is algebraically closed. Let $Z \subseteq W$ be a distinguished component of $W \subseteq V$ of dimension $\dim V - \dim \sigma(V) - 1$ and $C' \subseteq C_W V$ an associated irreducible component of $C_W V$, i.e., C' has base Z. Then $\operatorname{Span} C' = \sigma(V)$ and $j_Z \geq \deg \sigma(V)$.*

Proof. Let C'_Z be the fibre of $C' \to W$ over the generic point η_Z of Z. Set-theoretically we have

$$\mathbb{P}C'_Z \subseteq (\operatorname{Span} C') \otimes_K k(Z) \subseteq \sigma(V) \otimes_K k(Z), \qquad (*)$$

and so, using the assumption, we get

$$\dim \operatorname{Span} C' \geq \dim \mathbb{P}C'_Z = \dim V - \dim Z - 1 = \dim \sigma(V).$$

As $\sigma(V)$ and hence also $\sigma(V) \otimes_K k(Z)$ is irreducible, set-theoretically equality must hold in $(*)$. Hence $\deg \mathbb{P}C'_Z \geq \deg \sigma(V)$. Let $I \subseteq A := \mathcal{O}_{C_W V, Z}$ be the ideal of W. By 4.2, j_Z is the coefficient of Z in $v(\underline{\sigma^*}, C_W V)$, and the latter number is equal to $e(I, A)$ by 1.10. Since $\mathcal{O}_{Z',Z}$ is a quotient of A, it follows that

$$j_Z = e(I, A) \geq e(I\mathcal{O}_{C',Z}, \mathcal{O}_{C',Z}) = \deg \mathbb{P}C' \geq \deg \sigma(V),$$

as required. □

For the next proposition recall that every component of v has dimension $\geq \tau$ where $\tau := \dim V - \dim \sigma(V) - 1$.

Proposition 6.11. *Assume that K is algebraically closed and that $\tau < \dim W$. Then $\deg \sigma(V) \leq \sum j_Z \deg Z$, where the sum is taken over all τ-dimensional components of v which are not irreducible components of W.*

Proof. Write $W = W' \cup W''$, where W' is the union of irreducible components of dimension $> \tau$ and W'' is the union of irreducible components of dimension τ. By our assumption, $W' \neq \emptyset$. Let $H_i \subseteq \mathbb{P}^d_L$ be the i-th generic hyperplane given by $\sum_{j=0}^{d} U_{ij}\xi_j = 0$ where $[\xi_0 : \cdots : \xi_d]$ are homogeneous coordinates on $\mathbb{P}^d$. Then $B_{\tau+1} \backslash W_L = \sigma^{-1}(H_0 \cap \cdots \cap H_{\dim V - \tau - 2})$ and

$$\sigma(B_{\tau+1}) = H_0 \cap \cdots \cap H_{\dim V - \tau - 2} \cap \sigma(V_L)$$

has dimension 0. Consider now the algebraic closure $\bar{L}$ of L. Then $\sigma(B_{\tau+1}) \otimes_L \bar{L}$ consists of $\delta := \deg \sigma(V)$ points. Hence $B_{\tau+1} \otimes \bar{L}$ has δ irreducible components, say $A_1, \ldots, A_\delta$. By Grothendieck's connectedness theorem, $(B_{\tau+1} \otimes \bar{L}) \cup W_{\bar{L}}$ is connected in dimension τ. Thus $B_{\tau+1} \otimes \bar{L}$ and consequently some A_i must intersect $W'_{\bar{L}}$ in a set of dimension τ. By the generic Bertini theorem 1.2, $B_{\tau+1}$ is irreducible. Therefore the Galois group of $\bar{L} : L$ acts transitively on the A_i. Hence every A_i must intersect W' in a set of dimension τ. It follows that $B_{\tau+1} \otimes \bar{L}$ intersects $W'_{\bar{L}}$ in δ subvarieties of dimension τ, counted with multiplicity. By construction, these subvarieties are not irreducible components of W. This gives the result. □

6.12. Proof of 6.2. Consider the ruled join $V := J(X, Y)$ and the intersection $W := X \cap Y$, which we embed as usual diagonally into V. Now 6.7 implies the first part of 6.2 whereas the second and third part follow from 6.10 and 6.11.

We also mention the following variant of 6.3, where we assume that K is an algebraically closed field.

Proposition 6.13. *Let X, Y be subvarieties of $\mathbb{P}^n$. Assume that $X \cup Y$ is nondegenerate and that $\mathcal{O}_{X,C}$ and $\mathcal{O}_{Y,C}$ are Cohen–Macaulay for all irreducible components C of $X \cap Y$. Then the following conditions are equivalent:*

1. $\deg X \deg Y = \sum_{irr} \lg(\mathcal{O}_{X\cap Y,C}) \deg C$.

2. $\deg X \deg Y = \deg X \cap Y$.

3. X and Y intersect properly.

As before, this follows from the following estimate.

Theorem 6.14. *Let X and Y be as in 6.13. Then*

$$\deg X \deg Y \geq \sum\nolimits_{irr} \lg(\mathcal{O}_{X\cap Y,C}) \deg C + e.$$

In view of 6.4 it is sufficient to prove this in the case that XY has minimal dimension. Using as before the join construction this is reduced to the following statement.

Lemma 6.15. *Let $V \subseteq \mathbb{P}^n$ be a subvariety and $\sigma_0, \ldots, \sigma_d \in H^0(\mathcal{O}_{\mathbb{P}^n}(1))$ such that $\sigma(V)$ (equipped with its reduced structure) is a non-degenerate subvariety of $\mathbb{P}^d$ of dimension $\tau = \dim V - \dim W - 1$, where as usual W is the zero set of $\sigma_0, \ldots, \sigma_d$ in V. Assume moreover that $\mathcal{O}_{V,C}$ is Cohen–Macaulay for all irreducible components C of W. Then*

$$\deg V \geq \deg W + d - \tau.$$

Proof. If $\dim W > 0$, let H be a generic hyperplane of $\mathbb{P}^n$. Then $\deg V = \deg V \cap H$, $\deg W = \deg W \cap H$, and $\sigma(V \cap H) = \sigma(V)$. By Bertini's theorem $V \cap H$ is irreducible if $\dim V \geq 2$. Using induction, we can henceforth assume that W has dimension 0.

Suppose that $\dim V > 1$. Let H be a generic hyperplane in $\mathbb{P}^d$ and denote by D its proper transform in $\mathbb{P}^n$. Then $\sigma(V \cap D) = \sigma(V) \cap H$ is again non-degenerate in H. Using again induction we are reduced to the case that $\dim V = 1$ so that V is a connected curve.

Consider now a sufficiently generic hyperplane $H \subseteq \mathbb{P}^d$ with equation, say, $f \in E := \sum K\sigma_i$ and let D be as before the proper transform. By assumption, the map $E \to H^0(\mathcal{O}_{\sigma(V)}(1))$ and hence also $E \to H^0(\mathcal{O}_{V_{red}}(1))$ is injective. Using the diagram

$$\begin{array}{ccccc} 0 \to K & \to & E & \to & E/fK \to 0 \\ \cong \downarrow & & \downarrow & & \downarrow \\ 0 \to H^0(\mathcal{O}_{V_{red}}) \cong K & \to & H^0(\mathcal{O}_{V_{red}}(1)) & \to & H^0(\mathcal{O}_{V_{red}\cap D}(1)) \end{array}$$

it follows that $E/fK \hookrightarrow H^0(\mathcal{O}_{V_{red}\cap D}(1))$ and hence also $E/fK \hookrightarrow H^0(\mathcal{O}_{V\cap D}(1))$ is injective. The cokernel of the later map is just $H^0(\mathcal{O}_W(1))$. Thus we obtain

$$\deg V = h^0(\mathcal{O}_{V\cap D}(1)) = \dim E - 1 + h^0(\mathcal{O}_W(1)) = d + \deg W,$$

as required. □

We note that 6.14 is more generally true for schemes X, Y which are only connected in dimension $\dim X - 1$ resp. $\dim Y - 1$ (instead of being irreducible). For a proof, see [FVo1, FOV].

7 Applications to linear projections

Following [BHSS] we will apply the results of the preceding sections to linear projections. In this section we will assume that *the ground field K is algebraically closed.* Moreover we will use the notation introduced in 6.6.

We saw in 6.9 that either $\sigma(V) = \operatorname{Span} C$, or $\dim \sigma(V) = \dim V$ and $\dim \operatorname{Span} C = \dim V - 1$. In the following we will examine the structure of V and $\sigma(V)$ in the case that $\dim \sigma(V) < \dim V$ so that $\sigma(V) = \operatorname{Span} C$.

For the first result recall that a variety Z is called *uniruled* if there is a dominant rational map $\mathbb{P}^1 \times Z' \dashrightarrow Z$ with $\dim Z' = \dim Z - 1$.

Proposition 7.1. *Assume that $W \subseteq V$ is locally a complete intersection of codimension $k \geq 2$. If $\dim \sigma(V) < \dim V$ then $\sigma(V)$ is uniruled.*

Proof. With the notation of 6.8, we have that $\sigma(V) = \operatorname{Span} C = \sigma^*(E)$, see 6.9. As $k \geq 2$, the exceptional set E of the blowing up $B_W V \to V$ is a ruled variety, where the ruling is given by the projection $\pi : E \to W$. The fibres of this ruling are projective spaces of dimension $k-1$. Moreover, the restriction of σ to the fibres of π is an embedding. Hence there is a subvariety, say, W' of W such that $\sigma^* : \pi^{-1}(W') \to \sigma(V)$ is dominant and generically finite. It follows that $\sigma(V) = \sigma^*(\pi^{-1}(W'))$ is uniruled. □

For the next result recall that a variety is called *unirational* if there is a dominant rational map $\mathbb{P}^{\dim Z} \dashrightarrow Z$.

Proposition 7.2. *Let V be an algebraic variety and assume that $W \subseteq V$ is unirational and locally a complete intersection. If $\dim \sigma(V) < \dim V$ then $\sigma(V)$ is unirational.*

Proof. By assumption, E is a ruled variety over W and so is unirational too. Hence the result follows from the well-known fact that the image of a unirational variety under a rational map is again unirational. □

As $\sigma(V)$ contains the span of $C_W V$, it follows in particular that

$$\dim \sigma(V) \geq k - 1, \qquad (*)$$

where $k := \operatorname{codim}_W V$. In the next corollary we characterize as to when equality holds.

Corollary 7.3. *Let V be an algebraic variety and let $W \subseteq V$ be a subscheme which is generically a complete intersection of codimension k. Assume that equality holds in $(*)$. Then $\sigma(V) \subseteq \mathbb{P}^d$ is a linear subspace of dimension $k-1$, and $W \subseteq V$ is set-theoretically the intersection $H_1 \cap \cdots \cap H_k$, where H_i are sufficiently generic hyperplanes in the linear system $|\sum K\sigma_i|$.*

Proof. Let $x \in W$ be a generic point. Then σ^* maps $E_x \cong \mathbb{P}^{k-1}$ injective and linearly into $\mathbb{P}^d$. As $\sigma^*(E_x)$ is contained in $\sigma(V)$ and has the same dimension these spaces must be equal. This proves the first part. The second part is immediate as the image of $H_1 \cap \cdots \cap H_k \backslash W$ is contained in the intersection of $\sigma(V)$ with k generic hyperplanes and so is empty. □

It is interesting to note that the equality of $H_1 \cap \cdots \cap H_k$ and W also holds scheme-theoretically if W is a locally complete intersection in V. This is a consequence of the following simple lemma.

Lemma 7.4. *Assume that W is locally a complete intersection in V. Then, after replacing $\sigma_0, \ldots, \sigma_d$ by sufficiently general linear combinations, the ideal sheaf $\mathcal{I}_W$ is generated by $\sigma_0, \ldots, \sigma_s$, where $s := \dim \sigma(V)$.*

Proof. With the notation of 6.8, we have $E \cong \mathbb{P}(\mathcal{I}_W/\mathcal{I}_W^2(1))$, and $\sigma : E \to \mathbb{P}^d$ is given by the sections $\sigma_0^*, \ldots, \sigma_d^*$ of $H^0(E, \mathcal{O}_E(1)) \cong H^0(W, \mathcal{I}_W/\mathcal{I}_W^2(1))$. After replacing $\sigma_0, \ldots, \sigma_d$ by sufficiently general linear combinations, we may assume that $\sigma_0, \ldots, \sigma_s$ have no zero on $V \backslash W$ and that $\sigma_0^*, \ldots, \sigma_s^*$ have no common zero in $\mathcal{O}_E(1)$. This means that the residue classes of $\sigma_0, \ldots, \sigma_s$ generate the sheaf $\mathcal{I}_W/\mathcal{I}_W^2(1)$. Now Nakayama's lemma gives that $\sigma_0, \ldots, \sigma_s$ generate $\mathcal{I}_W$. □

We note that the condition that $W \subseteq V$ is a locally complete intersection is essential.

For a generalization of 7.3 to the case that W is not necessarily a locally complete intersection in V, see 6.10 (note that the inequality $j_Z \geq \deg \sigma(V)$ implies that $\sigma(V)$ is a linear subspace of $\mathbb{P}^d$ in case that $j_Z = 1$).

Proposition 7.5. *Let $V \subseteq \mathbb{P}^N$ be an n-dimensional algebraic subvariety and $W \subseteq V$ be a linear subspace which is locally a complete intersection of codimension k in V. Assume that $\mathcal{L} \cong \mathcal{O}_V(1)$. Then the following hold.*

(a) If $\dim \sigma(V) = k - 1$ then $V \cong \mathbb{P}^n$ is a linear subspace of $\mathbb{P}^N$.

(b) If $\dim \sigma(V) = k$ and $n - k \geq 2$ then V is a hypersurface in an $(n+1)$-dimensional linear subspace of $\mathbb{P}^N$.

Proof. We claim that in both cases we have

$$h^0(\mathcal{O}_V(1)) \leq n - k + \dim \sigma(V) + 2. \tag{1}$$

In case (a) this implies that $h^0(\mathcal{O}_V(1)) \leq n+1$, i.e., V is contained in an n-dimensional linear subspace of $\mathbb{P}^N$. Hence V must be equal to this subspace. Similarly, in case (b) the above claim gives that $h^0(\mathcal{O}_V(1)) \leq n+2$ and so V can be embedded into a linear subspace of dimension $n+1$.

It remains to prove (1). Let $\mathcal{I}_W \subseteq \mathcal{O}_V$ be the ideal sheaf of W. Using the exact sequence

$$0 \to \mathcal{I}_W(1) \to \mathcal{O}_V(1) \to \mathcal{O}_W(1) \to 0$$

and the fact that $h^0(\mathcal{O}_W(1)) = n - k + 1$ it is sufficient to show that

$$h^0(\mathcal{I}_W(1)) \leq \dim \sigma(V) + 1. \tag{2}$$

By the preceding lemma, $\mathcal{I}_W$ is generated by $s+1$ elements, where $s := \dim \sigma(V)$. Hence we obtain an exact sequence

$$0 \to \mathcal{K} := \ker \beta \to \mathcal{O}_W^{s+1} \xrightarrow{\beta} \mathcal{I}_W/\mathcal{I}_W^2(1) \to 0. \tag{3}$$

As $\mathcal{I}_W/\mathcal{I}_W^2(1)$ has rank k it follows that in case (a) $\mathcal{K} = 0$ whereas in case (b) $\mathcal{K}$ is a line bundle on $W \cong \mathbb{P}^{n-k}$ and $n-k \geq 2$. In any case, $H^1(W, \mathcal{K}) = 0$ and so

$$h^0(\mathcal{I}_W/\mathcal{I}_W^2(1)) \leq s+1.$$

Taking symmetric powers, (3) gives rise to exact sequences

$$0 \to \mathcal{K} \otimes S^{j-1}(\mathcal{O}_W^{s+1})(-j+1) \to S^j(\mathcal{O}_W^{s+1})(-j+1) \to \mathcal{I}_W^j/\mathcal{I}_W^{j+1}(1) \to 0.$$

The associated cohomology sequence shows that $h^0(\mathcal{I}_W^j/\mathcal{I}_W^{j+1}(1)) = 0$ for $j \geq 2$. Using the exact sequences

$$0 \to \mathcal{I}_W^j/\mathcal{I}_W^{j+1}(1) \to \mathcal{I}_W/\mathcal{I}_W^{j+1}(1) \to \mathcal{I}_W/\mathcal{I}_W^j(1) \to 0$$

and a simple induction argument it follows that $h^0(\mathcal{I}_W/\mathcal{I}_W^j(1)) \leq s+1$ for all j. Hence also $h^0(\mathcal{I}_W(1)) \leq s+1$ as required. □

For the next result we need the following observation due to Faltings (see [Fal]).

Lemma 7.6. *Let W be an algebraic variety and let*

$$0 \to \mathcal{F} \xrightarrow{\alpha} \mathcal{E} := \mathcal{L}_1 \oplus \ldots \mathcal{L}_t \xrightarrow{\beta} \mathcal{Q} \to 0$$

be an exact sequence of locally free sheaves on W, where $\mathcal{L}_1, \ldots, \mathcal{L}_t$ are line bundles. Set $r := \operatorname{rk} \mathcal{F}$ so that $\operatorname{rk} \mathcal{Q} = t - r$. Assume that for every pair of non-empty subvarieties A, B which can be defined locally by r resp. $t-r$ equations, the intersection $A \cap B$ is non-empty. Then the above sequence splits, and $\mathcal{Q}$ is a direct sum of some of the $\mathcal{L}_i$, the isomorphism being given by β.

In the case of linear projections this lemma has the following application.

Corollary 7.7. *Let the notation be as in 6.6. Let V be an algebraic variety of dimension n and let $W \subseteq V$ be a subvariety which is locally a complete intersection of codimension k in V. Assume that W satisfies the following property:*

($\mathbf{P}_m$) *If A, B are subvarieties of W with $\operatorname{codim}_W A \leq k$ and $\operatorname{codim}_W A + \operatorname{codim}_W B \leq m$ then $A \cap B \neq \emptyset$.*

Then $\dim \sigma(V) = k-1$, or $\dim \sigma(V) \geq m$.

Proof. By 7.4 there is a surjection $\mathcal{O}_W^{s+1} \to \mathcal{I}_W/\mathcal{I}_W^2 \otimes \mathcal{L}$ with $s := \dim \sigma(V)$. If $s \geq m$ then we are done. So assume that $s < m$, i.e., $t := s+1 \leq m$. Then (P_m) implies that the assumptions of the lemma of Faltings are satisfied. Thus we obtain that $\mathcal{I}_W/\mathcal{I}_W^2 \otimes \mathcal{L} \cong \mathcal{O}_W^k$ is trivial and so

$$h^0(E, \mathcal{O}_E(1)) = h^0(W, \mathcal{I}_W/\mathcal{I}_W^2 \otimes \mathcal{L}) = k.$$

Hence $s = \dim \sigma^*(E) = k - 1$ as desired. □

As a typical application we obtain the following result.

Corollary 7.8. *Let $V \subseteq \mathbb{P}^N$ be an n-dimensional subvariety and let $W \subsetneq V$ be a linear subspace which is locally a complete intersection of codimension k in V. Let $\sigma = \sigma_W$ be the linear projection from W. Then* $\dim \sigma(V) = k - 1$, *or* $\dim \sigma(V) \geq n - k$.

The proof follows from the fact that condition (P_{n-k}) is satisfied.

Remark 7.9. More generally, it is known that (P_m), $m := \dim W$, is satisfied for any projective manifold W for which the étale cohomology groups $H^{2i}_{et}(W, \mathbb{Q}_\ell)$ are isomorphic to $\mathbb{Q}_\ell$ for $0 \leq i \leq k$. Here ℓ has to be prime to the characteristic of K. Equivalently, if W is defined over $\mathbb{C}$ and W^{an} is the associated analytic space then (P_m) is satisfied if

$$H^{2i}(W^{an}, \mathbb{Q}) \cong \mathbb{Q}, \quad 0 \leq i \leq k.$$

In particular, this is the case for complete intersections and $2k < \dim W$, or for subvarieties of low codimension in the projective space (see e.g. [Har1]).

8 Generic Projections and Double Point Cycles

In this and the next section we will give a further interpretation of the intersection algorithm for the self-intersection of subvarieties X of $\mathbb{P}^N$. Assume that X is smooth of dimension n and

$$p_m : X \to \mathbb{P}^m$$

is a generic linear projection, i.e., $p_m = [\lambda_0 : \cdots : \lambda_m]$ where $\lambda_i = \sum_{j=0}^N u_{ij} x_j$ are sufficiently generic linear forms. Let $R(p_m)$ be the *ramification locus* of p_m which is by definition the set of points $x \in X$ where the tangent map $T_x X \to T_{p_m(x)} \mathbb{P}^m$ is not injective. Moreover let $D(p_m)$ be the closure of the set of all $(x, y) \in X \times X$ with $x \neq y$ and $p_m(x) = p_m(y)$, which is called the set of *double points* of f. Then by a result of Johnson [Joh]

$$[R(p_m)] = H.[D^*(p_m)] - [D^*(p_{m+1})]$$

where $[D^*(p_m)] := \pi_{i*}[D(p_m)]$, H being a generic hyperplane. Here π_i denotes the i-th projection $X \times X \to X$, $i = 1, 2$. Observe that this definition is independent of i because of the symmetric definition of $D(f)$. This formula looks strikingly similar to the intersection algorithm

$$v_k = H.\beta_{k+1} - \beta_k.$$

Our aim in this and the next section is to prove that indeed $v_{k-1} = [R(p_{2n-k})]$. Moreover, β_k is the cycle associated to the closure of the set of all $[x : y] \in J(X,X)\backslash\Delta_X$ with $p_{2n-k}(x) = p_{2n-k}(y)$. As an application we will prove formulae for the degree of the ramification and double point cycle in terms of Segre classes. In particular we derive various obstructions for when a subvariety of $\mathbb{P}^N$ can be embedded into a lower-dimensional space by a generic linear projections. Thereby we obtain a new approach to the classical results of Holme [Hol2, Hol1] and Johnson [Joh]. We follow partly the exposition given in [FMa1].

As an example, let $C \subseteq \mathbb{P}^N$ be a smooth curve of degree d and genus g. Then $\deg \beta_0$ is the number of double points of a generic projection $p_2 : C \to \mathbb{P}^2$. By the well-known Plücker formula

$$\deg \beta_0 = 2(\frac{(d-1)(d-2)}{2} - g).$$

We know already that $v_1 = [C]$. Using the refined Bezout theorem $d^2 = \deg v_1 + \deg v_0 + \deg \beta_0$ we obtain that $\deg v_0 = 2d - 2 + 2g$. By Hurwitz' formula [Har2, p. 301], this number is just the number of ramification points of a projection $p_1 : C \to \mathbb{P}^1$.

First we will give a somewhat technical proposition which will be used in the proofs of the main results in this and the next section.

Proposition 8.1. *Let V be an algebraic manifold and $W \subseteq V$ a submanifold of codimension r with defining ideal $(g_1, \ldots, g_N)$. For indeterminates $U_{\mu\nu}, 1 \le \mu,\ \nu \le N$, set*

$$L := K(U_{\mu\nu})_{1\le\mu,\nu\le N} \quad \text{and} \quad F_\mu := \sum_{\nu=1}^{N} U_{\mu\nu} g_\nu.$$

For $r \le m < N$ let $X_m \subseteq V_L := V \otimes_K L$ be the set of zeros of $F_1, \ldots, F_m$, and let B^m be the closure of $X_m \backslash W_L$ in X_m. Then the following hold.

(1) $V^{m+1} := B^m \cap W_L$ is a Cohen–Macaulay subvariety of W_L of codimension $m+1$ in V_L. Moreover, this variety is equal to the degeneracy locus of the map

$$\varphi : \mathcal{O}_{W_L}^m \cong \bigoplus_{\mu=1}^{m} \mathcal{O}_{W_L} f_\mu \to \mathcal{J}_{W_L}/\mathcal{J}_{W_L}^2 \quad \text{with} \quad f_\mu \mapsto \bar{F}_\mu = \sum_{\nu=1}^{N} U_{\mu\nu}\bar{g}_\nu,$$

i.e., V^{m+1} is given by the maximal minors of φ in a suitable basis of $\mathcal{J}_{W_L}/\mathcal{J}_{W_L}^2$.

(2) *There is a closed subset* Σ *of* V^{m+1} *with* $\operatorname{codim}_{V^{m+1}} \Sigma \geq m - r + 3$ *such that* $\operatorname{Sing} V^{m+1}$ *and* $\operatorname{Sing} B^m$ *are contained in* Σ. *Thus* $V^{m+1} \backslash \Sigma$ *is a smooth Cartier divisor in* $B^m \backslash \Sigma$.

Proof. By a repeated application of 1.2 we get that the set $B^m \backslash W$ is smooth of codimension m in V_L. Hence $\operatorname{codim}_{V_L} B^m = m$. For the remaining assertions we may assume that $\mathcal{J}_W$ is generated by the elements $g_1, \ldots, g_r$. For $j > r$ write $g_j = \sum_{\varrho=1}^r \alpha_{j\varrho} g_\varrho$ so that

$$F_\mu = \sum_{\varrho=1}^{r} U'_{\mu\varrho} g_\varrho \quad \text{with} \quad U'_{\mu\varrho} := U_{\mu\varrho} + \sum_{j>r} U_{\mu j} \alpha_{j\varrho}.$$

Let D^m_{t-1} be the subscheme of W_L given by the t-minors of the matrix $\mathbf{U}'_m := (U'_{\mu\varrho})_{\substack{1 \leq \mu \leq m \\ 1 \leq \varrho \leq r}}$. By Cramer's rule, the r-minors of $\mathbf{U}'_m$ are contained in $\mathcal{J}_{B^m}$. Thus $D^m_{r-1} \supseteq V^{m+1}$. Being a generic determinantal variety, D^m_{r-1} is Cohen–Macaulay of codimension $m+1$ in V_L, as follows from the well-known result of Hochster–Eagon (cf. [BHe, 7.3.1 (c)]). Moreover its singular set is $\Sigma := D^m_{r-2}$ (cf. [BHe, 7.3.4]). Note that $\operatorname{codim}_{D^m_{r-1}} \Sigma = m - r + 3$.

Let $x \in B^m \backslash \Sigma$ be a point. By Cramer's rule, the ideal $(F_1, \ldots, F_m)\mathcal{O}_{V,x}$ contains $r-1$ minimal generators of $\mathcal{J}_{W_L,x}$. We may assume that $(g_1, \ldots, g_{r-1}) = (F_1, \ldots, F_{r-1})$ near x. Write $F_i = H_i g_r \bmod (g_1, \ldots, g_{r-1})$, $i \geq r$, near x. Then B^m is given by the equations $F_1 = \cdots = F_{r-1} = H_r = \cdots = H_m = 0$. Hence B^m is a complete intersection of codimension m in V_L near x, and $V^{m+1} \subseteq B^m$ is given by $g_r = 0$. It is clear that the ideal of r-minors of $\mathbf{U}'_m$ is generated by $H_r, \ldots, H_m \bmod \mathcal{J}_{W_L}$. Hence $V^{m+1} = D^m_{r-1}$ is smooth at x, and this is a Cartier divisor in B^m. In particular B_m is smooth at x too. This proves (2).

To finish the proof it remains to show that $V^{m+1} = D^m_{r-1}$ everywhere. As we remarked above, $V^{m+1} \subseteq D^m_{r-1}$, and generically these varieties are equal. As D^m_{r-1} is reduced this gives the desired equality. □

Remark 8.2. (1) Using the theory of residual intersections one can show that B^m is Cohen–Macaulay also (see [FMa1] for details).

(3) If the characteristic of the ground field K is 0 then in 8.1 it is possible to specialize the indeterminates U_{ij} to generic elements $u_{ij} \in K$. This follows from Bertini's theorem.

Let $X, Y \subseteq \mathbb{P}^N$ be algebraic varieties of dimension n, m respectively. Applying our intersection algorithm, see Sections 1 and 2, we get cycles $v = v(X,Y) = \sum v_k$ and $\beta = \beta(X,Y) = \sum \beta_k$. In the rest of this section we will give a geometric interpretation of the cycle β_k as the cycle of double points of a generic linear projection. In the following we equip the projective space $\mathbb{P}^N = \mathbb{P}^N_K$ with homogeneous coordinates $[X_0 : \cdots : X_N]$. For simplicity we will assume in this section that K is *algebraically closed.*

Definition 8.3. For indeterminates U_{ij} consider the linear forms

$$\lambda_i := \sum_{j=0}^{N} U_{ij}\, X_j, \qquad 0 \le i \le N\ ,$$

on $\mathbb{P}_L^N$, where $L := K(U_{ij})_{0\le i,j\le N}$. We call the map

$$p := [\lambda_0 : \cdots : \lambda_r] : \mathbb{P}_L^N \dashrightarrow \mathbb{P}_L^r$$

the r-th *generic linear projection.*

We remark that for a subscheme $Z \subseteq \mathbb{P}^N$ with $\dim Z \le r$ the restriction of p to Z_L is a regular map. Moreover we note that the λ_i's above and the $s_i = \sum U_{ij}(X_j - Y_j)$ from the intersection algorithm are related via

$$s_i(\underline{X}, \underline{Y}) = \lambda_i(\underline{X}) - \lambda_i(\underline{Y}) \quad \text{for } 0 \le i \le r.$$

Definition 8.4. Let $f : X \cup Y \longrightarrow \mathbb{P}^r$ be a linear projection. Let

$$\tilde{f} : J := J(X,Y) \longrightarrow \mathbb{P} := J(\mathbb{P}^r, \mathbb{P}^r)$$

be the map induced by f. Denote by $\Delta_{\mathbb{P}} \subseteq \mathbb{P}$, $\Delta_{X\cap Y} \subseteq J$ the diagonal subspaces. Then the closure of $\tilde{f}^{-1}(\Delta_{\mathbb{P}})\backslash\Delta_{X\cap Y}$ is called the *double point scheme* of f and will be denoted by $\mathbb{D}_f(X,Y)$.

Usually the double point scheme is introduced in a slightly different way as the closure $\tilde{\mathbb{D}}_f(X,Y)$ of $(f\times f)^{-1}(\Delta)\backslash(X\cap Y)$ where the diagonal $X\cap Y \hookrightarrow X\times Y$ and $\Delta \cong \mathbb{P}^r \hookrightarrow \mathbb{P}^r \times \mathbb{P}^r$ are the diagonals. We remark that these two definitions coincide essentially. More precisely, the canonical map

$$\begin{aligned} \mathbb{D}_f(X,Y)\backslash(i(X)\cup j(Y)) &\longrightarrow \tilde{\mathbb{D}}_f(X,Y) \\ [x:y] &\longmapsto (x,y) \end{aligned}$$

is an isomorphism outside the diagonals, where $i : X \hookrightarrow J(X,Y)$, $j : Y \longrightarrow J(X,Y)$ are the maps $x \mapsto [x:0]$ resp. $y \mapsto [0:y]$.

Proof. The reader may easily verify that this map is surjective. For injectivity, consider

$$[x:y], [x':y'] \in \mathbb{D}_f(X,Y)\backslash(i(X)\cup j(Y))$$

such that $(x,y) = (x',y')$, i.e., $x = \lambda x'$ and $y = \tau y'$ for some $\lambda, \tau \in K^*$. As $[f(x):f(y)]$, $[f(x'):f(y')]$ are in the diagonal it follows that

$$f(x) = f(y), \quad f(x') = f(y')$$

(as points in the affine space). Thus by the linearity of f

$$\lambda f(x') = f(\lambda x') = f(x) = f(y) = \tau f(y'),$$

forcing $\lambda = \tau$. □

The next proposition shows the connection between the double point cycle and the intersection algorithm.

Proposition 8.5. *Let $X, Y \subseteq \mathbb{P}^N$ be algebraic varieties of dimension n, m respectively. Let $p : X_L \cup Y_L \longrightarrow \mathbb{P}^{n+m-k}$ be the generic linear projection, $0 \le k \le \dim X \cap Y$, and set $\mathbb{D} := \mathbb{D}_p(X_L, Y_L)$, $\Delta := \Delta_{X_L \cap Y_L} \subseteq J_L$. Then the following hold.*

1. *$\mathbb{D}$ has pure dimension k and $[\mathbb{D}] = \beta_k$.*
2. $\operatorname{codim}_{\mathbb{D}}(\operatorname{Sing}\mathbb{D}\backslash\Delta) \le \operatorname{codim}_J(\operatorname{Sing} J\backslash\Delta)$.

Proof. The equality of the cycles β_k and $[\mathbb{D}]$ follows immediately from the fact that $\mathbb{D}\backslash\Delta_{X_L\cap Y_L}$ is given by the vanishing of

$$s_i(\underline{X}, \underline{Y}) = \lambda_i(\underline{X}) - \lambda_i(\underline{Y}), \quad 0 \le i \le n+m-k\ ,$$

and the construction of β_k. Thus $\mathbb{D}\backslash\Delta$ has pure dimension k. We leave it as an exercise to the reader to deduce (2) from 1.2. □

Applying this proposition to the case $k = 0$ and specializing the indeterminates U_{ij} to generic elements u_{ij} in K we obtain the following result.

Corollary 8.6. *Assume that $n + m \le N$. Let $p : X \cup Y \longrightarrow \mathbb{P}^{n+m}$ be a generic linear projection. Then* deg β_0 *is the number of points in the set $\mathbb{D}_p(X, Y)$ of double points of p.*

In the case $X = Y$ this gives the following corollary.

Corollary 8.7. *Let $X \subseteq \mathbb{P}^N$ be a subvariety of dimension n with $N \ge 2n$. Let $p : X \longrightarrow \mathbb{P}^{2n}$ be a generic linear projection. Then $p(X)$ has at most ordinary double points, and their number, say δ, is given by*

$$2\delta = \deg\beta_0 = \deg \operatorname{Sec} X \deg(J/\operatorname{Sec} X)$$

where $J = J(X, X)$ is the ruled join.

We wish to give a more explicit result for $\deg\beta_0$ in terms of Segre classes. Let X, Y be subvarieties. We remind the reader that by the refined Bezout theorem 2.5

$$\deg X \deg Y = \sum_{i\ge 0} \deg v_i + \deg XY \deg(J/XY),$$

where $J := J(X, Y)$ is the ruled join of X, Y. Moreover by 4.4

$$v^k = \sum_{i=0}^{k} \binom{k-1}{i-1} c_1(\mathcal{O}(1))^{k-i} \cap s^i(C_{X\cap Y}J).$$

Using 5.10 we get that $s^i(C_{X\cap Y}J) = s^{i-1}(C_{X\cap Y}(X\times Y))$ and $s^0(C_{X\cap Y}J) = 0$. Hence

8.8. $$v^k = \sum_{i=0}^{k-1} \binom{k-1}{i} c_1(\mathcal{O}(1))^{k-i-1} \cap s^i(C_{X\cap Y}(X \times Y)).$$

Taking degrees and summing up we obtain

$$\begin{aligned} \deg v &= \sum_{k=0}^{n+m+1} \sum_{i=0}^{k-1} \binom{k-1}{i} \deg s^i(C_{X\cap Y}(X \times Y)) \\ &= \sum_{i\geq 0} \binom{n+m+1}{n+m-i} \deg s^i(C_{X\cap Y}(X \times Y)) \\ &= \sum_{i\geq 0} \binom{n+m+1}{n+m-i} \deg s_{n+m-i}(C_{X\cap Y}(X \times Y)) \end{aligned}$$

where we have used 4.6 and the fact that $\dim s^i(C_{X\cap Y}(X \times Y)) = n + m - i$. (Recall that, by our convention, $s_i(C)$ denotes the part of the Segre class of dimension i of a cone C while $s^i(C)$ denotes the part which is of codimension i in C.) Thus we have shown the following formula which was given by Ådlandsvik in [Adl].

Theorem 8.9. *For subvarieties X, Y of $\mathbb{P}^N$ of degree d_X, resp. d_Y and dimension n, resp. m, we have*

$$\deg XY \deg(J/XY) = d_X d_Y - \sum_{k\geq 0} \binom{n+m+1}{k} \deg s_k(C_{X\cap Y} X \times Y).$$

Applying 8.9 and 8.7 to the special case of a self-intersection, we obtain the following corollary which was first proved by Peters and Simonis [PSi].

Corollary 8.10. *For X an n-dimensional submanifold of $\mathbb{P}^N$, the number of double points δ of a generic linear projection $X \to \mathbb{P}^{2n}$ with $N \geq 2n$ is given by*

$$2\delta = (\deg X)^2 - \sum_{k\geq 0} \binom{2n+1}{k} \deg s_k(T_X).$$

Thus X can be embedded into $\mathbb{P}^{2n}$ by a generic linear projection if and only if

$$(\deg X)^2 = \sum_{k\geq 0} \binom{2n+1}{k} \deg s_k(T_X).$$

Let us consider some classical examples.

Example 8.11. (1) If X is an abelian variety, then $s_i(T_X) = 0$ for $i \neq n$ and $s_n(T_X) = [X]$. Hence the above condition is equivalent to

$$\deg X = \binom{2n+1}{n},$$

which is a classical result, see e.g. [Ful1, 3.2.15 a]. For example, an abelian surface can be embedded into $\mathbb{P}^4$ if and only if its degree is 10. It was shown by Comessatti (see [LBi, Chapter 10, 1.6]) and rediscovered by Horrocks and Mumford that such abelian surfaces exist. In contrast, by a result of Van de Ven [VdV] an abelian variety of dimension $n \geq 3$ does not admit an embedding into $\mathbb{P}^{2n}$.

(2) For a smooth plane curve we rediscover the classical Plücker formulas as discussed at the beginning of this section.

(3) Let $X \subseteq \mathbb{P}^N$ be a smooth surface with canonical divisor K_X and degree d. Let H denote a hyperplane section of X. Using 3.9 we get $\deg v_2 = d$ and

$$\begin{aligned}\deg v_1 &= K_X H + 3d\\ \deg v_0 &= -c_2 + K_X^2 + 4K_X H + 6d.\end{aligned}$$

Hence the generic projection of X into $\mathbb{P}^4$ has

$$d^2 - 10d - 5K_X H + c_2 - K_X^2$$

double points. This is the classical double point formula of Severi.

As before, let X be a subvariety of $\mathbb{P}^n$ and assume that the secant variety X^2 has maximal dimension. At the end of this section, let us discuss the question as to how one can obtain a formula for the degree of X^2 and not only for $\deg(J/X^2) \deg X^2$. Of course, if X^2 has dimension n then $X^2 = \mathbb{P}^n$ and so the degree is 1. Otherwise we have the following criterion.

Proposition 8.12. *Let K be an algebraically closed field of characteristic 0. Assume that $2 \dim X + 1 < n$ and that one of the following two conditions is satisfied.*

1. *X is smooth and the twisted normal bundle $\mathcal{N}_{X/\mathbb{P}^n}(-1)$ is ample.*

2. *X is a nondegenerate curve.*

Then $\deg(J(X,X)/\operatorname{Sec} X) = 1$.

Proof. Assume that this is not the case. Let $u \in X^2$ be a generic point lying on the the secant $\ell := \overline{x_0 y_0}$, where $x_0, y_0 \in X$. By assumption, the closure, say, A of the set

$$\{x \in X \mid \exists y \in X \text{ with } \overline{xy} \cap \ell \neq \emptyset\}$$

has positive dimension. By Terracini's lemma, for every $x \in A$ we have $T_{X,x} \subseteq L := T_{X^2,u}$. In the first case take a linear form, say ℓ, on $\mathbb{P}^n$ which vanishes on L. Then ℓ defines a non-vanishing section in the sheaf $\mathcal{N}^\vee_{X/\mathbb{P}^n}(1)$ as the reader may easily verify. This contradicts the assumption that $\mathcal{N}_{X/\mathbb{P}^n}(-1)$ is ample. In the case that X is a curve, we obtain $A = X$ and so $X \subseteq L$, which again is a contradiction. □

Combining 8.12 with the previous results, we obtain explicit formulas for the degree of the secant variety X^2. For instance, if $X \subseteq \mathbb{P}^n$, $n \geq 4$, is a nondegenerate curve then

$$\deg X^2 = \binom{d-1}{2} - g.$$

Similarly, if (1) in 8.12 is satisfied then applying 8.9 we get an explicit formula for $\deg X^2$.

9 Generic Projections and the Ramification Cycle

Let $X \subseteq \mathbb{P}^N$ be a submanifold of dimension n and $p : X \longrightarrow \mathbb{P}^{2n-k-1}$ a generic linear projection, where $0 \leq k \leq n-1$. In this section we will show that the ramification locus of p has dimension $\leq k$ and that its associated k-cycle is just v_k. Using this and the formula for v_k in terms of Segre classes, see Sections 2.3 and 2.4, we will derive various classical obstructions for when a subvariety $X \subseteq \mathbb{P}^N$ can be embedded into a smaller projective space via a generic linear projection. This very natural approach was first given by van Gastel [vGa3]; see also [FMa1].

For a map of locally free sheaves $\varphi : \mathcal{E} \longrightarrow \mathcal{F}$ we denote by $\mathbb{D}(\varphi)$ the *degeneracy locus* of φ, i.e., the subscheme of X given locally by the maximal minors of φ in a suitable local basis of $\mathcal{E}$ and $\mathcal{F}$.

Let X be a smooth algebraic K-scheme and $f : X \to M$ a morphism into an algebraic manifold with $\dim M \geq \dim X$. Then we denote by $R(f)$ the *ramification locus* of f. We will equip it with the structure $R(f) := \mathbb{D}(\varphi)$, where $\varphi : f^*(\Omega^1_M) \to \Omega^1_X$ is the induced map.

The basic result is the following theorem (see [vGa3, FMa1]). We give here a simplified treatment.

Theorem 9.1. *Let $X \subseteq \mathbb{P}^N$ be an algebraic variety of dimension n and k an integer satisfying $0 \leq k \leq n-1$ and $N \geq 2n-k-1$. Let $p : X_L \to \mathbb{P}^{2n-k-1}_L$ be the generic linear projection, see 8.3, and $R(p)$ its ramification locus on $\operatorname{Reg} X$. Then $\dim R(p) \leq k$, and the associated k-cycle $[R(p)]_k$ is just $v_k | \operatorname{Reg} X_L$.*

Proof. Recall that the generic linear projection p is given by

$$p = [\lambda_0 : \cdots : \lambda_{2n-k-1}]$$

where $\lambda_i := \sum_{i=0}^N U_{ij} x_j$ is as in 8.3. The join $J(\mathbb{P}^{2n-k-1}, \mathbb{P}^{2n-k-1})$, which is a projective space of dimension $4n-2k-1$, is denoted by $\mathbb{P}$ and there is an induced map of joins

$$\tilde{p} : J := J(X, X) \longrightarrow \mathbb{P}.$$

Let $\Delta_X \cong X \hookrightarrow J$, $\Delta \cong \mathbb{P}^{2n-k-1} \hookrightarrow \mathbb{P}$ be the diagonals and $\mathcal{J}_X \hookrightarrow \mathcal{O}_J$, $\mathcal{J}_\Delta \subseteq \mathcal{O}_\mathbb{P}$ their ideal sheaves.

Let B_{k+1} be the closure of $\tilde{p}^{-1}(\Delta_L)\backslash\Delta_{X_L}$ and set $V_k := B_{k+1} \cap \Delta_{X_L}$. Applying 8.1 to $V := \operatorname{Reg} J_L$, $W := \Delta_{X_L} \cap \operatorname{Reg} X_L$, we get the following:

(a) B_{k+1} is of dimension $k+1$ and smooth on $\operatorname{Reg} J_L \backslash \Delta_{X_L}$.

(b) $V_k \cap W$ is Cohen–Macaulay and generically smooth. Moreover this scheme is equal to the degeneracy locus of the canonical map

$$\varphi : p^*(\mathcal{J}_{\Delta_L}/\mathcal{J}^2_{\Delta_L})|W \to (\mathcal{J}_{X_L}/\mathcal{J}^2_{X_L})|W.$$

(c) The codimension of V_k in B_{k+1} is 1 in the points of $\operatorname{Reg} X_L$; generically V_k is a Cartier divisor in B_{k+1}.

Thus V_k is of pure dimension k on $\operatorname{Reg} X_L$. Using the fact that $\tilde{p}^{-1}(\Delta_L)$ is given by the equations

$$s_i = \lambda_i \otimes 1 - 1 \otimes \lambda_i = \sum_{j=0}^{N} U_{ij}(X_j \otimes 1 - 1 \otimes X_j), \quad 0 \le i \le 2n-k-1,$$

we get from the intersection algorithm that

$$\beta_{k+1} = [B_{k+1}] \quad \text{and} \quad v_k|\operatorname{Reg}(X_L) = [V_k]|\operatorname{Reg}(X_L). \tag{$*$}$$

Let $\hat{X} \subseteq \mathbb{A}^{N+1}$ be the affine cone of X. Then the corresponding differential module $\Omega^1_{\hat{X}/K}$ is well known to be canonically isomorphic to $\mathcal{J}_{\hat{X}}/\mathcal{J}^2_{\hat{X}}$, with $\mathcal{J}_{\hat{X}}$ the ideal of the diagonal $\hat{X}$ in $\hat{X} \times \hat{X}$. It follows that

$$\tilde{\Omega}^1_{\hat{X}/K} \cong \mathcal{J}_X/\mathcal{J}^2_X,$$

where $\tilde{}$ denotes the associated sheaf on the projective space. Similarly, as $\mathbb{A}^{2n-k} = (\mathbb{P}^{2n-k-1})\tilde{}$, on $\mathbb{P}^{2n-k-1}$

$$\tilde{\Omega}^1_{\mathbb{A}^{2n-k}/K} \cong \mathcal{J}_\Delta/\mathcal{J}^2_\Delta.$$

Under these two isomorphisms the map $d\hat{p}$ is identified with φ in (b), where $\hat{p} : \hat{X}_L \to \mathbb{A}^{2n-k}_L$ is the map of the affine cones induced by p. Consider the following diagram of vector bundles on $\operatorname{Reg} X_L$:

$$\begin{array}{ccccccc}
0 \longrightarrow & p^*\Omega^1_{\mathbb{P}^{2n-k-1}_L/L} & \longrightarrow & p^*\tilde{\Omega}^1_{\mathbb{A}^{2n-k}_L/L} & \longrightarrow & \mathcal{O}_{X_L} & \longrightarrow 0 \\
 & dp\downarrow & & d\hat{p}\downarrow & & \cong\downarrow & \\
0 \longrightarrow & \Omega^1_{X_L/L} & \longrightarrow & \tilde{\Omega}_{\hat{X}_L/L} & \longrightarrow & \mathcal{O}_{X_L} & \longrightarrow 0\,.
\end{array}$$

Here the rows are just the Euler sequences. It follows that

$$\mathbb{D}(dp) = \mathbb{D}(d\hat{p}),$$

and so the ramification locus $R(p)$ is given by the maximal minors of the induced map $\varphi : p^*(\mathcal{J}_{\Delta_L}/\mathcal{J}^2_{\Delta_L}) \to \mathcal{J}_{X_L}/\mathcal{J}^2_{X_L}$. Together with (b) and $(*)$ above this proves the theorem. □

Example 9.2. (1) Let $X \subseteq \mathbb{P}^N$ be a smooth subvariety of dimension n and consider the generic linear projection $p : X \to \mathbb{P}^n$. Then the ramification locus is given by the determinant of the morphism $dp : p^*(\Omega^1_{\mathbb{P}^n}) \to \Omega^1_X$. Thus it is represented by the divisor $K_X + (n+1)H$, where H is a hyperplane section of X. In particular, its degree is $K_X H^{n-1} + (n+1)H^n = K_X H^{n-1} + (n+1)d$ (cf. 3.9). For instance, if X is a smooth curve of genus g and degree d this gives $2g - 2 + 2d$, which is just Hurwitz' ramification formula.

(2) Let $X \subseteq \mathbb{P}^N$ be a smooth surface and consider a generic linear projection $p : X \to \mathbb{P}^3$. The degree of its ramification locus is the degree of v_0, which is $K_X^2 - c_2 + 4K_X H + 6d$ by 8.11 (3).

(3) For a 3-fold, the degree of the ramification locus of a generic linear projection into $\mathbb{P}^5$ is given by the degree of v_0 which is

$$-c_3 - 2c_2K_X + K_X^3 + 6(K_X^2 H - c_2 H) + 15K_X H^2 + 20d$$

(see 3.9). Similarly, projecting X to $\mathbb{P}^4$ gives a ramification locus of dimension 1 and degree

$$-c_2 H + K_X^2 H + 5K_X H^2 + 10d.$$

Example 9.3. Assume that $C \subseteq \mathbb{P}^2$ is a plane curve of degree d with a unique singular point $p \in C$. By the refined Bezout theorem

$$d^2 = d + j_p + \deg v_0'$$

where j_p is the coefficient of p in the v-cycle and where v_0' is the part of v_0 which is lying on $C \backslash \{p\}$. To compute j_p and $\deg v_0'$, consider the normalization $\tau : \bar{C} \to C$ and a linear projection $\pi : C \to \mathbb{P}^1$ from a generic point, say $a \in \mathbb{P}^2$. Let $p_1, \ldots, p_r \in \bar{C}$ be the points over p, so that C has r branches at p. By Hurwitz' formula

$$2d + (2g - 2) = \sum_{\varrho=1}^{r} (v_{p_\varrho} - 1) + \sum_q (v_q - 1), \tag{1}$$

where v_b denotes the ramification index of $\pi \circ \tau$ at a point b of $\bar{C}$ and where the second sum is taken over the smooth points of C. By 9.1 above, $\sum_q (v_q - 1) = \deg v_0'$. On the other hand, if $\lambda \in \mathcal{O}_{\mathbb{P}^2,p}$ is a local equation for the line $\overline{ap}$ then

$$\sum_{i=1}^{r} (v_{p_\varrho} - 1) = \left(\sum_{i=1}^{r} \lg \mathcal{O}_{\bar{C},p_\varrho} / \lambda \mathcal{O}_{\bar{C},p_\varrho} \right) - r = \lg \mathcal{O}_{C,p} / \lambda \mathcal{O}_{C,p} - r = m_p - 1,$$

where m_p is the multiplicity of C at p. Thus we derive from (1) that

$$2d + (2g - 2) = m_p - r + \deg v_0', \tag{2}$$

By Plücker's formula, $g = \binom{d-1}{2} - \delta_p$, where δ_p is the length of the cokernel of $\mathcal{O}_{C,p} \to \bigoplus_\varrho \mathcal{O}_{\bar{C},p_\varrho}$ (this number is also called the virtual number of double points at p). Inserting this into (2) it follows that

$$d^2 = d + \delta_p + (m_p - r) + \deg v_0'.$$

Comparing with the refined Bezout theorem gives that

$$j_p = \delta_p + (m_p - 1).$$

Since every isolated plane curve singularity $A = K[[X,Y]]/(f)$ can be embedded into a plane curve as above, we get in particular the Teissier formula

$$e((\partial f/\partial X, \partial f/\partial Y), A) = \delta_p + (m_p - 1);$$

observe that the left-hand side is just j_p by 2.13.

In the remaining part of this section we will apply these results to embeddings and geometric injections given by generic linear projections. We introduce geometric injections as follows. A morphism $f : X \to Y$ of algebraic varieties over a field K is said to be a *geometric injection* if the map $X \otimes_K \bar{K} \to Y \otimes_K \bar{K}$ is injective, where $\bar{K}$ is the algebraic closure of K. Equivalently, $f : X \to Y$ is injective and for all $x \in X$ the field extension $k(x) : k(f(x))$ is purely inseparable. The reader may easily verify that f is geometrically injective if and only if $(f \times f)^{-1}(\Delta_Y) = \Delta_X$ set theoretically, where $\Delta_X \subseteq X \times X$ and $\Delta_Y \subseteq Y \times Y$ are the diagonals. Similarly, if we consider linear projections $p : X \to \mathbb{P}^r$ then p is geometrically injective if and only if $\tilde{p}^{-1}(\Delta) = \Delta_X$, with the notation in the proof of 9.1 above.

Proposition 9.4. *Let $X \subseteq \mathbb{P}^N$ be an algebraic variety of dimension n and k an integer with $0 \le k \le n-1$ and $N \ge 2n-k-1$. Let $p : X_L \to \mathbb{P}_L^{2n-k-1}$ be the generic linear projection. Then the following hold.*

(1) *p is geometrically injective if and only if $\beta_{k+1} = 0$.*

(2) *Assume that $\beta_{k+1} = 0$. Then $p|\operatorname{Reg} X$ is an embedding.*

Proof. By 8.5 (1), and using the notation of the proof of 9.1, $\beta_{k+1} = [\mathbb{D}]$ with $\mathbb{D} := \mathbb{D}_p(X_L, X_L)$. By definition $\mathbb{D}$ is the closure $\tilde{p}^{-1}(\Delta)\backslash\Delta_X$ and so $\mathbb{D} = \emptyset$ if and only if p is geometrically injective. This proves (1).

By 9.1 $[R(p)] = v_k = 0$ on $\operatorname{Reg} X_L$. As $R(p)$ has pure dimension k, this implies $R(p) \cap \operatorname{Reg} X = \emptyset$. Thus $p|\operatorname{Reg} X$ is unramified and an embedding. □

As a corollary we obtain Holme's embedding obstructions for the embedding of smooth varieties $X \subseteq \mathbb{P}^N$, stated in terms of Segre classes of the tangent bundle T_X.

Corollary 9.5. *Let $X \subseteq \mathbb{P}^N$ be a smooth algebraic variety of dimension n and k an integer with $0 \le k \le n-1$ and $N \ge 2n-k-1$. Then X can be* *embedded into* $\mathbb{P}^{2n-k-1}$ *by a generic linear projection if and only if*

$$(\deg X)^2 = \sum_{l \geq k+1} \binom{2n-k}{l-k-1} \deg s_l(T_X).$$

Proof. By the preceding proposition, we need to show that

$$\deg \beta_{k+1} = (\deg X)^2 - \sum_{l \geq k+1} \binom{2n-k}{l-k-1} \deg s_l(T_X).$$

As by 2.6

$$(\deg X)^2 = \sum_{i \geq k+1} \deg v_i + \deg \beta_{k+1}.$$

It remains to verify that

$$\sum_{i \geq k+1} \deg v_i = \sum_{l \geq k+1} \binom{2n-k}{l-k-1} \deg s_l(T_X).$$

Applying 8.8 gives

$$\begin{aligned} \sum_{i \geq k+1} \deg v_i &= \sum_j \sum_{i \leq 2n-j-k-1} \binom{j+i}{i} \deg s_{2n-j}(T_X) \\ &= \sum_j \binom{2n-k}{2n-j-k-1} \deg s_{2n-j}(T_X), \end{aligned}$$

where for the last equality we have used 4.6. This proves the corollary. □

Since for an arbitrary variety X the generic projection $p : X \to \mathbb{P}^{2n-k-1}$ is geometrically injective if and only if $\beta_{k+1} = 0$, we also obtain with the same proof the following criterion due to Johnson [Joh].

Corollary 9.6. *For an algebraic variety* $X \subseteq \mathbb{P}^N$, *the generic linear projection* $X_L \to \mathbb{P}^{2n-k-1}$ *is geometrically injective iff*

$$(\deg X)^2 = \sum_{l \geq k+1} \binom{2n-k}{l-k-1} \deg s_l(C_X(X \times X)).$$

Note that by 9.6 and 9.5, we get that for a smooth variety the generic projection is geometrically injective if and only if it is an embedding.

Remark 9.7. In [Joh] the author defines more generally a ramification cycle $R(p)$ for a morphism $p : X \to Y$, where X is a singular variety. Roughly speaking it is the set of all $x \in X$ around which the morphism p is not geometrically injective. We remark that more generally, as in 9.1, one has the equality of $[R(p)]_k$ and v_k in the case of a generic projection of an arbitrary (possibly singular) variety $X \subseteq \mathbb{P}^N$. This follows for example from the relationship between the double point and ramification cycles stated after the theorem of 5.2 in [Joh], and the intersection algorithm.

Let X be an arbitrary algebraic variety. In the remaining part of this section we consider the question as to when the generic linear projection is not only geometrically injective but even an embedding. For this we will introduce a modified tangent variety $\operatorname{ZTan} X \subseteq \mathbb{P}^N$ which is informally speaking the union of all embedded Zariski tangent spaces $t_{X,x} \subseteq \mathbb{P}^N,\ x \in X$. Before giving the precise definition, we remind the reader of the embedded tangent space: if $x \in X$ is a closed point and $\hat{x}$ is a closed point in the affine cone $\hat{X}$ lying over x then the embedded tangent space is the linear subspace $t_{X,x} := \mathbb{P}(T_{\hat{X},\hat{x}})$ of $\mathbb{P}^n$. It is well known and easy to check that $t_{X,x}$ is the projective closure of the usual tangent space $T_{X,x}$.

To introduce $\operatorname{ZTan} X$ let $J := J(X,X) \subseteq \mathbb{P}^{2N+1}$ be the ruled join and $X \cong \Delta \hookrightarrow J$ the diagonal. Let $\pi : J \dashrightarrow \mathbb{P}^N$ be the map $[x:y] \mapsto [x-y]$. Blowing up the diagonal subspace of J gives a morphism $\tilde{\pi} : \tilde{J} \to \mathbb{P}^N$ and in particular a map on the exceptional set

$$\tilde{\pi}^{-1}(\Delta) = \mathbb{P}(\mathcal{I}/\mathcal{I}^2) \xrightarrow{\varphi} \mathbb{P}^N,$$

where $\mathcal{I} \subseteq \mathcal{O}_J$ is the ideal sheaf of Δ. We call

$$\operatorname{ZTan} X := \varphi(\mathbb{P}(\mathcal{I}/\mathcal{I}^2))$$

the *Zariski tangent variety.*

We will say that a morphism $f : X \to Y$ is a *local embedding* if every $x \in X$ has a neighbourhood U such that the restriction $f : U \to Y$ is a locally closed embedding. In the next result we show that the dimension of $\operatorname{ZTan} X$ controls when p_m is a local embedding.

Proposition 9.8. ([Llu]) *Let $X \subseteq \mathbb{P}^N$ be a closed subscheme and $p_m : X_L \to \mathbb{P}^m_L$ be a generic projection, see 8.3. Then p_m is a local embedding if and only if* $\dim \operatorname{ZTan} X \le m$.

Proof. We may assume that the base field K is algebraically closed. Let $\Lambda \subseteq \mathbb{P}^N_L$ be the linear subspace such that p_m is the linear projection from Λ. Then p_m is a local embedding iff for all closed points $x \in X$ the tangent map of Zariski tangent spaces $T_{X,x} \to T_{\mathbb{P}^m, p_m(x)}$ is injective. This is equivalent to

$$t_{X,x} \cap \Lambda = \emptyset$$

where $t_{X,x} \subseteq \mathbb{P}^N$ is the projective tangent space of X at x. Thus p_m is a local embedding iff $\operatorname{ZTan} X \cap \Lambda = \emptyset$. As Λ is generic this is equivalent with $\dim \operatorname{ZTan} X < \operatorname{codim} \Lambda = m+1$. □

In the following we will describe a numerical obstruction for when the m-th generic linear projection is a local embedding. With the notation as above, consider the cone $C := \mathbb{V}(\mathcal{I}/\mathcal{I}^2)$. In general, C is not equidimensional. Write $[C] = \sum n_j [C_j]$ with C_j the irreducible components of C. The sheaf $\mathcal{I}/\mathcal{I}^2$ is globally generated by the residue classes of the sections $\sigma_i := X_i \otimes 1 - 1 \otimes X_i$, where $X_0, \ldots, X_N$ are the coordinate functions of

$\mathbb{P}^N$. Performing the intersection algorithm with these sections on the cone C_j leads to cycles

$$v(C_j) = \sum_i v^i(C_j), \quad \beta(C_j) = \sum_i \beta^i(C_j)$$

on X. We set

$$v^i(C) := \sum_j v^i(C_j) \quad \text{and} \quad \beta^i(C) := \sum_j \beta^i(C_j).$$

Note that in general the cycles $v^i(C)$, $\beta^i(C)$ will have a mixed dimension!

Lemma 9.9. *$\beta^{m+1}(C) = 0$ if and only if $\dim \operatorname{ZTan} X \leq m$.*

Proof. The map $\varphi : \mathbb{P}(\mathcal{I}/\mathcal{I}^2) = \mathbb{P}C \to \mathbb{P}^N$ is just given by the initial forms of the elements σ_i. By 1.14, $\beta^{m+1}(C_j) = 0$ iff $m \geq \dim \varphi(\mathbb{P}C_j)$. Since by definition $\operatorname{ZTan} X = \varphi(\mathbb{P}C)$, the lemma follows. □

The bundle $\mathbb{V}(\mathcal{I}/\mathcal{I}^2)$ is closely related to the tangent bundle $T_X := \mathbb{V}(\Omega^1_X)$ of X. By 5.11 there is an exact sequence of cones

$$0 \to X \times \mathbb{A}^1 \to C = \mathbb{V}(\mathcal{I}/\mathcal{I}^2) \to T_X \to 0.$$

In particular,

$$s^{i+1}(C \otimes \mathcal{O}(1)) = s^i(T_X \otimes \mathcal{O}(1)).$$

Collecting the above results gives the following theorem.

Theorem 9.10. *Let $X \subseteq \mathbb{P}^N$ be an arbitrary closed subscheme and $p_m : X_L \to \mathbb{P}^m_L$ be the generic projection. Then p_m is a local embedding if and only if $s^i(T_X \otimes \mathcal{O}(1)) = 0$ for $i > m$.*

Proof. This follows from the results above since by 3.3

$$v^{i+1}(C) = s^{i+1}(C \otimes \mathcal{O}(1)) = s^i(T_X \otimes \mathcal{O}(1)). \qquad \square$$

Taking degrees it is easy to derive a numerical criterion. If γ is a cycle (possibly of mixed dimension), say, $\gamma = \sum_C \gamma_C[C]$, then we set

$$\deg \gamma := \sum_C \gamma_C \deg C.$$

Being a cycle with non-negative coefficients, $v^k(C)$ vanishes if and only if its degree vanishes. Thus p_m is a local embedding if and only if the degree of $\sum_{k>m} v^{k+1}(C)$ vanishes. Moreover, using 3.6

$$\deg v^{k+1}(C) = \deg s^k(T_X \otimes \mathcal{O}(1)) = \sum_{i=0}^{k} \binom{k-1}{i-1} \deg c_1(\mathcal{O}(1))^{k-i} \cap s^i(T_X),$$

and $v^0(C) = 0$. This leads to the following corollary.

Corollary 9.11. *The generic projection p_m is a local embedding if and only if*

$$\sum_{k>m}\sum_{i=0}^{k}\binom{k-1}{i-1}\deg c_1(\mathcal{O}(1))^{k-i}\cap s^i(T_X)=0.$$

For instance, if $m \geq \dim T_X$ then

$$\dim c_1(\mathcal{O}(1))^{k-i}\cap s^i(T_X) \leq \dim T_X - k \leq m-k < 0$$

for $k > m$. Thus p_m is a local embedding. In particular, if X is smooth of dimension n then p_{2n} is a local embedding (but of course not an embedding, in general).

The first criterion for when the generic projection p_m is an embedding for singular varieties X was given by A. Holme, see [Hol1]. He defines a whole sequence of invariants which vanish if and only if p_m is an embedding. We remark that his sequence is essentially given by pairs of Segre classes of T_X and $C_X(X \times X)$.

10 Joins of maximal dimension

Let V be an equidimensional algebraic K-scheme, $W \subseteq V$ a closed subscheme and $C \subseteq W$ a subvariety. In this section we will turn to the question as to when a subvariety C of W appears in the v-cycle. As before we denote by $C_W V$ the normal cone of W in V and by $(C_W V)_C$ the fibre of the projection $C_W V \to W$ over the generic point, say η_C, of C. In 4.8 we have seen that

$$\dim(C_W V)_C \leq \min\{\dim \mathcal{O}_{V,C}, \mu(I)\}, \qquad (*)$$

where $I \subseteq \mathcal{O}_{V,C}$ is the ideal of W. Moreover, C appears in the v-cycle if and only if $\dim(C_W V)_C = \dim \mathcal{O}_{V,C}$.

Thus we need to find criteria for when equality holds in $(*)$. Applying this to the embedding of an intersection $X \cap Y \hookrightarrow X \times Y$, we will in particular obtain conditions under which $\mathrm{LJoin}_C(X,Y)$ has maximal dimension and thus C appears in $v(X,Y)$.

As a first step, let us specialize to the case of join varieties. By Section 5 we know that for projective varieties $X, Y \subseteq \mathbb{P}^n$ the limit of join variety $\mathrm{LJoin}_0(\hat{X},\hat{Y})$ is just the affine cone $(XY)\hat{}$. Therefore the above problem is closely related to the question as to when the join variety has maximal dimension. Before giving the first result, let us study a simple example. Consider the subvariety M_r of $\mathbb{P}^{nm-1}$ given by the set of all $n\times m$-matrices A with $\mathrm{rk}\, A \leq r$. It is clear that the sum of two matrices A, B of rank $\leq r$ resp. $\leq s$ is a matrix of rank at most $r+s$, and that for generic matrices A, B of rank r, s the sum $A+B$ will have rank $r+s$. This easily implies that the join of D_r and D_s is just D_{r+s}. It is well known that the codimension of D_r in $\mathbb{P}^{nm-1}$ is given by $(n-r)(m-r)$. This gives in particular that for $r+s \leq \min\{n, m\}$

$$\dim D_{r+s} = \dim D_r D_s < \dim D_r + \dim D_s + 1,$$

i.e., the join can have non-maximal dimension even if the varieties in question are Cohen–Macaulay.

Using curvature conditions we get the following result.

Proposition 10.1. *Let X, $Y \subseteq \mathbb{P}^n$ be projective varieties satisfying the following properties:*

1. *X is locally a complete intersection, and the twisted normal bundle $\mathcal{N}(-1) = (\mathcal{I}_X/\mathcal{I}_X^2(1))^\vee$ is an ample vector bundle (here $\mathcal{I}_X$ is the ideal sheaf of X in $\mathbb{P}^n$);*
2. *$X \cap Y \subseteq \operatorname{Reg} X$.*

Then XY has maximal dimension $\min\{n, \dim X + \dim Y + 1\}$.

Proof. We may assume that $XY \neq \mathbb{P}^n$. For a generic closed point $a \in XY$ we will study the so-called entry point set X_a, which is the closure of the set

$$X_a^o := \{x \in X \mid \exists y \in Y \text{ with } a \in \overline{xy}\}.$$

By Terracini's lemma (see e.g. [FOV, 4.3.2]), $t_{X,x} \subseteq t_{XY,a}$ for all $x \in X_a^o$. Since $X_a \backslash X_a^o \subseteq X \cap Y \subseteq X_{\text{reg}}$ it follows that $t_{X,x} \subseteq t_{XY,a}$ for all $x \in X_a$. Let $f \in H^0(\mathcal{O}_{\mathbb{P}^n}(1))$ be a nonzero linear form vanishing on $t_{XY,a}$. Then f vanishes on $t_{X,x}$ for all $x \in X_a$ and thus defines a section, σ say, in the conormal sheaf

$$\sigma \in H^0(X_a, \mathcal{I}_X/\mathcal{I}_X^2(1) \otimes \mathcal{O}_{X_a}).$$

As $\mathcal{I}_X/\mathcal{I}_X^2(1)$ is negative, the restriction of this sheaf to a subvariety of positive dimension cannot have nonzero sections. Thus it follows that X_a has dimension 0. A simple counting argument shows that $\delta(X,Y) = \dim X + \dim Y + 1 - \dim XY$ is equal to $\dim X_a$, and so is zero. Thus XY has maximal dimension. □

Assume that X is a nondegenerate complete intersection given by equations $f_1, \ldots, f_r$ of degree $d_1, \ldots, d_r$, where $r := \operatorname{codim}_{\mathbb{P}^n} X$. Then the twisted normal bundle is isomorphic to $\mathcal{O}_X(d_1 - 1) \oplus \cdots \oplus \mathcal{O}_X(d_r - 1)$ and so is negative. Thus we get the following corollary.

Corollary 10.2. *Let $X \subseteq \mathbb{P}^n$ be a nondegenerate complete intersection and $Y \subseteq \mathbb{P}^n$ a subvariety such that $X \cap Y \subseteq X_{\text{reg}}$. Then the join XY has maximal dimension, i.e.,* $\dim XY = \min\{n, \dim X + \dim Y + 1\}$.

Let us now turn to the limit of join variety. Unfortunately, we no longer have a criterion like 10.1 at our disposal. Instead, we will describe another method based on the theory of residual intersections of Artin–Nagata,

Huneke, Ulrich and others. This idea was first explored in [SUV] and [FUV]. In order to be able to formulate the main technical criterion, we must introduce the somewhat technical notion of sliding depth.

Recall first the notion of Koszul cohomology. Let M be a finite module over a local ring A and I an ideal of A. For a minimal system of generators $\underline{x} = x_1, \dots, x_m$ of I consider the Koszul complex $K^\bullet(\underline{x}, M)$. Then the cohomology

$$H^*(I, M) := H(K^\bullet(\underline{x}, M))$$

is called the *Koszul cohomology* of M with respect to I. It is easy to see that up to (a non-canonical) isomorphism these modules do not depend on the choice of a minimal system of generators of I. Following [HVV], we will say that (M, I) has *sliding depth* if

(a) M is a Cohen–Macaulay module, and
(b) $\operatorname{depth} H^i(I, M) \geq \dim M/IM - i$ for $i \geq 0$.

The second condition in particular is satisfied if all the modules $H^i(I, M)$ are Cohen–Macaulay over A/I. Following C. Huneke [Hun2], a pair (M, I) satisfying (a) and the latter condition is called *strongly Cohen–Macaulay* or SCM in brief. In the case $M = A$ we also say that the ideal I has sliding depth (resp. is SCM) in A.

To get an impression of these conditions, let us consider the following examples.

Example 10.3. (1) An ideal I in a Cohen–Macaulay ring A generated by a regular sequence $\underline{x} = (x_1, \dots, x_n)$ is always SCM in A. This follows from the fact that $H^i(\underline{x}, A) = 0$ for $i \neq n$ and $H^0(\underline{x}, A) = A/I$ (see [Mat]).

(2) (Huneke) Assume that $\mathfrak{a} \subseteq J$ are ideals in a local Cohen–Macaulay ring R, which are both generated by R-regular sequences. Then the ideal $I := J/\mathfrak{a}$ is SCM in the ring $A := R/\mathfrak{a}$.

Proof. By assumption $A/I \cong R/(g_1, \dots, g_n)$ with $g_1, \dots, g_n \in R$ a regular sequence, where I is generated by the residue classes, say $\bar{g}_1, \dots \bar{g}_n$, in A. So, with $\underline{g} := (g_1, \dots, g_n)$,

$$H_p(\underline{g}, A) = H_p(K_\bullet(\underline{g}, R) \otimes A) = \operatorname{Tor}_p^R(A/I, A),$$

because the Koszul complex $K_\bullet(\underline{g}, R)$ is a R-free resolution of A/I. Suppose that $A = R/(h_1, \dots, h_s)$, with $h_1, \dots, h_s \in R$ a regular sequence. By the same argument

$$\operatorname{Tor}_p^R(A/I, A) = H_p(A/I \otimes K_\bullet(\underline{h}, R)) \cong A/I \otimes K_p(\underline{h}, R),$$

and the latter group is Cohen–Macaulay.

(3) Let me also mention the following remarkable result due to Huneke: *let R be a local Gorenstein ring and suppose that $\mathfrak{a} \subseteq R$ is an ideal in the linkage class of a complete intersection. Then $\mathfrak{a}$ is SCM in R.* For the proof see [Hun2] or [FOV, Section 7.3].

In particular, assume that I is an ideal of codimension 2 (resp. 3) in a Gorenstein local ring A such that A/I is Cohen–Macaulay (resp. Gorenstein). Then it is known that I is in the linkage class of a complete intersection, and so (A, I) is SCM.

(4) For us, the following interesting result of Buchweitz and Ulrich is of importance also: *let M be a smooth algebraic variety and $X \subseteq M$ a subscheme which is in the linkage class of a complete intersection. Let $\mathcal{I}_\Delta \subseteq \mathcal{O}_{X\times X}$ be the ideal of the diagonal $\Delta \cong X \hookrightarrow X \times X$. Then for every subvariety C of $X \cong \Delta$ the pair $(p_1^*(\omega_X)_C, \mathcal{I}_{\Delta,C})$ is SCM over the ring $A = \mathcal{O}_{X\times X,C}$, where p_1 denotes the first projection.* For a proof we refer the reader to [BUl] or [FOV, Section 7.3].

The following theorem is the main tool to get an estimate for the limit of join variety. If not otherwise stated, we will work in the remaining part of this section over an arbitrary field K. In the next results, let V be an algebraic K-scheme, $W \subseteq V$ a subscheme and $C \subseteq W$ a subvariety. Let $\mathcal{I} \subseteq \mathcal{O}_V$ denote the ideal belonging to W.

Theorem 10.4. *Assume that for an integer $k \leq \dim \mathcal{O}_{V,C}$ the following conditions are satisfied.*

(i) *$\mu(\mathcal{I}_D) \leq \dim \mathcal{O}_{V,D}$ for all subvarieties $D \subseteq W$ with $C \subseteq D$ and $\dim \mathcal{O}_{V,D} < k$. (Here $\mu(\mathcal{I}_D)$ denotes the minimal number of generators of $\mathcal{I}_D$ as an $\mathcal{O}_{V,D}$-module.)*

(ii) *There exists an $\mathcal{O}_V$-module $\mathcal{M}$ with $\operatorname{supp} \mathcal{M} = V$ and $\mathcal{I} \otimes_{\mathcal{O}_V} \mathcal{M} \cong \mathcal{I}\mathcal{M}$ such that $(\mathcal{M}_C, \mathcal{I}_C)$ satisfies sliding depth over the ring $\mathcal{O}_{V,C}$.*

Then $(C_W V)_C$ has dimension $\geq \min\{k, \mu(\mathcal{I}_C)\}$.

The proof relies on the theory of residual intersections of Artin–Nagata, Huneke and others. For a proof see [FUV] or, for a more detailed exposition, [FOV, Section 7.4].

In the remaining part of this section we will give some applications. Applying this theorem to the case $\mathcal{M} = \mathcal{O}_V$ and $k = \dim \mathcal{O}_{V,C}$ we get:

Corollary 10.5. *Assume that $\mathcal{I}_C \subseteq \mathcal{O}_{V,C}$ has sliding depth and that*

(*) *$\mu(\mathcal{I}_D) \leq \dim \mathcal{O}_{V,D}$ for all subvarieties $D \subseteq W$ with $C \subsetneq D$.*

Then $(C_W V)_C$ has maximal dimension, i.e., $\dim(C_W V)_C = \min\{\dim \mathcal{O}_{V,C}, \mu(I_C)\}$.

As ideals in the linkage class of a complete intersection are always SCM by 10.3 (3), we obtain in particular:

Corollary 10.6. *Assume that $\mathcal{I}_C \subseteq \mathcal{O}_{V,C}$ is in the linkage class of a complete intersection, that $\mathcal{O}_{V,C}$ is Gorenstein and that condition (*) in 10.5 is satisfied. Then $(C_W V)_C$ has maximal dimension.*

We will apply the preceding results to the intersection of schemes. Let X, Y be closed subschemes of an algebraic K-scheme, say Z, and let $C \subseteq X \cap Y$ be a subvariety. Let us consider the diagonal embedding $W := X \cap Y \hookrightarrow V := X \times Y$.

Corollary 10.7. *Assume that the following conditions are satisfied.*

(i) $\dim T_{X\cup Y,D} \leq \dim \mathcal{O}_{X,D} + \dim \mathcal{O}_{Y,D} + \dim D$ *for all subvarieties* $D \subseteq X \cap Y$ *with* $C \subsetneq D$.

(ii) *There is an* $\mathcal{O}_{X\times Y}$*-module* $\mathcal{M}$ *with* $\operatorname{supp}\mathcal{M} = X \times Y$ *and* $\mathcal{I} \otimes \mathcal{M} \cong \mathcal{I}\mathcal{M}$ *such that* $(\mathcal{M}_C, \mathcal{I}_C)$ *is SCM, where* $\mathcal{I} \subseteq \mathcal{O}_{X\times Y}$ *is the ideal of* $X \cap Y \hookrightarrow X \times Y$.

Then $\operatorname{LJoin}_C(X,Y)$ *has maximal dimension, i.e.,*

$$\dim \operatorname{LJoin}_C(X,Y) = \min\{\dim T_{X\cup Y,C}), \dim \mathcal{O}_{X,C} + \dim \mathcal{O}_{Y,C} + \dim C\}.$$

Notice that $X \cup Y$ always denotes the scheme theoretic union of X, Y, i.e., the ideal sheaf of $X \cup Y$ in $\mathcal{O}_Z$ is the intersection of the ideal sheaves of X, Y in $\mathcal{O}_Z$.

Proof. By definition, $(C_W V)_C$ is just $\operatorname{LJoin}_C(X,Y)$. We will apply 10.4. Clearly (ii) is just the condition 10.4 (ii). It remains to show that (i) implies 10.4 (i). The dimension of the tangent space $T_{X\cup Y,C}$ is given by $\mu(\Omega^1_{X\cup Y/K,C})$. As $\mathcal{I}/\mathcal{I}^2 \cong \Omega^1_{X\cup Y/K}$ by 5.5, the result follows. □

Remark 10.8. If the base field K is perfect then, by 5.3, for an algebraic K-scheme V and a subvariety D of V

$$\dim T_{V,D} = \operatorname{emdim} \mathcal{O}_{V,D} + \dim D.$$

Thus in this case the inequality in 10.7 (i) is equivalent to

$$\operatorname{emdim} \mathcal{O}_{X\cup Y,D} \leq \dim \mathcal{O}_{X,D} + \dim \mathcal{O}_{Y,D}.$$

(2) Note that 10.7 (i) is always satisfied in the case that $X = Y$ and $\mathcal{O}_{X,C}$ has an isolated singularity.

Corollary 10.9. *Let* $X, Y \subseteq Z$ *be closed subschemes and* $C \subseteq X \cap Y$ *a subvariety. Suppose that the following conditions are satisfied:*

(i) $\dim T_{X\cup Y,D} \leq \dim \mathcal{O}_{X,D} + \dim \mathcal{O}_{Y,D} + \dim D$ *for all subvarieties* $D \subseteq X \cap Y$ *with* $C \subsetneq D$.

(ii) $\mathcal{O}_{X,C}, \mathcal{O}_{Y,C}$ *and* $\mathcal{O}_{X\cap Y,C}$ *are complete intersection rings.*

Then $\dim \operatorname{LJoin}_C(X,Y)$ *is maximal, i.e.,*

$$\dim \operatorname{LJoin}_C(X,Y) = \min\{\dim T_{X\cup Y,C}, \dim \mathcal{O}_{X,C} + \dim \mathcal{O}_{Y,C} + \dim C\}.$$

The proof follows from 10.7 and 10.3 (2). For the case of self-intersections, the above result 10.9 can be improved.

Corollary 10.10. *Let X be an algebraic K-scheme and $C \subseteq X$ a subvariety. Suppose that the following conditions are satisfied:*

(i) $\dim T_{X,D} \leq 2 \dim \mathcal{O}_{X,D} + \dim D$ *for all subvarieties D of X with $C \subsetneq D$.*

(ii) $\mathcal{O}_{X,C}$ *is in the linkage class of a complete intersection, i.e., $\mathcal{O}_{X,C} = R/\mathfrak{a}$, where R is a smooth K-algebra essentially of finite type and $\mathfrak{a} \subseteq R$ is in the linkage class of a complete intersection.*

Then $\mathrm{LSec}_C(X)$ *has maximal dimension.*

Proof. After shrinking X as a neighbourhood of the generic point of C we may assume that X is embedded into a smooth K-scheme Z such that $\mathcal{O}_{Z,C} = R$. By 10.3 (4), the pair $(p_1^*(\omega_X)_C, \mathcal{I}_C)$ is SCM in a neighbourhood of C, where $p_i : X \times X \to X$, $i = 1, 2$, is the i-th projection and $\mathcal{I} \subseteq \mathcal{O}_{X \times X}$ is the ideal of the diagonal $\Delta \cong X \hookrightarrow X \times X$. Using 10.7 it remains to show that $\mathcal{I} \cdot p_1^*(\omega_X) \cong \mathcal{I} \otimes p_1^*(\omega_X)$. The sequence $0 \to \mathcal{I} \to \mathcal{O}_{X \times X} \to \mathcal{O}_X \to 0$ is a sequence of flat $\mathcal{O}_X$-modules with respect to the first projection p_1. Tensoring over $\mathcal{O}_X$ with ω_X leads to the exact sequence

$$0 \to \mathcal{I} \otimes p_1^*(\omega_X) \to p_1^*(\omega_X) \to \omega_X \to 0,$$

and so $\mathcal{I} p_1^*(\omega_X) \cong \mathcal{I} \otimes p_1^*(\omega_X)$ as required. □

Applying the preceding result to the affine cone over a projective variety we obtain the following corollary.

Corollary 10.11. *Let $X \subseteq \mathbb{P}^n_K$ a nondegenerate subscheme which is in the linkage class of a complete intersection. Assume that for all points $x \in X$ we have $\dim T_{X,x} \leq 2 \dim \mathcal{O}_{X,x}$. Then the secant variety* $\mathrm{Sec}\, X$ *has maximal dimension.*

As an example, consider the Veronese surface $V \subseteq \mathbb{P}^5$. It is well known that $\mathrm{Sec}\, V$ is a hypersurface in $\mathbb{P}^5$ and so it has not maximal dimension (see e.g. [FOV, Section 1.3]). Thus V is not in the linkage class of a complete intersection although it is arithmetically Cohen–Macaulay. In contrast, an arithmetically Gorenstein subscheme in $\mathbb{P}^n$ of codimension 3 belongs to the linkage class of a complete intersection; see for instance [Wat] or [BEi].

Remark 10.12. (1) For further applications of the method described here, we refer the reader to [SUV, FUV] or [FOV].

(2) In all our results we have used the condition that (∗) in 10.5 holds, resp. for the case of limits of joins that

$$\dim T_{X \cup Y, D} \leq \dim \mathcal{O}_{X,D} + \dim \mathcal{O}_{Y,D} + \dim D$$

holds for all subvarieties $D \subseteq X \cap Y$ with $C \subsetneq D$. Note that without this condition the results are no longer true, as is easily seen by the example of a product.

We know that a subvariety C of an intersection $X \cap Y$ of equidimensional subschemes of $\mathbb{P}^n$ belongs to the v-cycle if and only if $\mathrm{LJoin}_C(X,Y)$ has maximal dimension $\dim X + \dim Y$. Thus from the preceding results we get in particular criteria for when a subvariety belongs to the intersection cycle. However, these results do not give a formula for the j-multiplicity with which C appears. It would be interesting to have such a formula in terms e.g. of Koszul cohomology or to relate this coefficient to other more accessible invariants.

As a special case let me mention the following result, see [FMa3].

Theorem 10.13. *Let $H = \{f = 0\} \subseteq \mathbb{A}^n_K$ be a hypersurface and C an irreducible component of the singular locus of H with $\operatorname{codim}_H C = c$. Then C appears in the self-intersection cycle H^{c+1} with coefficient*

$$j_0 = e(\operatorname{jac} f, \mathcal{O}_{H,C}),$$

where $\operatorname{jac} f = (\partial f/\partial x_1, \ldots, \partial f/\partial x_n)$ is the Jacobian ideal of f.

If H has not an isolated singularity then it is much harder to decide whether C occurs in the self-intersection cycle H^c (cf. also the conjecture made in [ELN]).

We believe that some of the results of this section can be improved significantly. For instance, in view of 10.1 the following should be true.

Problem 10.14. *Let X, $Y \subseteq Z$ be subschemes of an algebraic K-scheme Z. Assume that for an arbitrary subvariety $C \subseteq X \cap Y$ the following conditions are satisfied:*

(1) $\operatorname{emdim} \mathcal{O}_{X,C} = \operatorname{emdim} \mathcal{O}_{Z,C}$*;*

(2) $\mathcal{O}_{X,C}$ is a complete intersection;

(3) $X \cap Y \backslash C \subseteq \operatorname{Reg} X$.

Is $\dim \mathrm{LJoin}_C(X,Y)$ then maximal?

References

[AHV] Achilles, R.; Huneke, C.; Vogel, W.: A criterion for intersection multiplicity one. *Nagoya Math. J.* **100** (1985), 49–63

[AMa1] Achilles, R.; Manaresi, M.: An algebraic characterization of distinguished varieties of intersection. *Rev. Roumaine Math. Pures Appl.* **38** (1993), 569–578

[AMa2] Achilles, R.; Manaresi, M.: Multiplicity for ideals of maximal analytic spread and intersection theory. *J. Math. Kyoto Univ.* **33** (1993), 1029–1046

[ATW] Achilles, R.; Tworzewski, P.; Winiarski, T.: On improper isolated intersection in complex analytic geometry. *Ann. Polon. Math.* **51** (1990), 21–36

[AVo] Achilles, R.; Vogel, W.: On multiplicities for improper intersections. *J. Algebra* **168** (1994), 123–142

[Adl] Ådlandsvik, B.: Joins and higher secant varieties. *Math. Scand.* **61** (1987), 213–222

[ANa] Artin, M.; Nagata, M.: Residual intersections in Cohen-Macaulay rings. *J. Math. Kyoto Univ.* **12** (1972), 307–323

[AMa] Atiyah, M.F.; Macdonald, I.G.: *Introduction to commutative algebra.* Reading: Addison–Wesley, 1969

[BHSS] Beltrametti, M.; Howard, A.; Schneider, M.; Sommese, A.J.: *Projections from subvarieties* Preprint 1998

[Ber] Berzolari, L.: Sulle intersezioni di tre superficie algebriche. *Ann. di Mat.* **24**:2 (1896), 165–191

[BHe] Bruns, W.; Herzog, J.: *Cohen–Macaulay Rings.* Cambridge: Cambridge University Press, 1993

[BVe] Bruns, W.; Vetter, U.: *Determinantal rings.* Lecture Notes in Mathematics, Volume 1327. Berlin–Heidelberg–New York: Springer, 1988

[BEi] Buchsbaum, D.A.; Eisenbud, D.: Algebra structures for finite free resolutions, and some structure theorems for ideals of codimension 3. *Amer. J. Math.* **99** (1977), 447–485

[BUl] Buchweitz, R.O.; Ulrich, B.: *Homological properties invariant under linkage.* Preprint

[Coo] Coolidge, J.L.: *A treatise on algebraic plane curves.* New York: Dover Publications, Inc., 1959

[ELN] Ein, L.; Lazarsfeld, R.; Nakamaye, M.: Zero-estimates, intersection theory, and a theorem of Demailly. In: *Higher-dimensional complex varieties.* Proceedings of the International Conference, Trento (1994), 183–207, Berlin–New York: de Gruyter, 1996

[Eis1] Eisenbud, D.: *Commutative algebra with a view toward algebraic geometry.* Graduate Texts in Mathematics 150, Springer-Verlag, Berlin, Heidelberg, New York, 1995

[Fal] Faltings, G.: Ein Kriterium für vollständige Durchschnitte. *Invent. Math.* **62** (1980), 393–401

[FGV] Flenner, H.; Gastel, L.J. van; Vogel, W.: Joins and intersections. *Math. Ann.* **291** (1991), 691–704

[FMa1] Flenner, H.; Manaresi, M.: Intersections of projective varieties and generic projections. *Manuscripta Math.* **92** (1997), 273–286

[FMa2] Flenner, H.; Manaresi, M.: Variation of ramification loci of generic projections. *Math. Nach.*, to appear

[FMa3] Flenner, H.; Manaresi, M.: *On distinguished components of the selfintersection of hypersurfaces.* Preprint Bochum/Bologna 1998

[FOV] Flenner, H.; O'Carroll, L.; Vogel, W.: *Joins and intersections.* In preparation.

[FUV] Flenner, H.; Ulrich, B.; Vogel, W.: On limits of joins of maximal dimension. *Math. Ann.* **308**, 291–318 (1996)

[FVo1] Flenner, H.; Vogel, W.: Improper intersections and a converse to Bezout's theorem. *J. Algebra* **159**, 460–476 (1993)

[FVo2] Flenner, H.; Vogel, W.: Limits of joins and intersections. In *Higher dimensional complex varieties.* Proceedings of the International Conference, Trento (1994), 209–220, Berline–New York: de Gruyter, 1996

[Ful1] Fulton, W.: Intersection theory. *Erg. Math. Grenzgeb.*, 3. Folge, Bd. 2, Berlin–Heidelberg–New York: Springer, 1984

[Ful2] Fulton, W.: *Algebraic refined intersections.* Notes for Math. 271, Providence, R.I.: Brown University 1986

[FHa] Fulton, W.; Hansen, J.: A connectedness theorem for projective varieties, with applications to intersections and singularities of mappings. *Annals of Math.* **110**, 159–166 (1979)

[FLa1] Fulton, W.; Lazarsfeld, R.: Connectivity and its applications in algebraic geometry. In: *Algebraic Geometry*, University of Illinois at Chicago Circle, 1980. 26–92, Lecture Notes in Mathematics, Volume 862. Springer-Verlag, Berlin, Heidelberg, New York, 1980

[FLa2] Fulton, W.; Lazarsfeld, R.: Positivity and excess intersections. In: *Enumerative and classical geometry*, 97–105, Nice 1981, Progress in Math. 24, Birkhäuser, Cambridge, MA, 1982,

[FLa3] Fulton, W.; Lazarsfeld, R.: Positive polynomials for ample vector bundles. *Annals of Math.* **118**, 35–60 (1983)

[GLa] Gaffney, T.; Lazarsfeld, R.: On the ramification of branched coverings of $\mathbb{P}^n$. *Invent. Math.* **59**, 53–58 (1980)

[vGa1] Gastel, L.J. van: *A geometric approach to Vogel's intersection theory.* Preprint Nr. 427, University of Utrecht, 1986 (A corrected version published in 1987)

[vGa2] Gastel, L.J. van: *Excess intersections.* PhD Thesis, University of Utrecht 1989

[vGa3] Gastel, L.J. van: Excess intersections and a correspondence principle. *Invent. Math.* **103**, 197–221 (1991)

[Gro] Grothendieck, A.: *Cohomologie locale des faisceaux cohérents et théorèmes des Lefschetz locaux et globaux* (SGA2). Amsterdam: North-Holland 1968

[Han] Hansen, J.: Double points of compositions. *J. reine angew. Math.* **352**, 71–80 (1984)

[Harr] Harris, J.: Algebraic geometry. Graduate Texts in Mathematics 133, Springer-Verlag, Berlin, Heidelberg, New York, 1992

[Har1] Hartshorne, R.: Varieties of small codimension in projective space. *Bull. Amer. Math. Soc.* **80**, 1017–1032 (1974)

[Har2] Hartshorne, R.: *Algebraic geometry.* Graduate Texts in Mathematics 52, Springer-Verlag, Berlin, Heidelberg, New York, 1977

[HVV] Herzog, J.; Vasconcelos, V.; Villareal, R.: Ideals with sliding depth. *Nagoya Math. J.* **99**, 159–172 (1985)

[Hol1] Holme, A.: Embedding-obstructions for singular algebraic varieties in $\mathbb{P}^N$. *Acta Math.* **135**, 155–186 (1975)

[Hol2] Holme, A.: Embedding-obstructions for smooth projective varieties I. In: *Studies in algebraic topology*, 39–67, Advances in Math. Suppl. Studies **5**, 1979

[Hun1] Huneke, C.: Linkage and the Koszul homology of ideals. *Amer. J. Math.* **104** (1982), 1043–1062

[Hun2] Huneke, C.: Strongly Cohen–Macaulay schemes and residual intersections. *Trans. AMS* **277** (1983), 739–763

[HUl] Huneke, C.; Ulrich, B.: Residual intersections. *J. reine angew. Math.* **390** (1988), 1–20

[Joh] Johnson, K.W.: Immersion and embedding of projective varieties. *Acta Math.* **140** (1978), 49–74

[Kle1] Kleiman, S.L.: The enumerative theory of singularities. In: *Real and Complex Singularities*, Proc. 9th Nordic Summer School/NAVF Symposia Math. Oslo 1976, 297–396, Alphen aan den Rijn: Sijthoff & Nordhoff, 1977

[Kle2] Kleiman, S.L.: Concerning the dual variety. In: *Proc. 18th Scandinavian congress of Mathematicians* Aarhus, 1980, 396–396, Progr. in Math., vol. 11, Birkhäuser, Cambridge, MA, 1981,

[LBi] Lange, H.; Birkenhake, C.: *Complex Abelian Varieties.* Grundlehren der Math. Wiss. 302, Springer-Verlag, Berlin, Heidelberg, New York, 1992

[Laz] Lazarsfeld, R.: Excess intersections of divisors. *Compositio Math.* **43** (1981), 281–296

[Llu] Lluis, E.: Sur l'immersion des variétés algébriques. *Ann. of Math.* **62** (1955), 120–127

[Mat] Matsumura, H.: *Commutative ring theory.* Cambridge University Press, Cambridge, UK, 1986

[Nag] Nagata, M.: *Local rings.* Interscience tracts in Pure and Appl. Math. 13. J. Wiley, New York, 1962

[PSi] Peters, C.A.M.; Simonis, J.: A secant formula. *Quart. J. Math. Oxford* **27** (1976), 181–189

[RVo] Renschuch, B.; Vogel, W.: Perfektheit und die Umkehrung des Bezoutschen Satzes. *Math. Nachr.* **148** (1990), 313–323

[SSt] Scheja, G.; Storch, U.: Differentielle Eigenschaften der Lokalisierungen analytischer Algebren. *Math. Ann.* **197** (1972), 137–170

[SUV] Simis, A.; Ulrich, B.; Vasconcelos, W.: Tangent star cones. *J. reine angew. Math.* **483** (1997), 23–59

[SVo] Stückrad, J.; Vogel, W.: An algebraic approach to the intersection theory. In: *The curves seminar at Queen's Vol. II*, 1–32. Queen's papers in pure and applied mathematics, No. 61, Kingston, Ontario, Canada, 1982

[Tei] Teissier, B.: Cycles évanescants, sections planes et conditions de Whitney. *Astérisque* **7–8** (1973), 285–362

[Ten] Tennison, B.R.: Intersection multiplicities and tangent cones. *Math. Proc. Cambridge Philos. Soc.* **85** (1979), 33–42

[Ulr] Ulrich, B.: Remarks on residual intersections. In: *Free resolutions in commutative algebra and algebraic geometry.* 133–138, Boston: Jones and Bartlett Publishers, 1992

[VdV] Van de Ven, A.: On the embedding of abelian varieties in projective spaces. *Ann. Mat. Pura Appl.* **103** (1975), 127–129

[Vas] Vasconcelos, W.: Koszul homology and the structure of low codimension Cohen–Macaulay ideals. *Trans. AMS* **301** (1987), 591–613

[Vog1] Vogel, W.: *Lectures on results on Bezout's theorem.* (Tata Lect. Notes, Bombay, vol. 74) Springer-Verlag, Berlin, Heidelberg, New York, 1984

[Vog2] Vogel, W.: Old and new problems about Bezout's theorem. *Rend. Sem. Mat. Univ. Politec. Torino* **43** (1985), 43–58

[Vog3] Vogel, W.: A converse to Bezout's theorem. *Ann. Univ. Ferrara Sez. VII (N.S.)* **36** (1990), 85–95

[Vog4] Vogel, W.: On multiplicities of non-isolated intersection components. *Publ. Res. Inst. Math. Sci.* **27** (1991), 845–860

[Wat] Watanabe, J.: A note on Gorenstein rings of embedding codimension three. *Nagoya Math. J.* (1973) **50**, 227–232

[Wei] Weil, A.: *Foundations of algebraic geometry.* Amer. Math. Soc. Colloq. Publ. 29. Providence: Amer. Math. Soc., 1946

[Whi1] Whitney, H.: Tangents to an analytic variety. *Ann. of Math.* **81** (1965), 496–549

[Whi2] Whitney, H.: *Complex analytic varieties.* Reading: Addison–Wesley, 1972

[Zar] Zariski, O.: Generalized weight properties of the resultant of $n+1$ polynomials in n indeterminates. *Trans. AMS* **41** (1937), 249–265

Fakultät für Mathematik
Ruhr-Universität Bochum
Universitätsstraße 150, Geb. NA 2/72
44780 Bochum
Germany
Hubert.Flenner@ruhr-uni-bochum.de

Equimultiplicity and Equidimensionality of Normal Cones

Hubert Flenner
Mirella Manaresi

1 Introduction

The purpose of this paper is to give a new approach to some classical results on equimultiplicity. Our proofs are based on [FVo] in which there was described in a very precise way the behaviour of Hilbert function and multiplicities in an exact sequence and, in particular, under hyperplane sections. In [FOV], Section 1.1, this approach was used to give a simplified account to the classical theory of multiplicities including Rees' theorem.

In this paper we wish to show that with these methods we also obtain new and simplified proofs of some further important results on multiplicities. For instance in Section 3 we will give a new and simple proof of the following theorem of Böger: *Let $I, J \subseteq \mathfrak{m}_A$ be ideals in a formally equidimensional local Noetherian ring with $l(I) = \operatorname{ht}(I)$, where $l(I)$ is the analytic spread of I. Then I is a reduction of J if and only if $\sqrt{I} = \sqrt{J}$ and $e(I_\mathfrak{p}, A_\mathfrak{p}) = e(J_\mathfrak{p}, A_\mathfrak{p})$ for all minimal primes of J.*

As another application of our method we obtain new proofs of results of Lipman, Schickhoff and Teissier. We describe the first of these results in brief in the analytic context. Let $f : X \to Y$ be a holomorphic map of complex spaces and $W \subseteq X$ be a closed subspace which is finite and surjective over Y. Then it is well known that the function $d : W \to \mathbb{N}$ with $d(w) := \dim \mathcal{O}_{X(f(w)),w}$ is upper semicontinuous on W, where $X(f(w)) := f^{-1}(f(w))$ denotes the fibre over $f(w)$. Moreover, if $\mathcal{M}$ is flat over Y and if the fibres $\mathcal{M}(f(w))_w$ are equidimensional for all $w \in W$, then the function d is locally constant on W.

This paper was prepared during stays at the University of Bologna, resp., Bochum financed by Indam, MURST and the University of Bochum. Moreover these investigations were supported by University of Bologna Fonds for selected research topics. The authors want to thank these institutions.

Now consider the function

$$y \mapsto e(y) := \sum_{w \in f^{-1}(y) \cap W} e(\mathcal{I}(y)_w, \mathcal{O}_{X(y),w}),$$

where $\mathcal{I} \subseteq \mathcal{O}_X$ denotes the ideal of W in X. Then by results of Lipman [Lip] and Teissier [Tei2], 5.5, the following hold. *(1) If f is flat and if all fibres of $C_W X \to W$ are of dimension d then the function $y \mapsto e(y)$ is locally constant on Y. (Here $C_W X$ denotes the normal cone of X along W.) (2) Conversely assume that all fibres $X(f(w))$ are equidimensional of dimension d in the points of W and that $y \mapsto e(y)$ is constant on Y. Then $C_W X \to W$ has equidimensional fibres of dimension d.*

We will also give an algebraic version of these two results similarly as stated in [Lip] with a simplified proof. The analytic case is deduced from this in Section 6 using Stein compact sets as e.g., in [BFl].

Finally in Section 5 we apply our methods to obtain a new proof of the following theorem due to Lipman, Schickhoff and Teissier. *Let $\mathfrak{p}$ be a prime ideal in the local ring $(A, \mathfrak{m})$ so that $A/\mathfrak{p}$ is regular. Then the following hold.*

(1) If $l(\mathfrak{p}) = \operatorname{ht}\mathfrak{p}$ then $e(A) = e(A_{\mathfrak{p}})$.

(2) If moreover $\hat{A}$ is equidimensional then the converse of (1) is true.

(3) Assume that $\hat{A}$ is equidimensional and that $l(\mathfrak{p}) = \operatorname{ht}\mathfrak{p}$. Consider elements $t_1, \ldots, t_k \in \mathfrak{m}$ whose residue classes form a regular parameter system in the regular ring $A/\mathfrak{p}$. Let $t_1^*, \ldots, t_k^*$ denote the initial forms of $t_1, \ldots, t_k$ in the associated graded ring $G_{\mathfrak{m}}(A)$. Then the canonical surjections

$$\alpha : G_{\mathfrak{p}}(A)/\mathfrak{m}G_{\mathfrak{p}}(A) \longrightarrow G_{\mathfrak{m}}(A)/(t_1^*, \ldots, t_k^*)$$
$$\beta : (G_{\mathfrak{p}}(A)/\mathfrak{m}G_{\mathfrak{p}}(A))[T_1, \ldots, T_k] \longrightarrow G_{\mathfrak{m}}(A)$$

have nilpotent kernel, where β is given by $T_i \mapsto t_i^*$.

Again in Section 6 we state an analytic version of this result which is deduced using Stein compact sets.

2 Review of some known results

2.1. Let A be a Noetherian ring and $I \subseteq A$ an arbitrary ideal. Consider the associated graded ring

$$G(A) := G_I(A) := \bigoplus_{n \in \mathbb{N}} (I^n/I^{n+1})T^n.$$

If M is a finite A-module then the associated graded module

$$G(M) := G_I(M) := \bigoplus_{n \in \mathbb{N}} (I^n M/I^{n+1} M)T^n$$

is a module over $G(A)$ in a natural way.

We recall the following useful result; see [FOV, 1.2.6] for a proof.

Theorem 2.2. *Let* $0 \to M_0 \to M_1 \to M_2 \to 0$ *be an exact sequence of* A*-modules. Then the cycles on* $\operatorname{Spec} G(A)$ *associated to*

$$\ker(G(M_0) \to G(M_1)) \quad \text{and} \quad \ker(G(M_1)/G(M_0) \to G(M_2))$$

are equal.

We will often use this result in the situation that M is a finitely generated A-module and x is a M-regular element. Assume that $x \in I^\alpha$, $\alpha \geq 0$, and let $x^* := \bar{x}T^\alpha$ denote the initial form of x. Applying 2.2 to the sequence

$$0 \to M \xrightarrow{x} I^\alpha M \to I^\alpha M/xM$$

gives immediately the following result.

Corollary 2.3. *Under the assumptions above, the kernels of the maps*

$$\ker(G(M) \xrightarrow{x} G(M)) \quad \text{and} \quad \ker(G(M)/x^*G(M) \to G(M/xM))$$

have the same support.

These results have the following important consequence. For a proof, see [FOV, 1.2.19].

Proposition 2.4. *Let* M, M' *be finitely generated* A*-modules. Then the following hold.*

(a) *If* $\operatorname{supp} M' \subseteq \operatorname{supp} M$ *then the support of* $G(M') := G_I(M')$ *(as module over* $G := G_I(A)$*) is contained in the support of* $G(M) := G_I(M)$*.*

(b) *Assume that* $G(M)$ *is equidimensional. Let* $x \in I^\alpha$, $\alpha \geq 0$, *be an element with* $\dim G(M)/x^*G(M) < d$, *where* $x^* = \bar{x}T^\alpha$ *denotes the initial form of* x *in* G*. Then the supports of the modules* $G(M/xM)$ *and* $G(M)/x^*G(M)$ *are equal.*

Using induction this gives immediately the following corollary.

Corollary 2.5. *Assume that* $x_j \in I^{\alpha_j}$, $1 \leq j \leq k$ *and* $\alpha_j \geq 0$, *are elements with* $\dim G(M)/(x_1^*, \ldots, x_k^*)G(M) \leq d-k$, *where* $x_j^* = \bar{x}T^{\alpha_j}$ *denotes the initial form of* x_j *in* $G := G_I(A)$*. Then the supports of the modules* $G(M/(x_1, \ldots, x_k)M)$ *and* $G(M)/(x_1^*, \ldots, x_k^*)G(M)$ *are equal.*

It is convenient to introduce the following notation.

Definition 2.6. Let $(A, \mathfrak{m})$ be a local ring, $I \subseteq \mathfrak{m}$ an ideal and M a finite A-module. We call $x \in I^\alpha$ a *G-parameter of degree* α for (M, I) if

$$\dim G_I(M)/x^*G_I(M) = d - 1,$$

where $d := \dim M$ and where $x^* = \bar{x}T^\alpha$ denotes the initial form of x in $G_I(A)$. Note that it then follows that $\dim M/xM = d-1$. In the case $\alpha = 1$ we will say in brief that x is a G-parameter.

The following lemma is well known.

Lemma 2.7. *Let $(A, \mathfrak{m})$ be a local ring with infinite residue field $A/\mathfrak{m}$. Assume that $J \subseteq I$ is an ideal with*

$$\dim G_I(M)/J^* G_I(M) < \dim G_I(M),$$

where J^* denotes the initial ideal of J, i.e., the ideal in $G_I(A)$ generated by all initial forms $x^* = \bar{x}T$ of degree 1 of elements in J. Then J contains a G-parameter for (M, I).

Generalizing the well-known notion of a reduction ideal we will say that an ideal $J \subseteq I$ forms a *reduction* of (M, I) if

$$\dim G_I(M)/(J^* G_I(M) + \mathfrak{m} G_I(M)) = 0.$$

Note that then J is generated by at least $l_M(I) := \dim G_I(M)/\mathfrak{m} G_I(M)$ elements. J will be called a *minimal reduction* for (M, I) if furthermore J is generated by $l_M(I)$ elements. If $A/\mathfrak{m}$ is infinite then this equivalent to saying that J is minimal among the reductions of (M, I). Thus J is a (minimal) reduction of (A, I) if and only if it is a (minimal) reduction of I in the usual sense.

Observe that by Nakayama's lemma J is a reduction of (M, I) if and only if

$$JI^n M = I^{n+1} M \quad \text{for } n \gg 0.$$

For later use we note the following simple facts.

Lemma 2.8. *(a) J is a reduction of I if and only if the Rees module $R_I(M) := \bigoplus_{n \geq 0} I^n M$ is finite over the Rees ring $R_J := \bigoplus_{n \geq 0} J^n$ of J.*

(b) If $\operatorname{supp} N \subseteq \operatorname{supp} M$ and J is a reduction of (M, I) then J is also a reduction of (N, I).

(c) (Transitivity) Let $K \subseteq J \subseteq I$ be ideals. Then K is a reduction of (M, I) if and only if K is a reduction of (M, J) and J is a reduction of (M, I).

Proof. (a) follows from the graded version of Nakayama's lemma. In order to prove (b), note that by 2.4 (a) the support of $G_I(N)$ is contained in the support of $G_I(M)$. Hence

$$\dim G_I(N)/J^* G_I(N) + \mathfrak{m} G_I(N) \leq \dim G_I(M)/J^* G_I(M) + \mathfrak{m} G_I(M).$$

Finally, (c) follows from the usual transitivity properties for finite extensions using (a). □

Later on we will need the behaviour of multiplicities under hyperplane sections. We recall the following very precise result from [FOV, 1.2.11]. In the next two results we assume that $I = \mathfrak{q}$ is an ideal with $\dim M/\mathfrak{q}M = 0$ so that we can consider the multiplicity $e(\mathfrak{q}, M)$. For any $d \geq \dim M$ we set $e_d(\mathfrak{q}, M) = e(\mathfrak{q}, M)$ if $d = \dim M$, and $e_d(\mathfrak{q}, M) = 0$ for $d > \dim M$.

Theorem 2.9. *Assume that* $x \in \mathfrak{q}^\alpha$ *is an element with* $\dim M/xM < \dim M =: d$. *Then*

$$e_{d-1}(\mathfrak{q}, M/xM) \geq \alpha e_d(\mathfrak{q}, M) + e_{d-1}(\mathfrak{q}, \operatorname{Ann}_M x)$$

with equality if and only if x *is a G-parameter of degree* α *for* $(M, \mathfrak{q})$.

We remind the reader also of the theorem of Rees, which we state in a module theoretic version. For a proof see e.g., [FOV, 1.2.18].

Theorem 2.10. *Let* $\mathfrak{q}' \subseteq \mathfrak{q}$ *be an ideal such that* $\dim M/\mathfrak{q}'M = 0$ *and* $e(\mathfrak{q}, M) = e(\mathfrak{q}', M)$. *Assume moreover that the associated graded module* $G(M) := G_\mathfrak{q}(M)$ *is equidimensional. Then* $\mathfrak{q}'$ *is a reduction of* $(M, \mathfrak{q})$.

3 A simple proof of Böger's theorem

3.1. In this section let $(A, \mathfrak{m})$ denote a local ring and $J \subseteq I \subseteq A$ ideals. Let M be a finite A-module.

In order to be able to formulate the main results in a efficient way, we need a generalization of the notions of height and analytic spread of an ideal to the case of modules.

Definition 3.2. We call the number

$$\operatorname{ht}_M I := \min\{\dim M_\mathfrak{p} \mid \mathfrak{p} \in \operatorname{supp} M \cap V(I)\}$$

the *M-height* of I. Moreover, the number

$$l_M(I) := \dim G_I(M)/\mathfrak{m}G_I(M)$$

is called the *analytic spread of* (M, I).

In the case $M = A$ we also write in brief $l(I)$, $\operatorname{ht} I$ instead of $l_A(I)$, $\operatorname{ht}_A I$, respectively.

These numbers can be interpreted geometrically as follows. For simplicity let us assume that $M = A$. Let $p : V' \to V$ be the blowing up of $V = \operatorname{Spec} A$ along the subscheme defined by the ideal I so that V' is the projective scheme associated to the Rees ring R_I. Then $l(I) - 1$ is the dimension of the fibre $p^{-1}(\mathfrak{m})$.

We need the following elementary properties of these numbers.

Proposition 3.3. *(1) If* N, M *are A-modules with* $\operatorname{supp} N \subseteq \operatorname{supp} M$ *then* $l_N(I) \leq l_M(I)$ *and* $\operatorname{ht}_N I \leq \operatorname{ht}_M I$.

(2) $\operatorname{ht}_M I \leq l_M(I) \leq \min\{\dim M, \mu(I)\}$, *where* $\mu(I)$ *denotes the minimal number of generators for the ideal* I.

(3) If J *is a reduction of* (M, I) *then* $l_M(I) = l_M(J)$ *and* $\operatorname{ht}_M I = \operatorname{ht}_M J$.

Proof. The first part of (1) follows easily from 2.4 (a). Moreover, the inequality $\mathrm{ht}_N I \le \mathrm{ht}_M I$ is immediate from the definition.

The remaining assertions (2), (3) are well known in the case that $M = A$ (see e.g., [FOV, 3.2.6]). In view of (1) this proves the proposition. □

The theorem of Böger will be deduced from the following more general result which is essentially due to B. Ulrich.

Proposition 3.4. *Assume that the completion $\hat{M}$ is equidimensional and that $J \subseteq I$ are ideals. Then either $\mathrm{ht}_M(JI^{n-1}M :_A I^nM) \le l_M(J)$ for all $n \ge 1$, or $JI^{n-1}M :_A I^nM = A$ for large n.*

In other words, this result says that the (closed!) set of all primes $\mathfrak{p} \subseteq A$ where $J_\mathfrak{p}$ is not a reduction of $(M_\mathfrak{p}, I_\mathfrak{p})$, has codimension $\le l_M(J)$ in $\mathrm{Spec}\, A$. To interpret this result geometrically, consider the case $M = A$ and let $p : V' \to V = \mathrm{Spec}\, A$ be as before the blowing up of V along $V(I)$. Then $I\mathcal{O}_{V'}$ is an invertible ideal whose set of zeros, say E, is a p-ample divisor in V'. The elements of J define sections in $\mathcal{O}_{V'}(E) = I\mathcal{O}_{V'}$. Let Z be their common set of zeros. Then for a point $x \in V(I)$ given by a prime ideal $\mathfrak{p}$, the ideal $J_\mathfrak{p}$ is a reduction of $I_\mathfrak{p}$ if and only if $x \notin p(Z)$. Thus the result above says that every irreducible component of $p(Z)$ has codimension at most $l(I)$ in V. In this form the result is a special case of a generalized principal ideal theorem due to F. Steffen, see [Ste, FOV].

We will give here another simple inductive proof of 3.4 as follows.

Proof. We may assume that A is complete and that $A/\mathfrak{m}$ is infinite. Using a simple induction argument we need to show the following claim:

$(*)$ Suppose that for all primes $\mathfrak{p} \ne \mathfrak{m}$ the ideal $J_\mathfrak{p}$ is a reduction of $(M_\mathfrak{p}, I_\mathfrak{p})$ and that $d := \dim M > l_M(J)$. Then J is a reduction of (M, I).

Using induction on the dimension we may suppose that $(*)$ is true for all modules of smaller dimension. Obviously J is a reduction for (M, I) if and only if it is a reduction for (I^nM, I). Moreover, by 3.3 (1) $l_{I^nM}(J) \le l_M(J)$. Hence we may replace M by I^nM, $n \gg 0$, and can therefore assume that $\dim M/IM < d$. Finally, we may suppose that $\mathrm{supp}\, M = \mathrm{Spec}\, A$ so that also $\dim A/J = \dim A/I < d$.

First suppose that $l_M(J) = 0$ so that $J^nM = 0$ for $n \gg 0$. From the assumption in $(*)$ we infer that $I^nM_\mathfrak{p} = 0$ for every minimal prime $\mathfrak{p}$ of M. Hence $I^{kn}M = 0$ for $k \gg 0$, and we are done.

So we can suppose for the rest of the proof that

$$0 < l_M(J) < d. \tag{1}$$

By the assumption in $(*)$ we know that the fibres of $G_I(M)/J^*G_I(M)$ over primes $\mathfrak{p} \in \mathrm{Spec}\, A \setminus \{\mathfrak{m}\}$ are 0-dimensional, where $J^* = \bar{J}T$ is the initial ideal of J. Moreover, by (1) the fibre over $\mathfrak{m}$ has dimension $< d$. Hence

$$\dim G_I(M)/J^*G_I(M) < d.$$

Then for a sufficiently generic element $f \in J$ we have that $\dim G_I(M)/f^*G_I(M) < d$. By 2.3

$$G_I(M/fM) \quad \text{and} \quad G_I(M)/f^*G_I(M) \tag{2}$$

have the same support. Moreover $l_{M/fM}(J) = l_M(J) - 1$ and so, by induction, J is a reduction of $(M/fM, I)$, i.e., $G_I(M/fM)/J^*G_I(M/fM) + \mathfrak{m}G_I(M/fM)$ has dimension zero. Using (2) we get that also $G_I(M)/J^*G_I(M) + \mathfrak{m}G_I(M)$ has dimension 0 and so J is a reduction of (M, I). □

Now we are able to give a proof Böger's theorem.

Corollary 3.5. *Assume that the completion $\hat{M}$ is equidimensional and $l_M(J) = \operatorname{ht}_M J$. Moreover assume that for every prime $\mathfrak{p} \supseteq J$ with $\operatorname{ht}_M \mathfrak{p} = \operatorname{ht}_M J$ also $I \subseteq \mathfrak{p}$ and $e(IA_\mathfrak{p}, M_\mathfrak{p}) = e(JA_\mathfrak{p}, M_\mathfrak{p})$. Then J is a reduction of (M, I).*

Proof. The proof is the same as in [FOV, Sect. 3.6]. For the convenience of the reader we repeat the simple argument. Because of 3.4 we need to to verify that for every prime $\mathfrak{p}$ with $\operatorname{ht}_M \mathfrak{p} \leq \operatorname{ht}_M J$ we have

$$JI^{n-1}M_\mathfrak{p} = I^n M_\mathfrak{p} \quad \text{for} \quad n \gg 0. \tag{$*$}$$

This is certainly true for primes with $J \not\subseteq \mathfrak{p}$. So assume that $J \subseteq \mathfrak{p}$ and $\operatorname{ht}_M \mathfrak{p} \leq \operatorname{ht}_M J$. To prove $(*)$ for such primes we may restrict to the case that $\operatorname{ht}_M \mathfrak{p} = \operatorname{ht}_M J$. By our assumption, we can then apply the theorem of Rees 2.10 and obtain that $J_\mathfrak{p}$ is a reduction of $(M_\mathfrak{p}, I_\mathfrak{p})$ proving $(*)$. □

4 Equimultiple families of modules

4.1. Let A be a Noetherian ring and B an A-algebra. Let $I \subseteq B$ be an ideal and M a finite B-module. We will always assume that

(1) M/IM is finite over A,
(2) $\operatorname{supp} M/IM \to \operatorname{Spec} A$ is surjective,
(3) A, B are universally Japanese and catenary.

Geometrically speaking, let $X = \operatorname{Spec} B$ and $Y = \operatorname{Spec} A$ be the associated schemes and $\mathcal{M}$ the $\mathcal{O}_X$-module given by M. Then by (1) the restriction $\mathcal{M}|W$ to the subscheme $W := V(I)$ of X is finite over Y. Moreover (2) says that the support of $\mathcal{M}|W$ maps surjectively to Y. Thus $\mathcal{M}$ may be viewed as a family of modules parameterized by Y, the fibres being given by

$$M(\mathfrak{p}) := M_\mathfrak{p}/\mathfrak{p}M_\mathfrak{p} \quad \text{for } \mathfrak{p} \in Y = \operatorname{Spec} A.$$

We first note the following standard fact; for a proof see e.g., [BFl] or, for a slightly different treatment, [Lip].

Proposition 4.2. *(1) The function*

$$W \to \mathbb{N} \quad \textit{with} \quad \mathfrak{P} \mapsto d(\mathfrak{P}) := \dim M(\mathfrak{p})_{\mathfrak{P}}, \quad \mathfrak{p} := \mathfrak{P} \cap A,$$

is upper semicontinuous on W.

(2) If M is a flat module over A and if the fibres $M(\mathfrak{p})_{\mathfrak{P}}$ are equidimensional for all $\mathfrak{P} \in W$, where $\mathfrak{p} := \mathfrak{P} \cap A$, then the function d is locally constant on W.

Proof. For the convenience of the reader we include the short proof. In a first step we show that $\mathfrak{P} \mapsto d(\mathfrak{P})$ is constructible on $\operatorname{Spec} B/I$. For this it is sufficient to show that every closed irreducible subset, say, $C = \operatorname{Spec} A'$ of $\operatorname{Spec} B/I$ has an open dense subset on which d is constant. We can perform base change $A \to A'$ and replace M by $M' = M \otimes_A A'$. Note that C embeds diagonally into $\operatorname{Spec} B \otimes_A A'$ and that the fibres remain unchanged up to a field extension. Thus we may suppose that $C \cong \operatorname{Spec} A$ the isomorphism being given by $\mathfrak{P} \mapsto \mathfrak{p} = \mathfrak{P} \cap A$. By the theorem on generic flatness the associated graded module $G_I(M)$ is flat over some open subset U of $C \cong \operatorname{Spec} A$. In particular the function $\mathfrak{P} \mapsto \dim G_I(M) \otimes_B B_{\mathfrak{P}}/\mathfrak{p}B_{\mathfrak{P}}$, $\mathfrak{p} = \mathfrak{P} \cap A$, is constant on $U \subseteq C$. Using the fact that the associated graded module has the same dimension as the module itself it remains to show that $G_I(M) \otimes_B B_{\mathfrak{P}}/\mathfrak{p}B_{\mathfrak{P}} \cong G_I(M(\mathfrak{p})_{\mathfrak{P}})$ for $\mathfrak{P} \in U$. To check this, we may replace U by some affine subset and so we may assume that $U = \operatorname{Spec} A$. More generally we will show that for every finite A-module N the canonical map

$$G_I(M) \otimes_A N \to G_I(M \otimes_A N)$$

is an isomorphism. For this it is sufficient to verify that the natural map $\alpha : I^n M \otimes_A N \to I^n(M \otimes_A N)$ is an isomorphism for all n. We need to show the injectivity of this map. By the flatness of $I^n M/I^{n+1}M$ there is an exact sequence

$$0 \to I^{n+1}M \otimes_A N \to I^n M \otimes_A N \to (I^n M/I^{n+1}M) \otimes_A N \to 0.$$

Using induction it follows that α is injective.

In a second step we prove that d is upper semicontinuous. For this it is sufficient to show that for any two primes $\mathfrak{Q} \subseteq \mathfrak{P}$ of W we have $d(\mathfrak{Q}) \leq d(\mathfrak{P})$. Using a base change argument as before, we may suppose that $A \cong B/I$ is a local domain of dimension 1 and $\mathfrak{q} = 0$ and $\mathfrak{p} = \mathfrak{m}_A$, where $\mathfrak{p} = \mathfrak{P} \cap A$ and $\mathfrak{q} = \mathfrak{Q} \cap A$. Moreover, as our rings are universally Japanese, we can perform a base change with the normalization and so are reduced to the case that A is a discrete valuation ring. Let t denote a uniformizing parameter in $\mathfrak{m}_A$. By our reductions, $I = \mathfrak{Q}$ and $\mathfrak{P} = \mathfrak{Q} + tB$. Now the semicontinuity follows from the chain of inequalities

$$\dim M(\mathfrak{q})_{\mathfrak{Q}} = \dim M_{\mathfrak{Q}} \leq \dim M_{\mathfrak{P}} - 1 \leq \dim M_{\mathfrak{P}}/tM_{\mathfrak{P}}. \qquad (*)$$

Finally, to prove (2) we need to show that in the situation above we have $d(\mathfrak{Q}) = d(\mathfrak{P})$. Using the assumptions that the fibre is equidimensional and that by the flatness assumption t is not a zerodivisor on $M_{\mathfrak{P}}$, it follows that the first resp. second inequality in $(*)$ is an equality. This proves the proposition. □

4.3. Assume now that the function $\mathfrak{P} \mapsto d := d(\mathfrak{P})$ is constant on W. In the following we will study for $\mathfrak{p} \in \operatorname{Spec} A$ the Hilbert–Samuel function of the fibre

$$H_{\mathfrak{p}}^{M}(n) := \lg_{A_{\mathfrak{p}}} M(\mathfrak{p})/I^n M(\mathfrak{p}).$$

Note that

$$H_{\mathfrak{p}}^{M}(n) = \sum_{\mathfrak{P}} \lg_{A_{\mathfrak{p}}} M_{\mathfrak{P}}/(I^n+\mathfrak{p})M_{\mathfrak{P}} = \sum_{\mathfrak{P}} [k(\mathfrak{P}) : k(\mathfrak{p})] \lg_{B_{\mathfrak{P}}} M_{\mathfrak{P}}/(I^n+\mathfrak{p})M_{\mathfrak{P}},$$

where the sum is taken over all primes $\mathfrak{P} \in W$ with $\mathfrak{P} \cap A = \mathfrak{p}$ and where $[k(\mathfrak{P}) : k(\mathfrak{p})]$ denotes the degree of the residue field extension. Therefore we see that $H_{\mathfrak{p}}^{M}(n)$ is a polynomial of degree d in n. Its leading term is given by $e_{\mathfrak{p}}(I, M)/d!$ with

$$e_{\mathfrak{p}}(I, M) := \sum_{\mathfrak{P}} [k(\mathfrak{P}) : k(\mathfrak{p})] e(IB(\mathfrak{p})_{\mathfrak{P}}, M(\mathfrak{p})_{\mathfrak{P}}),$$

where the sum is taken over all primes $\mathfrak{P} \in W$ with $\mathfrak{P} \cap A = \mathfrak{p}$. For a fixed n the function $\mathfrak{p} \mapsto H_{\mathfrak{p}}^{M}(n)$ is obviously upper semicontinuous on $\operatorname{Spec} A$. This immediately implies the following lemma.

Lemma 4.4. *Assume that 4.3 is satisfied. Then the function*

$$\mathfrak{p} \mapsto e_{\mathfrak{p}}(I, M)$$

is semicontinuous on $Y = \operatorname{Spec} A$.

The next two results, due to Lipman [Lip], give a criterion for when $e_{\mathfrak{p}}(I, M)$ is constant.

Theorem 4.5. *Assume that M is a flat A-module and that all fibres of $G_I(M)$ over A have dimension d, i.e., $\dim G_I(A) \otimes_B B(\mathfrak{p})_{\mathfrak{P}} = d$ for all $\mathfrak{P} \in W$. Then $d(\mathfrak{P}) = d$ for all $\mathfrak{P} \in W$. Moreover $\mathfrak{p} \mapsto e_{\mathfrak{p}}(I, M)$ is locally constant on $Y = \operatorname{Spec} A$.*

Proof. As was shown in the proof of 4.2, the natural map

$$\alpha_{\mathfrak{P}} : G_I(M) \otimes_B B(\mathfrak{p})_{\mathfrak{P}} \to G_I(M(\mathfrak{p})_{\mathfrak{P}})$$

is an isomorphism for $\mathfrak{P}$ in an open dense subset of W, where as usual $\mathfrak{p} = \mathfrak{P} \cap A$. As $d(\mathfrak{P}) = \dim G_I(M(\mathfrak{p})_{\mathfrak{P}})$ it follows that $d(\mathfrak{P}) = d$ generically. By semicontinuity $d(\mathfrak{P}) \geq d$ for all $\mathfrak{P} \in W$. As $\alpha_{\mathfrak{P}}$ is always surjective, we also get $d(\mathfrak{P}) \leq d$ and so $d(\mathfrak{P}) = d$ for all $\mathfrak{P} \in W$.

To prove the remaining assertion we need to show that for all primes $\mathfrak{p}$, $\mathfrak{q}$ in A with $\dim A_\mathfrak{p}/\mathfrak{q}A_\mathfrak{p} = 1$ we have $e_\mathfrak{p}(I, M) = e_\mathfrak{q}(I, M)$. Clearly we may replace A by $A_\mathfrak{p}$ and B, M by $B \otimes_A A_\mathfrak{p}$, $M \otimes_A A_\mathfrak{p}$, respectively. Henceforth we assume that $(A, \mathfrak{m})$ is local with $\mathfrak{p} = \mathfrak{m}$ and $\dim A/\mathfrak{q} = 1$.

Now consider $\bar{A} := A/\mathfrak{q}$, $\bar{B} := B/\mathfrak{q}B$ and $\bar{M} := M/\mathfrak{q}M$. Then $G_I(\bar{M})$ is a quotient of $G_I(M) \otimes_A \bar{A}$ and the fibres of $G_I(\bar{M})$ over $\bar{A}$ are again of dimension d. Thus we are reduced to the case that A is an integral domain of dimension 1.

Let $A \to A'$ be the normalization of A and consider $I' := IA'$, $M' := M \otimes_A A'$. Then $\operatorname{supp} G_{I'}(M') \to \operatorname{supp} G_I(M)$ is finite and surjective. Hence $G_{I'}(M')$ has equidimensional fibres over A'. Moreover, the fibres of M over A and M' over A' are the same up to a field extension. Hence we may replace A by a localization, say, A'' of A' with a respect to a maximal ideal and M by $M'' := M \otimes_A A''$. Thus we are reduced to the case that A is a discrete valuation ring.

Let $t \in \mathfrak{m}$ be a uniformizing parameter. Applying 2.3 to the exact sequence $0 \to M \xrightarrow{t} M \to M/tM \to 0$ we get that

$$\ker(G_I(M) \xrightarrow{t} G_I(M)) \quad \text{and} \quad \ker(G_I(M)/tG_I(M) \to G_I(M/tM))$$

define the same cycles. Taking multiplicities gives, with $N := G_I(M)$ and I^* the initial ideal of I in $G_I(B)$,

$$e_\mathfrak{m}(I^*, G_I(M/tM)) = e_\mathfrak{m}(I^*, N/tN) - e_\mathfrak{m}(I^*, \operatorname{Ann}_N t).$$

By definition we have $e_\mathfrak{m}(I^*, G_I(M/tM)) = e_\mathfrak{m}(I, M/tM) = e_\mathfrak{m}(I, M)$. It remains to prove that, with $(0) = \mathfrak{q}$ the 0-ideal of A,

$$e_{(0)}(I, M) = e_\mathfrak{m}(I^*, N/tN) - e_\mathfrak{m}(I^*, \operatorname{Ann}_N t). \tag{1}$$

Set $K := \bigcup_{n \geq 1} \operatorname{Ann}_N t^n$ and consider the diagram

$$\begin{array}{ccccccccc} 0 & \to & K & \to & N & \to & N/K & \to & 0 \\ & & \downarrow t & & \downarrow t & & \downarrow t & & \\ 0 & \to & K & \to & N & \to & N/K & \to & 0\,. \end{array}$$

Using the snake lemma we obtain that $\operatorname{Ann}_K t = \operatorname{Ann}_N t$ and $[N/K + tN] + [K/tK] = [N/tN]$ as cycles on $\operatorname{Spec} G_I(B)$. As $[K/tK] = [\operatorname{Ann}_K t] = [\operatorname{Ann}_N t]$ this implies $[N/K + tN] + [\operatorname{Ann}_N t] = [N/tN]$. Taking multiplicities gives

$$e_\mathfrak{m}(I^*, N/tN) - e_\mathfrak{m}(I^*, \operatorname{Ann}_N t) = e_\mathfrak{m}(I^*, N/K + tN). \tag{2}$$

All homogeneous components of N/K are flat A-modules and so

$$e_\mathfrak{m}(I^*, N/K + tN) = e_{(0)}(I^*, N/K) = e_{(0)}(I^*, N) = e_{(0)}(I, M).$$

Together with (2) this proves (1), as required. $\square$

Theorem 4.6. *Assume that for all $\mathfrak{P} \in W$ the fibres $M(\mathfrak{p})_{\mathfrak{P}}$ are formally equidimensional of dimension d and that $\mathfrak{p} \mapsto e_{\mathfrak{p}}(I, M)$ is constant for $\mathfrak{p} \in$ Spec A. Then the fibres of $G_I(M)$ over A are equidimensional of dimension d.*

Proof. We may assume that $A = (A, \mathfrak{m})$ is a local domain. We need to show that $l_M^A(I) := \dim G_I(M)/\mathfrak{m}G_I(M)$ is equal to d (the number $l_M^A(I)$ may be regarded as a relative version of the analytic spread). First suppose that $d = 0$ so that $\dim M = \dim A$. The condition $e_{\mathfrak{p}}(I, M) = e_{\mathfrak{m}}(I, M)$ implies that for $n \gg 0$ the function

$$\mathfrak{p} \mapsto \lg(M/I^{n+1}M)(\mathfrak{p})$$

on Spec A is constant and independent of n. Hence $M/I^{n+1}M$ is a free module over A and $I^n M/I^{n+1}M = 0$ for $n \gg 0$. Thus $G_I(M)$ has 0-dimensional fibres over A.

Now assume that $d > 0$. Let K be the submodule of all $m \in M$ with $\dim Bm < \dim M$ so that $\bar{M} := M/K$ has no embedded components. Then $G_I(M)$ and $G_I(\bar{M})$ have the same support, see 2.4 (a). The support of the generic fibre $K(0)$ over the zero ideal $(0) \subseteq A$ has codimension ≥ 1 in the support of $M(0)$ and so

$$e_{(0)}(I, M) = e_{(0)}(I, \bar{M}) \leq e_{\mathfrak{m}}(I, \bar{M}) \leq e_{\mathfrak{m}}(I, M),$$

where the first inequality follows from the semicontinuity of multiplicities, see 4.4. Since by assumption $e_{(0)}(I, M) = e_{\mathfrak{m}}(I, M)$ we get equality everywhere. Therefore we can replace M by $\bar{M}$ and henceforth assume that M has no embedded components.

Now we choose $t \in I$ sufficiently generic so that t is not a zerodivisor on M, $e_{\mathfrak{m}}(I, M/tM) = e_{\mathfrak{m}}(I, M)$ and $l_M^A(I) = l_{M/tM}^A(I) + 1$. Consider the following chain of inequalities:

$$e_{\mathfrak{m}}(I, M) = e_{(0)}(I, M) \leq e_{(0)}(I, M/tM) \leq e_{\mathfrak{m}}(I, M/tM) = e_{\mathfrak{m}}(I, M).$$

Thus in this chain all terms are equal. Using induction we obtain that $l_{M/tM}^A(I) = d - 1$. As $l_{M/tM}^A(I) = l_M^A(I) - 1$ the theorem follows. □

5 Equimultiplicity and equidimensionality of the normal cone

In this section we will give a short proof of a theorem due to Schickhoff and Teissier in the analytic case. The following algebraic version was proved by Lipman [Lip, Th. 5].

Theorem 5.1. *Let $\mathfrak{p}$ be a prime ideal in the local ring $(A, \mathfrak{m})$ and assume that $A/\mathfrak{p}$ is regular. Then the following hold.*

(1) If $l(\mathfrak{p}) = \operatorname{ht}\mathfrak{p}$ then $e(A) = e(A_\mathfrak{p})$.

(2) If moreover $\hat{A}$ is equidimensional then the converse of (1) is true.

(3) Assume that $\hat{A}$ is equidimensional and that $l(\mathfrak{p}) = \operatorname{ht}\mathfrak{p}$. Consider elements $t_1, \ldots, t_k \in \mathfrak{m}$ whose residue classes form a regular parameter system in the regular ring $A/\mathfrak{p}$. Let $t_1^*, \ldots, t_k^*$ denote the initial forms of $t_1, \ldots, t_k$ in $G_\mathfrak{m}(A)$. Then the canonical surjections

$$\alpha : G_\mathfrak{p}(A)/\mathfrak{m}G_\mathfrak{p}(A) \longrightarrow G_\mathfrak{m}(A)/(t_1^*, \ldots, t_k^*)$$
$$\beta : (G_\mathfrak{p}(A)/\mathfrak{m}G_\mathfrak{p}(A))[T_1, \ldots, T_k] \longrightarrow G_\mathfrak{m}(A)$$

have nilpotent kernel, where β is given by $T_i \mapsto t_i^*$.

Proof. With the standard reduction we may assume that A is complete and that $A/\mathfrak{m}$ is infinite. Moreover because of 2.4 (a) we can always assume that A has no embedded components.

ad (1): In the case $\operatorname{ht}\mathfrak{p} = 0$ we have $l(\mathfrak{p}) = 0$ and so $\mathfrak{p}^n = 0$ for $n \gg 0$. The associativity formula for multiplicities gives

$$e(A) = \lg A_\mathfrak{p} \cdot e(A/\mathfrak{p}) = \lg A_\mathfrak{p} = e(A_\mathfrak{p}),$$

as required. So assume that $\operatorname{ht}\mathfrak{p} > 0$. There is a canonical isomorphism

$$G_\mathfrak{m}(A)/\mathfrak{p}^* G_\mathfrak{m}(A) \cong (A/\mathfrak{m})[t_1^*, \ldots, t_k^*] = G_\mathfrak{m}(A/\mathfrak{p}).$$

Therefore for a sufficiently generic element $t \in \mathfrak{p}$ the quotient $G_\mathfrak{m}(A)/t^*G_\mathfrak{m}(A)$ has dimension $d - 1$, where $d := \dim A$. As A has no embedded components, we deduce from 2.9 that $e(A) = e(A/tA)$. By the generic choice of t the dimension of $G_\mathfrak{p}(A_\mathfrak{p})/t^*G_\mathfrak{p}(A_\mathfrak{p})$ is $\dim A_\mathfrak{p} - 1$. Thus with the same argument as before $e(A_\mathfrak{p}) = e(A_\mathfrak{p}/tA_\mathfrak{p})$. Moreover, $l(\mathfrak{p}) = l(\mathfrak{p}/tA) + 1$ and $\operatorname{ht}\mathfrak{p} = \operatorname{ht}\mathfrak{p}/tA + 1$. Hence (1) follows by induction.

In order to prove (2) we again proceed by induction on $\operatorname{ht}\mathfrak{p}$. If $\operatorname{ht}\mathfrak{p} = 0$ then by the associativity formula

$$e(A_\mathfrak{p}) = e(A) = \sum \lg(A_\mathfrak{q}) e(A/\mathfrak{q}) \geq \lg(A_\mathfrak{p}) e(A/\mathfrak{p}) = e(A_\mathfrak{p}),$$

where the sum is taken over the minimal primes of A; note that A is equidimensional by assumption. Hence $\mathfrak{p}$ is the only minimal prime and $\mathfrak{p}^n = 0$ for $n \gg 0$.

Now assume that $\operatorname{ht}\mathfrak{p} > 0$ and consider again $t \in \mathfrak{p}$ generic. As above $l(\mathfrak{p}/tA) = l(\mathfrak{p}) - 1$, $\operatorname{ht}\mathfrak{p}/tA = \operatorname{ht}\mathfrak{p} - 1$, $e(A/tA) = e(A)$ and $e(A_\mathfrak{p}/tA_\mathfrak{p}) = e(A_\mathfrak{p})$. Moreover, the ring A/tA is again equidimensional. Hence the result follows by induction.

In order to prove (3) note first that the ring $G_\mathfrak{p}(A)/\mathfrak{m}G_\mathfrak{p}(A) \cong G_\mathfrak{p}(A)/(t_1^*, \ldots, t_k^*)$ has dimension $l(\mathfrak{p}) = \operatorname{ht}\mathfrak{p} = \dim R - k$. By 2.5 the rings

$$G_\mathfrak{p}(A)/(t_1^*, \ldots, t_k^*) \quad \text{and} \quad G_\mathfrak{p}(A/(t_1, \ldots, t_k))$$

have the same support. Hence in the diagram

$$\begin{array}{ccc} G_{\mathfrak{p}}(A)/\mathfrak{m}G_{\mathfrak{p}}(A) \cong G_{\mathfrak{p}}(A)/(t_1^*,\ldots,t_k^*) & \xrightarrow{\alpha} & G_{\mathfrak{m}}(A)/(t_1^*,\ldots,t_k^*) \\ \downarrow & & \downarrow \\ G_{\mathfrak{p}}(A/(t_1,\ldots,t_k)) & \xrightarrow{\sim} & G_{\mathfrak{m}}(A/(t_1,\ldots,t_k)) \end{array}$$

all the rings must have the same support. Thus the kernel of α is a nilpotent ideal.

Finally, consider β which is a surjective map of equidimensional rings of the same dimension. The support of $G_{\mathfrak{m}}(A)$ is a union of irreducible components of $\mathrm{supp}(G_{\mathfrak{p}}(A)/\mathfrak{m}G_{\mathfrak{p}}(A))[T_1,\ldots,T_k]$. But the components of the latter ring are in 1–1 correspondence with the irreducible components of $G_{\mathfrak{p}}(A)/\mathfrak{m}G_{\mathfrak{p}}(A)$. As α gives a bijection of the irreducible components of $G_{\mathfrak{p}}(A)/\mathfrak{m}G_{\mathfrak{p}}(A)$ and $G_{\mathfrak{m}}(A)/(t_1^*,\ldots,t_k^*)$ also the rings $(G_{\mathfrak{p}}(A)/\mathfrak{m}G_{\mathfrak{p}}(A))[T_1,\ldots,T_k]$ and $G_{\mathfrak{m}}(A)$ have the same support. Thus the kernel of β is nilpotent, as required. □

6 The analytic case

In this section we will give analytic versions of the results of Sections 4 and 5. We will indicate how the analytic case can be deduced from the algebraic one using Stein compact sets as in [BFl].

Recall that a subset K of a complex space X is called *Stein compact* if it is a compact semianalytic subset of X which is Stein, i.e., it admits arbitrary small open Stein neigbourhoods. By a fundamental theorem of Frisch [Fri], then the ring $A_K := \Gamma(K, \mathcal{O}_X)$ is Noetherian. Similarly, if $\mathcal{M}$ is a coherent $\mathcal{O}_X$-module then $M_K := \Gamma(K, \mathcal{M}_X)$ is a finitely generated module over A_K, and the functor $\mathcal{M} \mapsto M_K$ is exact. The maximal ideals of A_K are in 1–1 correspondence with the points of K, where one associates to $x \in K$ the ideal $\mathfrak{m}_x \subseteq A_K$ of functions vanishing at x. Moreover the completions of the modules $\mathcal{M}_x$ and $M_{\mathfrak{m}_x}$ are isomorphic.

We also remind the reader of the following fact. If $f : W \to Y$ is a finite holomorphic mapping and $L \subseteq Y$ is a Stein compact set then $f^{-1}(L)$ is a Stein compact subset of W. Similarly, if $W \subseteq X$ is a closed subspace and K is a Stein compact subset of W then K also is a Stein compact subset of X.

In the following a subset U of a complex space X is called *Zariski open* (Z-open in brief) if $X \setminus U$ is an analytic subset of X. Similarly, a function $f : X \to \mathbb{R}$ is called Z-upper semicontinuous if it is upper semicontinuous with respect to the Zariski topology on X, i.e., for all $a \in \mathbb{R}$ the set $f^{-1}(]-\infty, a[)$ is Z-open.

6.1. Let $f : X \to Y$ be a holomorphic map of complex spaces and $W \subseteq X$ a closed complex subspace given by the coherent ideal $\mathcal{I} \subseteq \mathcal{O}_X$. In the following we will assume that

$$f : W \to Y$$

is finite and surjective. For a point $y \in Y$ let $X(y)$ denote the fibre $f^{-1}(y)$ (with the structure given by $\mathfrak{m}_y \mathcal{O}_X$). Similarly, if $\mathcal{M}$ is a coherent $\mathcal{O}_X$-module then we set $\mathcal{M}(y) := \mathcal{M} \otimes_{\mathcal{O}_X} \mathcal{O}_{X(y)}$.

First let us give an analytic version of 4.2.

Proposition 6.2. *(1) The function*

$$W \to \mathbb{N} \quad \textit{with} \quad w \mapsto d(w) := \dim \mathcal{M}(f(w))_w$$

is Z-upper semicontinuous on W.

(2) If $\mathcal{M}$ is a flat module over Y and if the fibres $\mathcal{M}(f(w))_w$ are equidimensional for all $w \in W$, then the function d is locally constant on W.

Proof. A proof can be found e.g., in [BFl]. For the convenience of the reader we repeat the argument which consists in a reduction to the algebraic case. Let L be a Stein compact neighbourhood of $f(w)$ in Y and consider $K := f^{-1}(L) \cap W$. As $W \to Y$ is finite, K is a Stein compact set in W and then also in X. Consider the ring extension

$$A := \Gamma(L, \mathcal{O}_Y) \longrightarrow B := \Gamma(K, \mathcal{O}_X).$$

The rings A and B are Noetherian, and $M := \Gamma(L, \mathcal{M})$ is a finitely generated A-module. The ideal $I := \Gamma(L, \mathcal{I})$ of B has the property that $B/I \cong \Gamma(L, \mathcal{O}_W)$ is a finite extension of A. Let $w \in W$ be a point and $\mathfrak{m}_w \subseteq B$ and $\mathfrak{m}_{f(w)} \subseteq A$ be the maximal ideals corresponding to $w \in X$ resp. $f(w) \in Y$. Note that $\mathfrak{m}_{f(w)} = \mathfrak{m}_w \cap A$. As the completions of $\mathcal{M}_w$ and $M_{\mathfrak{m}_w}$ are isomorphic we have that

$$d(w) = \dim \mathcal{M}(f(w))_w = \dim(M/\mathfrak{m}_{f(w)}M)_{\mathfrak{m}_w}.$$

By 4.2 we obtain that the function

$$\mathfrak{P} \mapsto \dim M(\mathfrak{p})_{\mathfrak{P}}, \quad \mathfrak{p} := \mathfrak{P} \cap A,$$

is upper semicontinuous on W. This easily implies that also $d(w)$ is Z-upper semicontinuous on $W \cap K$. Hence (i) follows. Similarly, if all fibres $\mathcal{M}(f(w))_w$, $w \in W$, are equidimensional then also all fibres $M(\mathfrak{p})$, $\mathfrak{p} \in \operatorname{Spec} A$, are equidimensional. Applying 4.2 (ii) we obtain as above that (ii) holds. □

6.3. Assume now that the function $w \mapsto d := d(w)$ is constant on W. In the following we will study the Hilbert–Samuel function of the fibre

$$H_y^{\mathcal{M}}(n) := \sum_{w \in X(y) \cap W} \dim_{\mathbb{C}} \mathcal{M}(y)_w / \mathcal{I}^n \mathcal{M}(y)_w, \quad y \in Y.$$

For large n, $H_y^M(n)$ is a polynomial of degree d in n. Its leading term is given by $e_y(\mathcal{I}, \mathcal{M})/d!$ with

$$e_y(\mathcal{I}, \mathcal{M}) := \sum_{w \in X(y) \cap W} e(\mathcal{I}(y)_w, \mathcal{M}(y)_w),$$

where $\mathcal{I}(y)$ denotes the ideal $\mathcal{I}\mathcal{O}_{X(y)}$. For fixed n the function $y \mapsto H_y^{\mathcal{M}}(n)$ is obviously Z-upper semicontinuous on Spec A. This immediately implies the following lemma.

Lemma 6.4. *Assume that 6.3 is satisfied. Then the function*

$$y \mapsto e_y(\mathcal{I}, \mathcal{M})$$

is Z-upper semicontinuous on Y.

The following result is the analytic analogue of 4.5.

Theorem 6.5. *Assume that $\mathcal{M}$ is coherent $\mathcal{O}_X$-module which is flat over Y. Moreover assume that all fibres of $G_{\mathcal{I}}(\mathcal{M}) = \bigoplus_{n \geq 0} \mathcal{I}^n \mathcal{M}/\mathcal{I}^{n+1}\mathcal{M}$ over W are of dimension d. Then the function $y \mapsto e_y(\mathcal{I}, \mathcal{M})$ is locally constant on Y.*

Remark 6.6. The most important case is when $\mathcal{M} = \mathcal{O}_X$. In this case the condition that the fibre of $G_{\mathcal{I}}(\mathcal{O}_X)$ over W are of dimension d is equivalent to saying that all fibres of the projection $C_W X \to W$ are of dimension d.

Proof. Again let us consider K, L, $I \subseteq A$, B, M as in the proof of 6.2. Then we have

$$\Gamma(K, G_{\mathcal{I}}(\mathcal{M})) = G_I(M).$$

Hence the condition that all fibres of $G_{\mathcal{I}}(\mathcal{M})$ over W have dimension d implies that all fibres of $G_I(M)$ over Spec B/I are of dimension d. Applying 4.5 the result follows. □

The next result is the analytic version of 4.6.

Theorem 6.7. *Assume that for all $w \in W$ the fibre $\mathcal{M}(f(w))_w$ is equidimensional of dimension d and that*

$$y \mapsto e_y(\mathcal{I}, \mathcal{M})$$

is constant on Y. Then the fibres of $G_{\mathcal{I}}(\mathcal{M})$ over Y are equidimensional of dimension d.

The proof is easily deduced from the algebraic case 4.6 with the same arguments as above. Let us finally turn to an analytic version of the result in Section 5, which one can again prove by reducing to the algebraic case; we leave the details to the reader. This result is due to Schickhoff [Sch]; the equivalence of (1) and (2) was also shown by Teissier [Tei2, Th. 5.5].

Theorem 6.8. *Let X be a complex space and let W be a closed smooth subspace of X defined by the ideal $\mathcal{I} \subseteq \mathcal{O}_X$. Then the following hold.*

(1) If $C_W X \to W$ has equidimensional fibres then the function

$$W \longrightarrow \mathbb{Z} \quad \text{with} \quad w \mapsto e(\mathcal{O}_{X,w})$$

is locally constant.

(2) If moreover X is equidimensional then the converse of (1) holds.

(3) Assume that X is equidimensional and that $C_W X \to W$ has equidimensional fibres. Consider elements $t_1, \dots, t_k \in \mathcal{O}_{X,w}$ which are local coordinates on W with $t_i(w) = 0$ and denote by $t_1^*, \dots, t_k^*$ the initial forms of $t_1, \dots, t_k$ in $G_{\mathcal{I}}(\mathcal{O}_X)$. Then $V(t_1^*, \dots, t_k^*) \subseteq C_W X$ has the same support as the fibre $(C_W X)_w$ of $C_W X \to W$ over w. Similarly, the map

$$(\beta, t_1^*, \dots, t_d^*) : C_{\{w\}} X \to (C_W X)_w \times \mathbb{C}^d$$

is bijective on the underlying topological spaces, where β denotes the canonical map $C_{\{w\}} X \to (C_W X)_w$.

References

[Bö] Böger, E.: Einige Bemerkungen zur Theorie der ganzalgebraischen Abhängigkeit von Idealen. *Math. Ann.* **185**, 303–308 (1970)

[BFl] Bingener, J.: Flenner, H.: On fibres of analytic mappings. In: *Complex analysis and geometry* (V. Ancona and A. Silva, eds.) New York–London: Plenum Publishers, 1993, 45–103

[Da] Dade, E.C.: *Multiplicity and monoidal transformations.* Thesis, Princeton University, 1960

[Ful] Fulton, W.: Intersection theory. *Erg. Math. Grenzgeb.*, 3. Folge, Bd. 2. Berlin–Heidelberg-New York: Springer-Verlag, 1984.

[FVo] Flenner, H.: Vogel, W.: On multiplicities of local rings. *Manuscripta Math.* **78**, 85–97 (1993)

[FOV] Flenner, H.: O'Carroll, L.; Vogel, W.: *Joins and intersections.* Springer Monographs in Math., Springer, Berlin, 1999.

[Fri] Frisch, J.: Points de platitude d'un morphisme d'espaces analytiques complexes. *Invent. Math.* **4**, 118–138 (1967)

[Lip] Lipman, J.: Equimultiplicity, reduction, and blowing up. In: *Commutative Algebra, analytic methods* (R.N. Draper, ed.) Math. sciences regional research conf. (Mason Univ. 1979); Lect. notes in pure and applied Math. **68**, Marcel Dekker 1982

[Ste] Steffen, F.: A generalized principal ideal theorem with an application to Brill-Noether theory. *Invent. Math.* **132**, 73–89 (1998)

[Tei1] Teissier, B.: The hunting of invariants in the geometry of discriminants. In: *Real and complex singularities* (Oslo 1976), Sijthoff and Noordhoff, Alphen aan den Ridn, 1977, 567–677

[Tei2] Teissier, B.: Variétés polaires. II: Multiplicités polaires, sections planes et conditions de Whitney. In: *Algebraic geometry*, Proc. int. Conf., La Rabida/Spain 1981, Springer Verlag, Lect. Notes Math. 961, 314–491 (1982)

[Sch] Schickhoff, W.: *Whitneysche Tangentenkegel, Multiplizitätsverhalten, Normal-Pseudoflachheit und Äquisingularitätstheorie für Ramissche Räume.* Schriftenreihe des Math. Instituts der Univ. Münster, 2. Serie, Heft 12, 1977

Hubert Flenner
Fakultät für Mathematik
Ruhr-Universität Bochum
Universitätsstraße 150, Geb. NA 2/72
44780 Bochum
Germany
Hubert.Flenner@ruhr-uni-bochum.de

Mirella Manaresi
Dipartimento di Matematica
Università Bologna
P.za di Porta S. Donato 5
40126 Bologna
Italy
manaresi@dm.unibo.it

Intersection Products for Spaces of Algebraic Cycles

Eric M. Friedlander

This is a slightly expanded version of three lectures given by the author in Bologna in December 1997. Our purpose is to introduce to algebraic geometers some constructions developed in recent years for spaces of cycles that relate to classical intersection theory. The work discussed appears primarily in joint papers of the author and H. Blaine Lawson as well as a joint paper of the author and Ofer Gabber. We recommend the reader consult the expository article [L2] which not only provides further details concerning spaces of algebraic cycles but also provides a somewhat different point of view.

Unless mention is made to the contrary, all varieties considered will be quasi-projective varieties over the complex field **C**. By a closed subvariety of such a variety X, we shall mean (implicitly) closed in the Zariski topology on X, but otherwise when we speak of the topology on X we shall mean the analytic topology. As mentioned in the lectures, certain aspects of what we discuss hold over an arbitrary field (e.g., moving lemmas as in [FL-3]) or over a field which admits resolution of singularities (e.g., duality theorems as in [F-V]).

1 Topological abelian groups of algebraic cycles

In this section, we survey some of the basic definitions and properties of Lawson homology, a homology theory defined in terms of homotopy groups

---

This research was partially supported by the NSF and the NSA.

of topological abelian groups of algebraic cycles.

Let X be a (complex, quasi-projective) variety. Then the effective 0-cycles on X of a given degree d (i.e., formal sums of points of X with nonnegative integer coefficients whose sum equals d) admit the natural structure of an algebraic variety $SP^d(X)$, the d-fold symmetric power of X. The formal sum of 0-cycles determines a topological monoid structure on

$$\mathcal{C}_0(X) \equiv \coprod_{d\geq 0} SP^d(X),$$

where we set $SP^0(X)$ equal to a point (the set of effective 0-cycles containing only the empty- or zero-cycle). The naïve group completion of this topological monoid is the topological abelian group

$$Z_0(X) \equiv \mathbf{Z}[X] = \mathcal{C}_0(X)^{\times 2}/\sim$$

of all 0-cycles on X (or, equivalently, the free abelian group on X), where the equivalence relation $\sim$ consists of pairs of pairs $(\sum x_i, \sum y_j) \sim (\sum z_k, \sum w_\ell)$ such that $\sum x_i + \sum w_\ell = \sum y_j + \sum z_k$ and where $Z_0(X)$ is provided with the quotient topology with respect to $\mathcal{C}_0(X)^{\times 2} \to Z_0(X)$.

Theorem 1.1. (Dold–Thom [D-T]) *The integral singular homology of the complex projective variety X can be computed as*

$$H_*(X) = \pi_*(\mathbf{Z}[X]) = \pi_*(\Omega B\mathcal{C}_0(X)),$$

where $B\mathcal{C}_0(X)$ is the classifying space of the abelian topological monoid $\mathcal{C}_0(X)$ and $\Omega B\mathcal{C}_0(X)$ is its loop space.

Now we proceed with a very similar construction for positive dimensional cycles on algebraic varieties. We begin by assuming that our variety is projective so that we can use Chow varieties as recalled below.

Theorem 1.2. ([C-W]) *Let X be a closed subvariety of some projective space $\mathbf{P}^N$. Then the set of effective, degree d dimension r-cycles on X admits a natural structure of a projective algebraic variety $C_{r,d}(X)$.*

Moreover, the Chow monoid

$$\mathcal{C}_r(X) \equiv \coprod_{d\geq 0} C_{r,d}(X)$$

is an abelian topological monoid which is independent up to algebraic isomorphism of the chosen embedding $X \subset \mathbf{P}^N$.

The independence of projective embedding of $\mathcal{C}_r(X)$ is proved in [B].

Definition 1.3. Let X be a projective variety. The topological abelian group of algebraic r-cycles is defined to be the naïve group completion of the Chow monoid $\mathcal{C}_r(X)$

$$Z_r(X) \equiv \mathcal{C}_r(X)^{\times 2}/\sim,$$

where the equivalence relation $\sim$ consists of pairs of pairs $(Z_1, Z_2) \sim (W_1, W_2)$ such that $Z_1 + W_2 = Z_2 + W_1$ and where $Z_r(X)$ is provided with the quotient topology.

The (bi-indexed) Lawson homology groups are defined to be the homotopy groups of these topological abelian groups (pointed at 0)

$$L_rH_n(X) \equiv \pi_{n-2r}(Z_r(X)), \quad n \geq 2r$$

(cf. Remark 1.11 for an explanation of this indexing).

The reader consulting the literature should be forewarned that early papers on Lawson homology used a somewhat different (less convenient, but essentially equivalent) definition. From the point of view of getting sensible algebraic invariants, it seemed evident that one wanted to use a homotopy theoretic group completion of $\mathcal{C}_r(X)$. Thus, the first formulations used $\Omega B\mathcal{C}_r(X)$ or the group completion of the simplicial monoid Sing.$(\mathcal{C}_r(X))$. However, the underlying discrete group of $Z_r(X)$ is precisely the group of all r-cycles on X, so that this more concrete object has definite advantages. It was proved by Paulo Lima-Filho [Li2] and by Ofer Gabber and the author [F-G] that the natural map $\mathcal{C}_r(X) \to Z_r(X)$ is a homotopy theoretic group completion: The induced map in integral homology can be identified with the following localization:

$$H_*(\mathcal{C}_r(X)) \to \mathbf{Z}[\pi_0^+] \otimes_{\mathbf{Z}[\pi_0]} H_*(\mathcal{C}_r(X)),$$

where $\pi_0 = \pi_0(\mathcal{C}_r(X))$ is the discrete abelian monoid of connected components of $\mathcal{C}_r(X)$. Another reassuring property of the topological abelian groups $Z_r(X)$ is that they are "nice" as topological spaces; namely, $Z_r(X)$ has the homotopy type of a (countable) C.W. complex (cf. [F4]).

We shall require constructions on these topological abelian groups which require working modulo homotopy equivalences which are continuous group homomorphisms. We recall that this context is a familiar one in algebraic geometry.

Formalism. The category whose objects are topological abelian groups of the homotopy type of C.W. complexes and whose maps are homotopy equivalence classes of continuous group homomorphisms is equivalent to a full subcategory of the derived category of abelian groups $\mathcal{D}^+(Ab)$ defined as the the triangulated category of chain complexes of abelian groups (with differential of degree +1) which are bounded above and localized with respect to quasi-isomorphisms.

This equivalence can be realized by sending the topological abelian group Z to the normalized chain complex $\widetilde{Z}$ associated to the simplical abelian group Sing.(Z). Under this correspondence, $H_i(Z)$ is naturally isomorphic to $\pi_i(\widetilde{Z})$.

When working with the cycle spaces, we sometimes consider these as topological abelian groups $Z_r(X)$ and sometimes as chain complexes $\widetilde{Z}_r(X)$.

The following theorem of Blaine Lawson is the foundation for all that follows. We recall that $\mathbf{P}^{N+1}$ can be viewed as the union of all lines from a fixed hyperplane $\mathbf{P}^N \subset \mathbf{P}^{N+1}$ to a fixed point $x_\infty \in \mathbf{P}^{N+1}$ not on the hyperplane, $\mathbf{P}^{N+1} = \mathbf{P}^N \# x_\infty$. The algebraic suspension of a closed subvariety $X \subset \mathbf{P}^N$ is defined to be the subvariety $\Sigma X \subset \mathbf{P}^{N+1}$ given as the union of all lines from X to x_∞, where once again $\mathbf{P}^{N+1} = \mathbf{P}^N \# x_\infty$. Stated more algebraically, if the subvariety $X \subset \mathbf{P}^N$ is given by homogeneous equations $F_i(T_0, \ldots, T_N)$, then $\Sigma \subset \mathbf{P}^{N+1}$ is given by the same set of homogeneous equations now viewed as equations in the variables $T_0, \ldots, T_{N+1}$.

Theorem 1.4. (Lawson suspension theorem [L1]) *Let $X \subset \mathbf{P}^N$ be a closed subvariety. Then sending an irreducible subvariety $Y \subset X$ of dimension r to $\Sigma Y \subset \mathbf{P}^{N+1}$ determines a continuous group homomorphism*

$$\Sigma : Z_r(X) \to Z_{r+1}(\Sigma X),$$

which is a homotopy equivalence. In other words,

$$\Sigma : \widetilde{Z}_r(X) \to \widetilde{Z}_{r+1}(\Sigma X)$$

is a quasi-isomorphism in $\mathcal{D}^+(Ab)$.

Remarks. The essence of this theorem is a moving lemma. Namely, let

$$\mathcal{C}_{r+1,d}(\Sigma X, X) \subset \mathcal{C}_{r+1,d}(\Sigma X)$$

denote the submonoid of those effective $r+1$-cycles on ΣX which intersect $X \subset \Sigma X$ properly (i.e., in pure dimension r). Then it is not difficult to see that

$$\Sigma : \mathcal{C}_{r,d}(X) \to \mathcal{C}_{r+1,d}(\Sigma X, X)$$

is a deformation retract. The content of the theorem is the assertion that

$$C_{r+1,d}(\Sigma X, X) \subset C_{r+1}(\Sigma X)$$

induces a homotopy equivalence of homtopy theoretic group completions. One can view this as the assertion that after group completion any "bounded family" of $r+1$-cycles on ΣX can be moved so that each member of the family meets X properly.

As Lawson observed, this theorem plus Theorem 1.1 immediately gives the following computation (since $\mathbf{P}^N = \underbrace{\Sigma \circ \cdots \circ \Sigma}_{r \text{ times}} \mathbf{P}^{N-r}$).

Corollary 1.5. *For any $N \geq r \geq 0$,*

$$L_r H_n(\mathbf{P}^N) = H_n(\mathbf{P}^{N-r}),$$

where $H_n(\mathbf{P}^{N-r})$ denotes the integral singular homology of the analytic space $\mathbf{P}^{N-r}$.

Indeed, the possibility of such a result was a major motivation of Lawson, who sought to find an analogue of Theorem 1.1 for higher-dimensional cycles along the lines of a theorem of F. Almgren which asserted that integral homology of X could be computed as the homotopy groups of topological abelian groups of "integral cycles" (i.e., rectifiable currents on X with trivial boundary topologized with the flat norm topology) [A].

From the outset, we have taken a different point of view: Lawson homology groups are often algebraic invariants which have no classical algebraic topology representation. For example, we have the following elementary computation.

Proposition 1.6. ([F1]) *Let X be a projective algebraic variety. Then $L_rH_{2r}(X)$ equals the group of algebraic r-cycles modulo algebraic equivalence.*

We now extend our consideration to quasi-projective varieties, following the lead of Lima-Filho [Li1].

Definition 1.7. Let $X \subset \mathbf{P}^N$ be a closed subvariety, let $X_\infty \subset X$ be a closed embedding, and let $U = X - X_\infty$. Then we define

$$Z_r(U) \equiv Z_r(X)/Z_r(X_\infty), \quad L_rH_n(U) = \pi_{n-2r}(Z_r(U)).$$

Justification of this definition is a result of Lima-Filho that $Z_r(U)$ is independent (i.e., the isomorphism class of $\widetilde{Z}_r(U)$) of the choice of projective closure X of U and of the embedding $X \subset \mathbf{P}^N$ (cf. [Li1], [F-G, 2.2]).

The following localization theorem (due to Lima-Filho and Friedlander–Gabber) provides a useful computational tool.

Theorem 1.8. ([Li1], [F-G]) *Let X be a projective algebraic variety with closed subvariety $X_\infty \subset X$ and let $U = X - X_\infty$. Then*

$$Z_r(X_\infty) \to Z_r(X) \to Z_r(U)$$

is a fibration sequence; in other words, we have the following natural distinguished triangle in $\mathcal{D}^+(Ab)$:

$$\widetilde{Z}_r(X_\infty) \to \widetilde{Z}_r(X) \to \widetilde{Z}_r(U) \to \widetilde{Z}_r(X_\infty)[1].$$

We next consider operations on Lawson homology introduced by the author and Barry Mazur in [FM-1]. These operations are appropriate to discuss in a meeting on intersection theory, for their definition is essentially intersection-theoretic.

Consider two projective spaces $\mathbf{P}^m$ and $\mathbf{P}^n$ linearly embedded in $\mathbf{P}^{m+n+1}$ with disjont images. Then we may view $\mathbf{P}^{m+n+1}$ as the union of all lines from points on $\mathbf{P}^m$ to points on $\mathbf{P}^n$ and we write

$$\mathbf{P}^{n+m+1} = \mathbf{P}^m \# \mathbf{P}^n.$$

For any closed subvarieties $X \subset \mathbf{P}^m, Y \subset \mathbf{P}^n$, we define the algebraic join

$$X\#Y \subset \mathbf{P}^m \# \mathbf{P}^n = \mathbf{P}^{n+m+1}$$

as the union of all lines from points of X to points of Y. Taking $n = 0$ and $Y = x_\infty$, we obtain the algebraic suspension discussed earlier. If X is given by the set of homogeneous equations $\{F_i(S_0, \dots, S_m)\}$ and Y by the set of homogeneous equations $\{G_j(T_0, \dots, T_n)\}$, then $X\#Y$ is given by the union of these two sets of homogeneous equations viewed as functions in $S_0, \dots, S_m, T_0, \dots, T_n$. Observe that

$$X\#P^n = \Sigma^{n+1} X \subset \mathbf{P}^{m+n+1}.$$

This join construction extends by linearity to a continuous bilinear pairing of topological groups

$$\# : Z_r(X) \times Z_s(Y) \;\rightarrow\; Z_{r+s}(X\#Y).$$

Definition 1.9. ([FM-1]) Let $X \subset \mathbf{P}^N$ be a closed subvariety and let $r > 0$. We consider the chain of continuous group homomorphisms

$$Z_r(X) \times Z_0(\mathbf{P}^1) \xrightarrow{\#} Z_{r+1}(X\#\mathbf{P}^1) \xleftarrow{\Sigma^2} Z_{r-1}(X), \tag{1.9.1}$$

where the first is given by join of cycles and the second is two-fold algebraic suspension (or join with $x_\infty \in \mathbf{P}^1$). We define

$$h : Z_r(X) \to Z_{r-1}(X), \quad h : \widetilde{Z}_r(X) \to \widetilde{Z}_{r-1}(X)$$

to be the map (i.e., isomorphism class of maps in $\mathcal{D}^+(Ab)$) obtained by restricting the first map of (1.9.1) to $Z_r(X) \times \{x_\infty\}$ and composing this restriction with the homotopy inverse of Σ^2.

Consider the map $\mathbf{P}^1 \to Z_0(\mathbf{P}^1)$ given by sending a point x to $x - x_\infty$. We define

$$s : Z_r(X) \to \Omega^2 Z_{r-1}(X), \quad s : \widetilde{Z}_r(X) \to \widetilde{Z}_{r-1}(X)[-2]$$

as the map induced by adjunction by the composition of

$$Z_r(X) \times \mathbf{P}^1 \to Z_r(X) \times Z_0(\mathbf{P}^1) \xrightarrow{\#} Z_r(X \times \mathbf{P}^1)$$

and the homotopy inverse of Σ^2.

We view the operation h on $Z_r(X)$ as taking an r-cycle on X and intersecting it with a fixed hyperplane. Indeed, this operation does "cover" the operation in homology given by sending the homology class of a cycle (as defined below) to its intersection with the class of a hyperplane. Thus h evidently depends on the choice of embedding $X \subset \mathbf{P}^N$.

What is at first disturbing is that the operation h enables us to "continuously intersect" all r-cycles on X with a fixed hyperplane, whereas if r is less than the dimension of X then any hyperplane contains some r-cycles of X so that h cannot possibly be represented on such cycles as the intersection with the given hyperplane. Since the construction of h requires a choice of homotopy inverse for Σ, we see that h is not canonically defined. Indeed, if one fixes a given hyperplane L, then a representative for the homotopy class of h can be chosen so that on those effective r-cycles intersecting L properly this representative takes an effective cycle Z to $Z \bullet L$.

The operation s is more interesting. We view this geometrically as associating to any r-cycle on X the family of $r-1$-cycles on X parametrized by $\mathbf{P}^1$ given by intersecting the cycle with a fixed Lefschetz pencil of hyperplanes. Once again, we see that this can only be literally correct if we fix the pencil of hyperplanes and consider effective cycles which intersect properly each member of the pencil.

The following proposition mentions some of the good properties of the s operation.

Proposition 1.10. ([F-G], [F2]) *Let X be a projective algebraic variety with closed subvariety $X_\infty \subset X$ and let $U = X - X_\infty$.*

(a) *The s operation on $Z_*(X)$ restricts to an operation on $Z_*(X_\infty)$ and thereby determines operations*

$$s : Z_r(U) \to \Omega\, Z_{r-1}(U), \quad \widetilde{Z}_r(U) \to \widetilde{Z}_{r-1}(U)[-2].$$

(b) *The s operation can be represented as the composition of the map $Z_r(U) \times Z_0(\mathbf{P}^1) \to Z_r(U \times \mathbf{P}^1)$ sending an r-cycle ζ on U and a 0-cycle τ on $\mathbf{P}^1$ to $\zeta \times \tau$ and the map $Z_r(U \times \mathbf{P}^1) \to Z_{r-1}(U)$ given by intersection with $U \times x_\infty$. In particular, s is independent of the choice of projective closure $U \subset X$ and projective embedding $X \subset \mathbf{P}^N$.*

(c) *The map*

$$s^r : Z_r(U) \to \pi_0(Z_r(U)) \to \pi_{2r}(Z_0(U))$$

sends an algebraic equivalence class of r-cycles to its fundamental homology class in integral Borel–Moore homology $H_{2r}^{BM}(U) = H_{2r}(X, X_\infty)$.

Remark 1.11. Proposition 1.10 enables us to conclude that $s^r : Z_r(U) \to \Omega^r Z_0(U)$ determines a natural map (independent of projective closure $U \subset X$ and projective embedding $X \subset \mathbf{P}^N$)

$$L_r H_{2r+i}(U) \;\; \to \;\; H_{2r+i}^{BM}(U).$$

Using the s operation, Barry Mazur and the author introduced filtrations on singular homology and on algebraic cycles. The "topological filtration on homology" is defined by

$$T_i H_n(X) \equiv \mathrm{im}\{s^i : L_i H_n(X) \to H_n(X)\}$$

and investigated in [FM-1], [FM-2]. In particular, this topological filtration was shown to equal the "correspondence filtration" which is formulated in purely algebro-geometric terms in terms of images of homological correspondences. The S-filtration on cycles considered in [F4], [F5] is given by

$$S^j Z_r(X) \equiv \ker\{s^j : Z_r(X) \to \pi_0(Z_r(X)) \to L_{r-j}H_{2r}(X)\}.$$

This filtration can also be described in purely algebro-geometric terms in terms of images under correspondences of cycles homologically equivalent to 0 on smaller-dimensional subvarieties. Questions about these filtrations are directly related to the (generalized) Hodge Conjecture and Grothendieck's standard conjectures. For abelian varieties, the topological filtration on homology is investigated in a recent paper by Salman Abdulali [Ab]. A recent paper of C. Peters [P] investigates the stability of the S-filtration on cycles in certain cases.

2 Intersection pairings

In this section, we discuss various aspects of the our joint paper with Ofer Gabber [F-G]. The reader familiar with intersection theory as developed in William Fulton's book [Fu] will find that we are presenting some aspects of that theory in such a way that we treat all cycles of a given dimension at the same time. In particular, an intersection pairing

$$Z_r(U) \times Z_s(U) \to Z_{r+s-n}(U), \quad n = \dim(U)$$

is constructed for any smooth quasi-projective variety by defining Gysin maps for regular immersions and considering the Gysin map $\Delta^! : Z_{r+s}(U \times U) \to Z_{r+s-n}(U)$. We also discuss another approach to intersecting spaces of cycles based on the Moving Lemma developed by the author and Blaine Lawson [FL-3]. This approach has the advantage that it applies over an arbitrary field, providing interesting structure to motivic cohomology (cf. [F-V]).

As usual, all varieties discussed are quasi-projective complex algebraic varieties unless explicit mention to the contrary.

We begin with the following "homotopy invariance" property.

Proposition 2.1. ([F-G, 2.3]) *Let E be a locally free rank e sheaf on a quasi-projective variety U and let*

$$\pi : \mathbf{V}(E) = \mathrm{SpecSym}_{\mathcal{O}_X} E^* \to U$$

be the associated algebraic vector bundle. Then

$$\pi^* : Z_r(U) \to Z_{r+e}(\mathbf{V}(E))$$

is a homotopy equivalence, where π^* denotes flat pull-back of cycles.

The proof of this important property is surprisingly easy. Using the localization property (Theorem 1.8), we reduce to the case in which the bundle is trivial. We can similarly localize the context of the Lawson suspension theorem (trivializing the line bundle $\mathcal{O}_X(1)$ determining ΣX), thereby obtaining the desired homotopy equivalence.

Proposition 2.1 enables the following natural definition of the operational first Chern class of a line bundle.

Definition 2.2. Let L be a line bundle on the quasi-projective variety U. For any $r \geq 0$, we define

$$c_1(L) : Z_{r+1}(U) \to Z_r(U)$$

as the composition of $Z_{r+1}(U) \to Z_{r+1}(\mathbf{V}(L))$ induced by the 0-section $0 : X \to \mathbf{V}(L)$ followed by a choice of homotopy inverse for $\pi^* : Z_r(U) \to Z_{r+1}(\mathbf{V}(L))$.

Of course, $c_1(L)$ is only well defined up to homotopy; in other words, the isomorphism class of

$$c_1(L) : \widetilde{Z}_{r+1}(U) \to \widetilde{Z}_r(U)$$

is well defined in $\mathcal{D}^+(Ab)$. Observe that $c_1(L)$ is a generalization of the h operation: take U to be the suspension ΣW of some variety W and take L to be $\mathcal{O}_U(1)$.

From our perspective of looking at the entire topological group of algebraic cycles on a variety, it is natural to formulate intersection with the group of all divisors (not simply one divisor at a time). We state such a formulation in the next theorem. Only in the case of divisors (as in Theorems 2.3 and 2.4) can we successfully define intersection product on cycle spaces of a variety which is not necessarily smooth.

Theorem 2.3. ([F-G, 3.1]) *Let X be an irreducible projective variety. Let*

$$Div(X) = \coprod_{\alpha \in NS(X)^+} \mathrm{Div}_\alpha(X)$$

be the disjoint union of projective varieties whose points are effective Cartier divisors on X. Let $\mathrm{Div}(X)^+$ *denote the homotopy-theoretic group completion of* $\mathrm{Div}(X)$, *constructed as the infinite mapping telescope of the self-map given by addition with a chosen very ample divisor. Then there is a natural (in $\mathcal{D}^+(Ab)$) pairing*

$$Z_{r+1}(X) \times \mathrm{Div}(X)^+ \to Z_r(X)$$

satisfying the following properties:

(a) *For any effective Cartier divisor L, $c_1(L)$ is represented by restricting this pairing to $Z_{r+1}(X) \times \{L\}$. In particular, the homotopy class of $c_1(L)$ depends only upon the equivalence class of L in $\pi_0\mathrm{Div}(X)^+ = NS(X)$, the Neron–Severi group of X.*

(b) *There is a fibration sequence*

$$\mathbf{P}^\infty \to \mathrm{Div}(X)^+ \to \mathrm{Pic}(X);$$

in particular, $\pi_2 \mathrm{Div}(X) = \mathbf{Z}$.

(c) *Restricting this pairing to* $Z_{r+1}(X) \times S^2$ *(where* $S^2 = \mathbf{P}^1 \subset \mathbf{P}^\infty \subset$ *Div*$(X)^+$*) determines the s operation*

$$Z_{r+1}(X) \wedge S^2 \quad \to \quad Z_r(X).$$

We next observe that $c_1(L)$ can be refined to take values in spaces of cycles on the support of the Weil divisor associated to the Cartier divisor L. As one might expect, we do not need to assume we are working with smooth varieties when intersecting with (Cartier) divisors.

Theorem 2.4. ([F-G, 2.5]) *Let* D, D' *be effective Cartier divisors on a quasi-projective variety* U *with support on the closed subvarieties* $i_D : |D| \to U$, $i_{D'} : |D'| \to U$. *There exists a Gysin map (well defined up to homotopy, or up to isomorphism in* $\mathcal{D}^+(Ab)$*) of*

$$i_D^! : Z_{r+1}(U) \to Z_r(|D|), \quad i_D^! : \widetilde{Z}_{r+1}(U) \to \widetilde{Z}_r(|D|)$$

satisfying the following properties:

(a) *A representative of* $i_D^!$ *can be chosen whose restriction to effective cycles which meet* $|D|$ *properly is the usual intersection with a divisor.*

(b) *The composition of* $i_D^!$ *with the map induced by* i_D *equals* $c_1(\mathcal{O}(D))$: $Z_{r+1}(U) \to Z_r(U)$.

(c) $i_D! + i_{D'}^! = i_{D+D'}^! : \widetilde{Z}_{r+1}(U) \to \widetilde{Z}_r(|D| \cup |D'|)$.

(d) $i_D^! \circ i_{D'}^! = i_{D'}^! \circ i_D^! : \widetilde{Z}_{r+1}(U) \to \widetilde{Z}_{r-1}(|D| \cap |D'|)$.

(e) $i_D^!$ *depends naturally as a map in* $\mathcal{D}^+(Ab)$ *upon the pair* (U, D).

The proof of Theorem 2.4 utilizes the trivialization of $\mathbf{V}(\mathcal{O}(-D)) \to U$ when restricted to $U - |D|$ plus the localization sequence to construct a lifting of $Z_{r+1}(U) \to Z_{r+1}(\mathbf{V}(\mathcal{O}(-D))$ to $Z_r(|D|) \simeq Z_{r+1}(\mathbf{V}(i_D^* \mathcal{O}(-D))$.

Once one has a good formulation of intersection with a divisor, one can easily conclude the following projective bundle computation. As shown to us by Grothendieck, this enables us to define Chern classes of vector bundles in this theory.

We denote by $p_E : \mathbf{P}(E) = \mathrm{Proj}(\mathrm{Sym}_{\mathcal{O}_U} E^*) \to U$ the projectivization of $\mathbf{V}(E) \to U$; thus if E has rank e, then the fibres of p_E are projective spaces of dimension $e - 1$.

Corollary 2.5. ([F-G, 2.5]) *Let E be a rank e vector bundle over a quasi-projective variety U. Then the map*

$$\sum_{0\leq j<e} c_1(\mathcal{O}_{\mathbf{P}(E)}(1))^j \circ p_E^* : \prod_{0\leq j<e} Z_{r+j}(U) \to Z_{r+e-1}(\mathbf{P}(E))$$

is a homotopy equivalence.

Our next task is to extend the definition of the Gysin maps of Theorem 2.4 to a regular closed embedding $W \subset U$ of codimension greater than 1. As one might expect, the construction of this Gysin map uses the technique of deformation to the normal cone of a regular embedding.

Theorem 2.6. ([F-G, 3.4]) *Let U be a quasi-projective variety and let $i_W : W \to U$ be a regular closed embedding of codimension c. Then there exists a Gysin map (well defined up to homotopy)*

$$i_W^! : Z_{r+c}(U) \to Z_r(W)$$

satisfying the following properties:

(a) *A representative of $i_W^!$ can be chosen whose restriction to effective $r+c$-cycles Z on U which meet W properly is given by*

$$i_W^!(Z) = W \bullet Z,$$

where $\bullet$ denotes the intersection product on the smooth variety U.

(b) *If W is of codimension 1 in U, then $i_W^!$ equals the Gysin map of Theorem 2.4.*

(c) *If $i_V : V \to W$ is a regular embedding of codimension c', then*

$$(i_W \circ i_V)^! = i_V^! \circ i_W^! : Z_{r+c+c'}(U) \to Z_r(V).$$

(d) *If $g : U' \to U$ is flat, then*

$$i_{W\times_X X'}^! \circ g^* = g'^* \circ i_W^!;$$

if $f : X' \to X$ is proper, then

$$i_W^! \circ f_* = f'_* \circ i_{W\times_X X'}^!.$$

If U is smooth, then the diagonal $\Delta : U \to U \times U$ is a regular closed embedding. Thus Theorem 2.6 immediately provides the following intersection pairing on cycle spaces of a smooth variety.

Corollary 2.7. *Let U be a smooth variety of pure dimension n. We define the intersection product on cycle spaces (well defined up to homotopy) as the map*

$$\Delta^! \circ (- \times -) : Z_r(U) \times Z_s(U) \to Z_{r+s}(U \times U) \to Z_{r+s-n}(U)$$

sending a pair (ζ_1, ζ_2) to $\Delta^!(\zeta_1 \times \zeta_2)$. A representative of this map can be chosen so that it sends a pair of effective cycles (ζ_1, ζ_2) of dimensions r, s on U which meet properly to their intesection product $\zeta_1 \bullet \zeta_2$.

We conclude this section with a sketch of a different approach to intersection of cycle spaces following joint work of the author and Blaine Lawson. Here is a formulation of the Moving Lemma of [F3]. We remind the reader that a "continuous algebraic map" from X to Y is a morphism from the weak normalization of X to Y.

Theorem 2.8. *Let X be a projective variety of pure dimension n, let $r, s,$ be nonnegative integers with $r + s \geq n$, and let e be a positive integer. Then there exists a Zariski open neighborhood $\mathcal{O} \subset \mathbf{A}^1$ of $0 \in \mathbf{A}^1$ and a continuous algebraic map*

$$\Psi : \mathcal{C}_s(X) \times \mathcal{O} \to \mathcal{C}_s(X)^2$$

satisfying the following conditions:

(a) *$\psi_0^+ - \psi_0^- : \mathcal{C}_s(X) \to Z_s(X)$ is the natural inclusion.*

(b) *If Z_1, Z_2 are effective cycles on X of degree $\leq e$ of dimension r, s, respectively, then for any $0 \neq t \in \mathcal{O}$ Z_1 intersects $(\psi_t^+ - \psi_t^-)(Z_2)$ properly outside of the singular locus of X.*

Here we have denoted by $\psi_t^+ \times \psi_t^-$ the restriction of Ψ to $\mathcal{C}_s(X) \times \{t\}$ for any $t \in \mathcal{O}$.

In other words, once one bounds the degree of the effective cycles under consideration, Ψ gives a means of uniformly moving via rational moves all effective s-cycles of degree $\leq e$ so that each intersects properly all effective r-cycles of degree $\leq e$ outside of the singular locus of X. For X smooth, this implies that given compact subsets $R \subset Z_r(X), S \subset Z_s(X)$, we can find a rational move of all of $Z_s(X)$ which moves each s-cycle in S to a rationally equivalent cycle meeting properly every r-cycle in R.

In particular, we conclude that for X projective and smooth the intersection product on homotopy groups induced by the pairing of Corollary 2.7 can be realized more geometrically as follows.

Corollary 2.9. *Let X be projective and smooth of pure dimension n, and let $r, s,$ be nonnegative integers with $r + s \geq n$. Then the bilinear map on homotopy groups*

$$L_rH_m(X) \otimes L_sH_p(X) \ \to \ L_{r+s-n}H_{m+p-2n}(X)$$

induced by the pairing $\Delta_X^! \circ (- \times -)$ of Corollary 2.7 can be realized as follows: represent a class $\alpha \in L_r H_m(X)$ by a continuous map $a : S^{m-2r} \to Z_r(X)$ and a class $\beta \in L_s H_p(X)$ by a continuous map $b : S^{p-2s} \to Z_s(X)$; using an appropriate Ψ as in Theorem 2.8, move each s-cycle in the image of b so that it meets properly each r-cycle in the image of a; then choose some $0 \neq t \in \mathcal{O}$ and take the homotopy class of the map sending a point $x \wedge y \in S^{m+p-2r-2s} = S^{m-2r} \wedge S^{p-2s}$ to $a(x) \bullet (\psi_t^+ - \psi_t^-)(y) \in Z_{r+s-n}(X)$.

The reason that Theorem 2.8 does not provide intersection of cycle spaces on a smooth variety U which is not projective is that the moving achieved in Theorem 2.8 on a projective closure X of U might move certain effective s-cycles of degree $\leq e$ supported on $X_\infty = X - U$ to cycles with components meeting U. In other words, what Theorem 2.8 does not permit us to do is move cycles on X with a specified behaviour on $X_\infty \subset X$. A different approach to moving cycles which does permit such conditions on the moving but which applies in a much restricted situation has been given by Andrei Suslin [S].

3 Cocycles and duality

As we saw in the last section, we have a suitable intersection product on cycle spaces when our underlying quasi-projective variety is smooth. We now ask how one might proceed for singular varieties. Although we have no "magic answers", we do see that cocycle spaces admit a cup product pairing without hypotheses of smoothness. With this in mind, we can speculate how one might be able to refine intersection products using the stratification of a variety by singular loci.

In this section we survey "morphic cohomology" as introduced by the author and Blaine Lawson. As we shall see, morphic cohomology has a "join product" which is compatible with cup product in singular cohomology. One interesting feature of morphic cohomology is that it naturally maps (via a cap product with a fundamental class) to Lawson homology and this map is an isomorphism for smooth varieties. Moreover, this duality isomorphism for smooth varieties sends the join product in morphic cohomology to the intersection product in Lawson homology. Thus, in some sense, we may view join product in morphic homology as providing some sort of "intersection theory" on singular varieties.

We briefly mention a possible refinement of this theory which might more fully reflect the geometry of singular varieties. We conclude by mentioning motivic analogues of the theory we have discussed.

As mentioned in Remark 3.6, a somewhat more sophisticated theory which we call "topological cycle cohomology" has been developed (partly in response to a question of Robert Lateveer following the original lecture) which has particularly pleasing properties [F6].

Definition 3.1. An algebraic s-cocycle on a variety U with values in an equidimensional variety Y is a cycle ζ on $U \times Y$ each irreducible component of which meets each $\{u\} \times Y$ in codimension s.

The following proposition immediately implies that such cocycles are contravariant in U for U normal, covariant in Y for Y projective (with a shift in codimension).

Proposition 3.2. ([FL-1, 1.5]) *If U is a normal variety and Y is a projective variety of pure dimension n, then the abelian monoid of effective s-cocycles on U with values in Y can be naturally identified with the monoid of morphisms* $\mathrm{Hom}(U, \mathcal{C}_{n-s}(Y))$.

To find a suitable topology on $Hom(U, \mathcal{C}_{n-s}(Y))$ is somewhat subtle, for we want the topology to be as "algebraic" as possible yet to provide invariants which have good formal properties (cf. [F4] for a discussion of relative merits of other possible topologies).

Definition 3.3. Assume that U is a normal variety of pure dimension m and that Y is a projective variety of pure dimension n. Sending $f : U \to \mathcal{C}_{n-s}(Y)$ to its graph Γ_f determines a natural monomorphism

$$\mathrm{Hom}(U, \mathcal{C}_{n-s}(Y)) \to \mathcal{C}_{m+n-s}(U \times Y).$$

We provide $\mathrm{Hom}(U, \mathcal{C}_{n-s}(Y))$ with the subspace topology for this embedding and define $Z^s(U, Y)$ to be the topological abelian group given as the the naïve group completion of the topological abelian monoid $\mathrm{Hom}(U, \mathcal{C}_{n-s}(Y))$.

Similarly, if U is a normal variety of pure dimension m, then this graph construction determines a natural monomorphism

$$\mathcal{C}^s(U) \equiv \mathrm{Hom}(U, \mathcal{C}_0(\mathbf{P}^s))/\mathrm{Hom}(U, \mathcal{C}_0(\mathbf{P}^{s-1})) \to \mathcal{C}_m(U \times \mathbf{P}^s)/\mathcal{C}(U \times \mathbf{P}^{s-1}).$$

Once again, we provide $\mathcal{C}^s(U)$ with the subspace topology for this embedding and define $Z^s(U)$ to be the topological abelian group given as the the naïve group completion of the topological abelian monoid $\mathcal{C}^s(U)$.

The following proposition assures us that these topological abelian groups are reasonably well behaved from a topologist's point of view.

Proposition 3.4. ([FL-2, C.3], [F4, 1.5, 1.7]) *Let U be a normal equidimensional variety and Y a projective equidimensional variety. The topology on $Hom(U, \mathcal{C}_{n-s}(Y))$ is characterized by the property that a sequence of maps $\{f_i; i \in \mathbf{N}\}$ converges if and only if both* (a) *the sequence converges in the compact open topology and* (b) *for some Zariski locally closed embedding $U \times Y \subset \mathbf{P}^N$, the graphs of the maps f_i have bounded degree. In particular, if $X = U$ is compact, then* $\mathrm{Hom}(X, \mathcal{C}_{n-s}(Y))$ *has the compact-open topology.*

Furthermore, the topological abelian groups $Z^s(U,Y)$ and $Z^s(U)$ have the homotopy type of C-W complexes and are homotopy-theoretic group completions of the topological abelian monoids $\mathcal{C}^s(U,Y), \mathcal{C}^s(U)$.

The algebraic join $\# : C_{r,d}(Y) \times C_{r',d'}(Y') \to C_{r+r'+1,dd'}(Y\#Y')$ considered in section 1 induces an external pairing on cocycles

$$\begin{aligned}&\mathrm{Hom}(U,\mathcal{C}_{n-s}(Y)) \times \mathrm{Hom}(U',\mathcal{C}_{n'-s'}(Y'))\\ &\quad\to \mathrm{Hom}(U \times U',\mathcal{C}_{n-s+n'-s'+1}(Y\#Y')).\end{aligned}$$

Taking U equal to U', $Y = \mathbf{P}^s$, and $Y' = \mathbf{P}^{s'}$, we obtain pairings

$$Z^s(U)\otimes Z^{s'}(U) \to [\mathrm{Hom}(U,\mathcal{C}_1(\mathbf{P}^{s+s'+1}))/\mathrm{Hom}(U,\mathcal{C}_1(\mathbf{P}^{s+s'}))]^+ \simeq Z^{s+s'}(U),$$

where the indicated equivalence is proven using the observation that the equivalence $Z_0(\mathbf{P}^{s+s'}) \to Z_1(\mathbf{P}^{s+s'+1})$ in the Lawson suspension theorem (Theorem 1.4) is determined by retractions which are algebraic.

Definition 3.5. Let U be a normal equidimensional variety. Then the morphic cohomology groups of U are the homotopy groups of $Z^s(U)$, denoted as follows:

$$L^sH^{2s-j}(U) \equiv \pi_j(Z^s(U)).$$

As above, the join pairing induces an associative pairing

$$\# : L^sH^{2s-j}(U) \otimes L^{s'}H^{2s'-j'}(U) \to L^{s+s'}H^{2s+2s'-j-j'}(U).$$

Remark 3.6. In [F6], we introduce a more sophisticated formulation of such a cohomology theory (which we call topological cycle cohomology) which agrees with morphic cohomology for smooth varieties. Topological cycle cohomology is defined for all varieties. The important property that it possesses which fails for the morphic cohomology of normal but not smooth varieties is excision. Topological cycle cohomology in conjunction with Lawson homology constitute a "Poincaré duality theory with supports" as axiomitized by S. Bloch and A. Ogus [B-O].

In Remark 1.11, we observed that there is a natural map from Lawson homology to Borel–Moore homology. We have a similarly natural map from morphic cohomology to (singular) cohomology. Recall that $Z_0(\mathbf{A}^s)$ has the homotopy type of the Eilenberg–Mac Lane space $K(\mathbf{Z}, 2s)$ and thereby represents singular cohomology.

Proposition 3.7. ([FL-1, 6.3]) *Let U be a normal equidimensional variety. The composition of natural homomorphisms of topological monoids*

$$\mathrm{Hom}(U,\mathcal{C}_0(\mathbf{P}^s)) \to \mathrm{Hom}_{\mathrm{cont}}(U, Z_0(\mathbf{P}^s)) \to \mathrm{Hom}_{\mathrm{cont}}(U, Z_0(\mathbf{A}^s))$$

induces a continuous group homomorphism

$$\Phi : Z^s(U) \to \mathrm{Hom}_{\mathrm{cont}}(U, Z_0(\mathbf{A}^s))$$

whose map on homotopy groups in degree j is denoted

$$\Phi^{s,2s-j} : L^sH^{2s-j}(U) \to H^{2s-j}(U).$$

So defined, $\Phi^{*,*}$ is a ring homomorphism, where the product on morphic cohomology is the join pairing of (3.5) *and the product on integral singular cohomology is a cup product.*

As to be expected, algebraic vector bundles have Chern classes with values in morphic cohomology. Namely, if E is a rank e algebraic vector bundle on U generated by $N+1$ global sections, then E determines

$$U \to Grass^e(\mathbf{P}^N) \to \mathcal{C}_{N-e}(\mathbf{P}^N).$$

If we define the total Segre class (equal to the inverse of the total Chern class) of E to be the class of this map

$$s(E) \in \pi_0 Z^e(U, \mathbf{P}^N) \simeq \oplus_{i=0}^N L^i H^{2i}(U).$$

Mark Walker and the author have recently proved the desired Whitney sum formula for these Chern and Segre classes [FW-1], sharpening the original verification of the author and Blaine Lawson in [FL-1].

We proceed to relate cocycle spaces to cycle spaces.

Definition 3.8. Let U be a normal variety of pure dimension m and Y a projective variety of pure dimension n. Then the duality maps are the continuous homomorphisms

$$\mathcal{D} : Z^s(U,Y) \to Z_{m+n-s}(U \times Y),$$
$$\mathcal{D} : Z^s(U) \to Z_m(U \times A^s) \simeq Z_{m-s}(U).$$

determined by sending $f : U \to \mathcal{C}_{n-s}(Y)$ to its graph Γ_f

The author and Blaine Lawson prove that in the case $U = X$ is projective this duality map is compatible with cap product with the fundamental homology class $[X]$ of X in the sense that the following square commutes:

$$\begin{array}{ccc} L^sH^{2s-j}(X) & \xrightarrow{\mathcal{D}_*} & L_{m-s}H_{2m-2s+j}(X) \\ \Phi\downarrow & & \downarrow\Phi \\ H^{2s-j}(X) & \xrightarrow{\cap[X]} & H_{2m-2s+j}(X). \end{array}$$

A proof of this can be found in the original duality paper [FL-2], whereas a more "geometric" proof is presented in [FL-4].

The following duality theorem was proved by Blaine Lawson and the author for smooth projective varieties in [FL-2] and for general smooth varieties by the author in [F4]. The proof is an application of Theorem 2.8

that shows that a family of cycles parametrized by a sphere or the product of a sphere and an interval can be moved to a new family consisting of cocycles (i.e., each member of the new family intersects properly each fibre of the projection $X \times Y \to X$).

Theorem 3.9. *Let U be a smooth variety of dimension m and let Y be a smooth projective variety of dimension n. Then the duality maps*

$$\mathcal{D} : Z^s(U, Y) \to Z_{m+n-s}(U \times Y),$$
$$\mathcal{D} : Z^s(U) \to Z_m(X \times \mathbf{A}^s) \simeq Z_{m-s}(X)$$

are homotopy equivalences, inducing isomorphisms from morphic cohomology to Lawson homology groups

$$L^s H^{2s-j}(U, Y) \simeq L_{2m+2n-2s+j}(U \times Y),$$
$$L^s H^{2s-j}(U) \simeq L_{2m-2s+j}(U).$$

To complete the compatibility picture of join product in morphic cohomology and intersection product in Lawson homology, we state the following.

Proposition 3.10. ([FW-1], 3.5) *If U is a smooth connected variety of dimension m, then the follow square of topological abelian groups commutes up to homotopy for any s, t with $s + t \leq m$:*

$$\begin{array}{ccc} Z^s(U) \times Z^t(U) & \xrightarrow{\#} & Z^{s+t}(U) \\ \mathcal{D}\downarrow & & \downarrow\mathcal{D} \\ Z_{m-s}(U) \times Z_{m-t}(U) & \xrightarrow{\bullet} & Z_{m-s-t}(U). \end{array}$$

Let us briefly speculate how one might find intermediate theories analogous to intersection homology groups in our context. A different approach by Pawel Gajer can be found in [G]. Our idea is to relax the condition on a cocycle on X with values in Y that it meet each fibre $\{x\} \times Y$ properly. Of course, if we impose no condition on this intersection, we get arbitrary cycles which determine our (Lawson) homology theory.

Challenge. Define a "middle perversity Lawson homology theory" for normal varieties X for which there is an intersection pairing.

For example, one might try to proceed as follows. Let X be a normal variety of dimension m and let

$$X \supset X^{(1)} \supset X^{(2)} \supset \cdots$$

be a stratification of X with $X^{(i)} - X^{(i+1)}$ smooth of codimension i in X. Consider the subgroup

$$P^s(X) \subset Z_{2m-s}(X \times \mathbf{A}^m)$$

consisting of $(2m-s)$-cycles ζ on $X \times \mathbf{A}^m$ with the property that $|\zeta| \cap (X^{(i)} \times \mathbf{A}^m) \subset X^{(i)} \times \mathbf{A}^m$ has codimension greater or equal to $s + i/2$, where $|\zeta|$ denotes the support of the cycle ζ. The challenge is to provide a singular moving lemma which would enable us to define a pairing from $P^s(X) \times P^t(X)$ to $P^{s+t}(X)$.

We conclude our paper by mentioning that the constructions we have presented have "motivic" analogues. For these analogues, the role of algebraic equivalence of cycles is replaced by rational equivalence. One important feature of these analogues is that they can be formulated for varieties over an arbitrary field k.

Once one is familiar with Lawson homology and morphic cohomology, then the "naïve" motivic versions are not difficult to formulate. We utilize Chow varieties as before. Rather than impose the analytic topology and view cycle spaces as topological abelian groups, we consider the "Suslin complex" of the algebraically defined monoid of effective cycles and group complete the resulting simplicial abelian monoid. This approach was briefly mentioned by Ofer Gabber and the author in [F-G], then taken somewhat more seriously in [F2]. For example, a naïve formulation of motivic cohomology $H^*(U, \mathbf{Z}(s))$ of a normal variety U over a field k is given as the cohomology of the chain complexes associated to the simplicial abelian groups

$$A^s(U) = [\frac{\mathrm{Hom}(\Delta^\bullet \times U, \mathcal{C}_0(\mathbf{P}^s))}{\mathrm{Hom}(\Delta^\bullet \times U, \mathcal{C}_0(\mathbf{P}^{s-1}))}]^+$$

where $\Delta^\bullet = \{n \mapsto \Delta^n\}$ is the cosimplicial variety with $\Delta^n = \mathrm{Spec} k[x_0, \ldots, x_n]/\Sigma x_i = 1$.

The reader interested in such constructions is invited to consult [F-V] where Vladimir Voevodsky and the author formulate motivic homology and cohomology using this point of view and prove a duality theorem analogous to Theorem 3.9. The usefulness of this latter duality theorem can be seen in the proof given in [F-V] that for smooth varieties motivic cohomology groups are isomorphic to the higher Chow groups of Spencer Bloch.

To prove various properties of motivic cohomology and homology, one needs to follow the sophisticated point of view of Andrei Suslin and Vladimir Voevodsky (cf. [F7]). The reader somewhat familiar with the constructions in these notes might find that [F6] serves as an introduction to some of the derived category formalism used by Suslin and Voevodsky.

References

[Ab] S. Abdulali, Filtrations on the cohomology of abelian varieties, preprint

[A] F.J. Almgren, Jr., Homotopy groups of the integral cyclic groups, *Topology* **1** (1962), 257–299

[B] D. Barlet, Espace analytique réduit des cycles analytiques complexes compacts d'un espace analytique complexe de dimension finite, in *Fonctions de Plusieurs Variables* II, Lecture Notes in Math. 482, Spring-Verlag, Berlin, New York, Heidelberg, 1975, pp. 1–158

[B-O] S. Bloch and A. Ogus, Gersten's conjecture and the homology of schemes, *Ann. École Norm. Sup.* **7** (1974), 181–202

[C-W] W.L. Chow and B.L. van der Waerden, Zur algebraisches Geometrie IX, *Math. Ann.* **113** (1937), 692–704

[D-T] A. Dold and R. Thom, Quasifaesrungen und unendliche symmetrische produckte, *Ann. Math.* **67** (1958), 239–281

[F1] E. Friedlander, Algebraic cocycles, Chow varieties, and Lawson homology, *Compositio Math.* **77** (1991), 55–93

[F2] E. Friedlander, Some computations of algebraic cycle homology, *K-Theory* **8** (1994), 271–285

[F3] E. Friedlander, Filtrations on algebraic cycles and homology, *Ann. École Norm. Sup.* **28** (1995), 317–343

[F4] E. Friedlander, Algebraic cocycles on normal, quasi-projective varieties, *Compositio Math.* **110** (1998), 127–162

[F5] E. Friedlander, Relative Chow correspondences and the Griffiths group, preprint

[F6] E. Friedlander, Bloch–Ogus properties for topological cycle theory, to appear in *Ann. Ec. Norm. Sup.*

[F7] E. Friedlander, Motivic complexes of Suslin and Voevodsky, *Sém. Bourbaki*, *Astérisque* **245** (1997), 355–378

[F-G] E. Friedlander and O. Gabber, Cycles spaces and intersection theory, *Topol. Methods Mod. Math.* 1993, 325–370

[FL-1] E. Friedlander and H.B. Lawson, A theory of algebraic cocycles, *Ann. Math.* **136** (1992), 361–428

[FL-2] E. Friedlander and H.B. Lawson, Duality relating spaces of algebraic cocycles and cycles *Topology* **36** (1997), 533–565

[FL-3] E. Friedlander and H.B. Lawson, Moving algebraic cycles of bounded degree, *Invent. Math.* **132** (1998), 91–119

[FL-4] E. Friedlander and H.B. Lawson, Graph mappings and Poincaré duality, preprint

[FM-1] E. Friedlander and B. Mazur, *Filtration on the Homology of Algebraic Varieties*, Mem. Amer. Math. Soc. 529, American Mathematical Society, Providence, RI, 1994

[FM-2] E. Friedlander and B. Mazur, Correspondence homomorphisms for singular varieties, *Ann. Inst. Fourier* **44** (1994), 702–727

[F-V] E. Friedlander and V. Voevodsky, Bivariant cycle cohomology, in *Cycles, Transfers, and Motivic Homology Theories*, V. Voevodsky, A. Suslin, and E. Friedlander, eds., Ann. Math. Stud., Princeton University Press, Princeton, NJ

[FW-1] E. Friedlander and M. Walker, Function spaces and continuous algebraic pairings of spaces. Preprint

[FW-2] E. Friedlander and M. Walker, Semi-topological K-theory using function complexes. Preprint

[Fu] W. Fulton, *Intersection Theory*, Ergeb. Math., Springer-Verlag, Berlin, New York, Heidelberg, 1984

[G] P. Gajer, Intersection Lawson homology, *Trans. Amer. Math. Soc.* **349** (1997), 1527–1550

[L1] H.B. Lawson, Algebraic cycles and homotopy theory, *Ann. Math.* **129** (1989), 253–291

[L2] H.B. Lawson, Spaces of algebraic cycles, in *Surveys in Differential Geometry*, Vol. 2, International Press, Somerville, MA, 1995, pp. 137–213

[Li1] P. Lima-Filho, Lawson homology for quasi-projective varieties, *Compositio Math.* **84** (1992), 1–23

[Li2] P. Lima-Filho, Completions and fibrations for topological monoids, *Trans. Amer. Math. Soc.* **340** (1993), 127–147

[P] C. Peters, Lawson homology for varieties with small Chow groups and the induced filtration on the Griffiths group

[S] A. Suslin, Higher Chow groups and etale cohomology, in *Cycles, Transfers, and Motivic Homology Theories*, V. Voevodsky, A. Suslin, and E. Friedlander, Ann. Math. Stud., Princeton University Press, Princeton, NJ

[S-V] A. Suslin and V. Voevodsky, Chow sheaves, in *Cycles, Transfers, and Motivic Homology Theories*, V. Voevodsky, A. Suslin, and E. Friedlander, Ann. Math. Stud., Princeton University Press, Princeton, NJ

Department of Mathematics
Northwestern University
Evanston, IL 60208-2730
eric@math.nwu.edu

Schur Q-Functions and Degeneracy Locus Formulas for Morphisms with Symmetries

Alain Lascoux
Piotr Pragacz

To Maria

ABSTRACT We give closed-form formulas for the fundamental classes of degeneracy loci associated with vector bundle maps given locally by (not necessary square) matrices which are symmetric (resp. skew-symmetric) w.r.t. the main diagonal. Our description uses essentially Schur Q-polynomials of a bundle and is based on a push-forward formula for these polynomials in a Grassmann bundle, established in [P4].

"Something which is not testable is not scientific."
Popper's criterium

Introduction

The goal of the present paper is to state and prove new closed-form formulas for the fundamental classes of some degeneracy loci.

We will be here interested in degeneracy loci associated with vector bundle maps given locally by (not necessary square) matrices that are symmetric (resp. skew-symmetric) w.r.t. the main diagonal. To be more precise, let $\alpha : E \twoheadrightarrow F$ be a surjection of two vector bundles of respective ranks e and f on a variety X. We denote by $E \vee F$ (resp. $E \wedge F$) the kernel of the surjection

$$E \otimes F \xrightarrow{\alpha\otimes 1} F \otimes F \twoheadrightarrow \wedge^2 F$$

This research was supported by grant No. 5031 of French–Polish cooperation C.N.R.S.–P.A.N., and KBN grant No. 2P03A 05112.
AMS Mathematics Subject Classification: 14M12, 14C17, 14M15.

(resp.

$$E \otimes F \xrightarrow{\alpha \otimes 1} F \otimes F \twoheadrightarrow S^2F).$$

In the present paper, we will call a morphism $\varphi : E^* \to F$ *symmetric* (resp. *skew-symmetric*) provided it is induced by a section of the subbundle $E \vee F$ (resp. $E \wedge F$) of $E \otimes F$. (Note that for $E = F$ these notions coincide with the usual notions of "symmetric" and "skew-symmetric" morphisms.)

Our goal is to describe the fundamental classes of the loci

$$D_r(\varphi) = \{x \in X : \text{rank } \varphi(x) \leqslant r\}$$

with the help of some explicitly given polynomials in the Chern classes of E and F. Set $n = e - f$. More precisely, using the terminology of [F-P], the degeneracy locus $D_r(\varphi)$ is represented, in the symmetric case, by the polynomial

$$\sum_I Q_{(f-r,f-r-1,\dots,1)+I}(F) \cdot s_{\complement \widetilde{I}}(E - F), \tag{$*$}$$

where the sum is over all partitions I in $(n)^{f-r}$, and for the conjugate partition $\widetilde{I} = (\widetilde{i}_1, \dots, \widetilde{i}_n) \subset (f-r)^n$ we write $\complement \widetilde{I} = (f - r - \widetilde{i}_n, \dots, f - r - \widetilde{i}_1)$ for the "complement" of $\widetilde{I}$ in $(f-r)^n$. Here, $Q_J(\bullet)$ are Schur Q-polynomials whose definition is recalled in Section 1. The connection of Schur Q-polynomials to geometry was originally established in [P2] in two contexts. Firstly, in [P2, Sect. 7], the ideal of polynomials universally supported on the rth symmetric degeneracy locus was described in terms of Schur Q-polynomials. This is closely connected with the present paper, see below. Secondly, in [P2, (8.7)] the connection of Schur Q-functions to the cohomology rings of isotropic Grassmannians was given. (This subject was then developed in [P3, Sect. 6].) Consult also [P4] and [F-P] for more about geometric properties of Q-polynomials, and especially for their applications to intersection theory and enumerative geometry.

By virtue of [P2, Sect. 7], the form of polynomials (*) representing $[D_r(\varphi)]$ is not at all surprising. Observe that for $\varphi' = \varphi \circ \alpha^* : F^* \to E^* \to F$, we have $D_r(\varphi) \subset D_r(\varphi')$, so the polynomials representing $D_r(\varphi)$ are universally supported on the rth (symmetric) degeneracy locus associated with symmetric bundle maps from rank f vector bundles to their duals, in the sense of [P2, Sect. 7] (see also [F-P, Sect. 4.4]). Therefore, by [P2, Th.7.2 and Prop. 7.17], the polynomial representing $[D_r(\varphi)]$ is a $\mathbb{Z}[c.(E), c.(F)]$-combination of $Q_{(f-r,f-r-1,\dots,1)+I}(F)$ for I contained in $(r)^{f-r}$. For $n \leqslant r$, this is exactly the combination (*). For $n > r$, this suggests that at the cost of replacing in (*) the coefficients from $\mathbb{Z}[c.(E-F)]$ by the coefficients from a larger ring $\mathbb{Z}[c.(E), c.(F)]$, one can make the sum (*) smaller. We also note that to get (*), we use exactly the same instance of the Gysin map formula, recalled in (1.8) below, as to get the just-mentioned results from [P2, Sect. 7].

Similar formulas and discussion hold true for the degeneracy loci associated with skew-symmetric morphisms; see Section 3.

When $E = F$, our formulas specialize to the ones given by Józefiak and the authors in [J-L-P], and Harris and Tu in [H-T] (see also [P1]). Recall that all the three last-mentioned papers gave a "modern treatment" à la "Thom-Porteous" of the formulas for the degree of determinantal varieties in the spaces of symmetric and skew-symmetric matrices of forms, given classically by Giambelli in [G]. Some special cases of the formulas from [H-T], [J-L-P] and [P1] were given independently by Barth [Ba], Damon and Tyurin.

When $E = \bigoplus_{i=1}^{e} \mathcal{O}(p_i)$, $F = \bigoplus_{i=1}^{f} \mathcal{O}(p_i)$ are vector bundles over a projective space, our formulas give the degrees of determinantal varieties defined by matrices of forms, satisfying the above symmetry conditions. This problem was considered, in the symmetric case, by Bottasso in [Bo], generalizing Giambelli's study to the case of not necessarily square matrices. Bottaso did not give closed-form formulas for these degrees, but established some recursions for their computations. The present paper offers a modern treatment and closed-form version of [Bo].

There are several motivations to do such computations in intersection theory. These computations combine intersection theory on homogeneous spaces (notably Grassmannians and flag varieties) and algebra of different types of symmetric functions. The obtained formulas have applications to the enumerative theory of singularities, Brill–Noether theory etc. Again, we refer the reader to [P4] and to [F-P] for more information on these matters.

We give in the present paper formulas for the "Thom-Porteous" case. It would be interesting to generalize them to the "quiver variety" case, similarly to the work of Buch and Fulton [B-F].

Since the degeneracy loci studied here are intimately connected with some Schubert varieties in symplectic and orthogonal Grassmannians (see Section 1), the formulas given in the present paper should be gotten "in principle" from the divided difference operator approach developed in [F2-F3] from one side, and in [P-R] and [L-P] from another. We do not see, however, how to do it conceptually using divided differences.

The article is organized as follows.

In Section 1, we recall the definitions and properties of two families of symmetric polynomials that we need in the present paper: Schur polynomials in a difference of bundles and Schur Q-polynomials. We also recall a formula for Gysin push-forward of Schur Q-polynomials in a Grassmann bundle from [P4], which is basic to the present paper. Usually, we do not state results in their full generality, but only in the range needed in this paper. Moreover, we give some preliminary properties and examples of the degeneracy loci associated with symmetric and skew-symmetric morphisms.

In Section 2, we give some formulas for the top Chern classes of the bundles $E \vee F$ and $E \wedge F$, generalizing the formulas from [L].

In Section 3, we prove the main formula of the present paper. The proof follows a well-known pattern: it uses essentially a certain desingularization of $D_r(\varphi)$ for a "universal" φ and the above-mentioned push-forward formula for Schur Q-polynomials. Some examples are also discussed.

In Section 4, we discuss some variations of computation of the polynomials representing $D_r(\varphi)$, based on the technique of "constructions with a nontrivial generic fibre" presented in [P1, Sect. 2]. This material is recalled in Proposition 4.1, where, in fact, a certain straightening of this last-mentioned method is presented. As an application, we get some algebraic equations involving Gysin maps on their LHS's and some closed-form expressions involving symmetric polynomials on their RHS's.

1 Preliminaries: recollection of some definitions and results

In the present paper, we will need two families of symmetric polynomials. We recall now their definitions and needed properties. Let $A = (a_1, \dots, a_n)$ and $B = (b_1, \dots, b_m)$ be two sequences of commuting elements in a ring.

1.1 Let $I = (i_1, \dots, i_k)$ be a partition. Recall that by a *partition* (of some natural number) we understand a sequence of integers $I = (i_1, \dots, i_k)$, where $i_1 \geqslant i_2 \geqslant \dots \geqslant i_k \geqslant 0$ and the sum $\sum i_p$ is the partitioned number. We define

$$s_I(A - B) = \det\Big[s_{i_p - p + q}(A - B)\Big]_{1 \leqslant p,q \leqslant k},$$

where $s_i(A - B)$ is defined by

$$\sum_{i=-\infty}^{\infty} s_i(A - B) = \prod_{p=1}^{n}(1 - a_p)^{-1} \prod_{q=1}^{m}(1 - b_q).$$

Moreover, we put $s_I(A) = s_I(A - B)$ for $B = (0, \dots, 0)$. When A and B are sequences of variables, then the polynomials $s_I(A)$ are called *Schur polynomials* or *S-polynomials* and the polynomials $s_I(A-B)$ are often called *Schur polynomials in a difference of alphabets.* For example, the "resultant" can be expressed in a closed form as

$$\prod_{i,j}(a_i - b_j) = s_{(m,\dots,m)}(A - B), \tag{1.2}$$

where the part m appears n times.

1.3 Let $Q_i(A)$ be defined by the expansion

$$\sum_{i=-\infty}^{\infty} Q_i(A) = \prod_{p=1}^{n}(1+a_p)(1-a_p)^{-1}.$$

Given positive integers $i > j$, we set

$$Q_{(i,j)}(A) = Q_i(A) \cdot Q_j(A) + 2\sum_{p=1}^{j}(-1)^p Q_{i+p}(A) \cdot Q_{j-p}(A).$$

Finally, if $I = (i_1 > \ldots > i_k > 0)$ is a *strict* partition, then for odd k we put

$$Q_I(A) = \sum_{p=1}^{k}(-1)^{p-1} Q_{i_p}(A) \cdot Q_{(i_1,\ldots,i_{p-1},i_{p+1},\ldots,i_k)}(A),$$

and for even k,

$$Q_I(A) = \sum_{p=2}^{k}(-1)^p Q_{(i_1,i_p)}(A) \cdot Q_{(i_2,\ldots,i_{p-1},i_{p+1},\ldots,i_k)}(A).$$

These define the $Q_I(A)$'s by recurrence on k. When A is a sequence of variables, then the polynomials $Q_I(A)$ are called *Schur Q-polynomials* after [Sch]. For example, one has

$$Q_{(n,n-1,\ldots,1)}(A) = \prod_{i \leqslant j}(a_i + a_j) = 2^n s_{(n,n-1,\ldots,1)}(A). \tag{1.4}$$

1.5 Let E and F be two vector bundles. Then $s_I(E-F)$ is defined to be $s_I(A-B)$, where A and B are the sequences of the Chern roots of E and F respectively. Similarly, set $s_I(E) = s_I(A)$ and $Q_I(E) = Q_I(A)$.

1.6 For partitions I and J, we write $I \supset J$ if $i_1 \geqslant j_1$, $i_2 \geqslant j_2$, ... Moreover, we denote by $I + J$ the partition $(i_1 + j_1, i_2 + j_2, \ldots)$.

The "rectangular" partition $(i, \ldots, i)$ (r-times) is denoted by $(i)^r$ and the "triangular" partition $(k, k-1, \ldots, 2, 1)$ is denoted by ρ_k.

For a given partition I, $l(I) = \operatorname{card}\{p : i_p > 0\}$ denotes its length and $\widetilde{I}$ denotes the partition conjugate of I, i.e., $\widetilde{I} = (\widetilde{i}_1, \widetilde{i}_2, \ldots)$ where $\widetilde{i}_p = \operatorname{card}\{q : i_q \geqslant p\}$.

As is common, we will often omit brackets writing partitions in lower indices.

1.7 Recall that for every strict partition $I = (i_1 > \cdots > i_k > 0)$ and A a sequence of variables, one has

$$Q_I(A) = 2^k P_I(A)$$

for some polynomial $P_I(A)$ with integer coefficients. These polynomials are called *Schur P-polynomials.* For example,

$$P_{\rho_{n-1}}(A) = \prod_{i<j}(a_i + a_j) = s_{\rho_{n-1}}(A).$$

Given a vector bundle E, we set $P_I(E) = P_I(A)$, where A is specialized to the sequence of the Chern roots of E. For another expression of $P_I(A)$ in the form of a quadratic polynomial in the $s_J(A)$'s, see [L-L-T].

1.8 We recall the following two *factorization formulas.* Let I be a partition such that $l(I) \leqslant n$. Then

$$s_{(m)^n+I}(A - B) = s_{(m)^n}(A - B) \cdot s_I(A) \tag{1.9}$$

and

$$Q_{\rho_{n-1}+I}(A) = Q_{\rho_{n-1}}(A) \cdot s_I(A). \tag{1.10}$$

We refer to [B-R] [L-S] and [St] (see also [M], [P3] and [L-L-T]) for more about these formulas. In particular, the proof of (1.9) given in [L-S] is valid for any elements A and B of ranks n and m in a λ-ring.

1.11 We recall the following formulas for the top Chern classes of some tensor operations. We write, in the present paper, $c_{\mathrm{top}}(E)$ for the top Chern class of a bundle E. Let E and F be two vector bundles of ranks e and f respectively. We have

$$c_{\mathrm{top}}(E \otimes F) = \sum_I s_I(E) \cdot s_{\mathrm{C}\,\widetilde{I}}(F), \tag{1.12}$$

where the sum is over all partitions in $(f)^e$ and for the conjugate partition $\widetilde{I} \subset (e)^f$ we write $\mathrm{C}\,\widetilde{I} = (e - \widetilde{i}_f, \ldots, e - \widetilde{i}_1)$ for the "complement" of $\widetilde{I}$ in $(e)^f$. Also,

$$c_{\mathrm{top}}(S^2E) = 2^e s_{\rho_e}(E) = Q_{\rho_e}(E), \tag{1.13}$$

and

$$c_{\mathrm{top}}(\wedge^2 E) = s_{\rho_{e-1}}(E) = P_{\rho_{e-1}}(E). \tag{1.14}$$

We refer to [L] (see also [M, pp. 47–48 and p. 67]) for more details.

1.15 We recall the following push-forward formula from [P4]. Let $\pi : G = G^q(E) \to X$ be the Grassmann bundle parametrizing q-quotients of a vector bundle E on a variety X. Write $r = e - q$. Let

$$0 \to R \to E_G \to Q \to 0$$

be the tautological sequence on G with $\operatorname{rank} R = r$ and $\operatorname{rank} Q = q$. Let $I = (i_1 > \ldots > i_k > 0)$ be a strict partition with $k \leqslant q$. Then for $\alpha \in A_*(X)$, we have

$$\pi_*\Big[c_{\text{top}}(R \otimes Q) \cdot P_I(Q) \cap \pi^*\alpha\Big] = d \cdot P_I(E) \cap \alpha,$$

where d is zero if $(q-k)r$ is odd, and

$$d = \binom{[(e-k)/2]}{[(q-k)/2]}$$

in the opposite case. Here, the symbol $[\ \bullet\ ]$ means the integer part of a rational number. For a proof, we refer to [P4, App. 1].

We will need three special instances of this formula. First, suppose that I is a strict partition with $l(I) = q$. Then

$$\pi_*\Big[c_{\text{top}}(R \otimes Q) \cdot Q_I(Q) \cap \pi^*\alpha\Big] = Q_I(E) \cap \alpha. \tag{1.16}$$

Secondly, assume that I is a strict partition with $l(I) = q-1$ and r is even. Then

$$\pi_*\Big[c_{\text{top}}(R \otimes Q) \cdot P_I(Q) \cap \pi^*\alpha\Big] = P_I(E) \cap \alpha. \tag{1.17}$$

Thirdly, suppose that I is a strict partition with $l(I) = q-1$ and r is odd. Then

$$\pi_*\Big[c_{\text{top}}(R \otimes Q) \cdot P_I(Q) \cap \pi^*\alpha\Big] = 0. \tag{1.18}$$

1.19 Recall that the Gysin map in a Grassmann bundle admits the following explicit description. Let $(a_1, \ldots, a_q)$ be the sequence of the Chern roots of Q and $(a_{q+1}, \ldots, a_e)$ be the sequence of the Chern roots of R. Then, writing $A = (a_1, \ldots, a_e)$ for the sequence of the Chern roots of E, the Gysin map in question is induced by the following symmetrizing operator. Let S_e be the group of permutations of $(1, \ldots, e)$, S_q the group of permutations of $(1, \ldots, q)$, and S_r the group of permutations of $(q+1, \ldots, e)$. For $P \in \mathbb{Z}[A]^{S_q \times S_r}$, the symmetrizing operator in question acts as follows:

$$P \longmapsto \sum_{\overline{\sigma} \in S_e/S_q \times S_r} \sigma\left(P / \prod_{i \leqslant q < j} (a_i - a_j)\right).$$

For more on this, see, e.g., [P4, Sect. 4].

1.20 We now switch to the setup of the Introduction. In the present paper, to be on the safe side, we assume that the ground field k is algebraically closed of characteristic different from 2. This is because in Sections 1 and 4 we make use of isotropic symplectic and orthogonal Grassmannians. Let $U \twoheadrightarrow V$ be vector spaces of dimensions e and f, respectively. Let X denote the affine space $U \vee V$ (resp. $U \wedge V$). In this situation, there exists a tautological morphism $\varphi : (E = U_X)^* \to (F = V_X)$. For this φ, $D_r(\varphi)$ is the restriction to an appropriate open set of some Schubert variety in the symplectic (resp. orthogonal) Grassmannian of f-dimensional isotropic subspaces in k^{2e}. More precisely, let $\beta : U^* \to V$ be a linear map and $\gamma : V^* \to U$ its dual. Consider $U^* \oplus U$ equipped with canonical nondegenerate bilinear forms

$$\langle (v_1, u_1), (v_2, u_2) \rangle = v_1(u_2) \pm v_2(u_1) \qquad u_i \in U, v_i \in U^*,$$

with the + sign giving a symmetric form and the – sign a skew-symmetric form. The assignment $\beta \mapsto \text{graph}(\gamma)$ embeds $U \vee V$ (resp. $U \wedge V$) as an open subset A of the Grassmannian G of f-dimensional isotropic subspaces of $U^* \oplus U$ w.r.t. the just defined skew-symmetric (resp. symmetric) form on $U^* \oplus U$. Then $D_r(\varphi)$ is the restriction to A of the "determinantal" Schubert subvariety of G parametrizing those f-dimensional isotropic subspaces of $U^* \oplus U$, which intersect the maximal isotropic subspace $U^* \oplus 0$ in dimension $\geqslant f - r$. (When φ is skew-symmetric and $e = f$, we assume r to be even.) Hence $D_r(\varphi)$ is irreducible, normal and Cohen–Macaulay (by results of De Concini and Lakshmibai [DC-L]); moreover its codimension $c(r)$ equals

$$(e-f)(f-r)+(f-r)(f-r+1)/2 \quad (\text{resp. } (e-f)(f-r)+(f-r)(f-r-1)/2).$$

Perhaps the easiest way to remember this number is to set $n = e - f$ and $q = f - r$ (we will keep this notation throughout the rest of the paper), and note that $c(r)$ equals

$$nq+q(q+1)/2 = q(2n+q+1)/2 \quad (\text{resp. } nq+q(q-1)/2 = q(2n+q-1)/2).$$

In general, for $\varphi : E^* \to F$ as in the Introduction, the scheme structure on $D_r(\varphi)$ is defined as follows. Set $\tilde{X} = \text{Spec } S^\bullet(E \vee F)^*$ (resp. $\tilde{X} = \text{Spec } S^\bullet(E \wedge F)^*$). Observe that φ induces a section $s : X \to \tilde{X}$. There exists the tautological bundle homomorphism $\tilde{\varphi} : \tilde{E}^* \to \tilde{F}$ where $\tilde{E} = E_{\tilde{X}}$, $\tilde{F} = F_{\tilde{X}}$ such that $s^*(\tilde{\varphi}) = \varphi$. Then, defining first the scheme structure on $D_r(\tilde{\varphi})$ as above with the help of an appropriate Schubert bundle in an isotropic Grassmann bundle, we define the scheme structure on $D_r(\varphi)$ as the schematic preimage via s of the one on $D_r(\tilde{\varphi})$.

To the best of our knowledge, the above determinantal varieties have not been studied algebraically like the "ordinary" determinantal varieties were studied by Eagon and Hochster or using the Hodge algebra technique. Probably such a study would allow us to formulate the main results of the present paper with less restrictive assumptions on the ground field.

1.21 We pass now to some examples.

1.21.1 Let $\varphi : E^* \to F$ be symmetric, where $e \geqslant f$ are arbitrary and $r = f - 1$. Then the expected codimension of $D_r(\varphi)$ is $e - f + 1$, so we are in the situation of the Giambelli–Thom-Porteous formula for the locus defined by maximal minors. We infer that $D_r(\varphi)$ is represented by $s_{e-f+1}(F - E^*)$. For example, for $e = 4, f = 3$, writing formally by the splitting principle $c(E) = c(F)(1 + d)$, we get

$$\begin{aligned} s_2(F - E^*) &= s_2(F) - s_1(F)s_1(E^*) + s_{1,1}(E^*) \\ &= s_2(F) + s_1(F)s_1(E) + s_{1,1}(E) \\ &= s_2(F) + s_1(F)\big(s_1(F) + d\big) + s_{1,1}(F) + s_1(F)d \\ &= 2\big(s_2(F) + s_{1,1}(F) + s_1(F)d\big) \\ &= 2\big(s_2(F) + s_{1,1}(E)\big) \\ &= 2s_1(E)s_1(F). \end{aligned}$$

For $e = 5, f = 3$, we get the representing polynomial

$$s_3(F - E^*) = 2\big(s_1(E)s_2(F) + s_{1,1,1}(E)\big) = 2\big(s_3(F) + s_{1,1}(E)s_1(F)\big).$$

1.21.2 In this example, X is a projective space, $F = \mathcal{O}(a) \oplus \mathcal{O}(b) \oplus \mathcal{O}(c)$ and $r = 2$.

Assume first that $E = \mathcal{O}(a) \oplus \mathcal{O}(b) \oplus \mathcal{O}(c) \oplus \mathcal{O}(d)$ and $\varphi : E^* \to F$ is given by a matrix of forms

$$\begin{pmatrix} 0 & a_{12} & a_{13} & x \\ -a_{12} & 0 & a_{23} & y \\ -a_{13} & -a_{23} & 0 & z \end{pmatrix}.$$

Then $D_2(\varphi)$ is of codimension 1, defined by

$$a_{12}z - a_{13}y + a_{23}x = 0.$$

The degree of this equation is $a + b + c + d = s_1(E)$.

Assume now that $E = \mathcal{O}(a) \oplus \mathcal{O}(b) \oplus \mathcal{O}(c) \oplus \mathcal{O}(d) \oplus \mathcal{O}(e)$ and $\varphi : E^* \to F$ is given by a matrix of forms

$$\begin{pmatrix} 0 & a_{12} & a_{13} & x & w \\ -a_{12} & 0 & a_{23} & y & v \\ -a_{13} & -a_{23} & 0 & z & u \end{pmatrix}.$$

Then $D_2(\varphi)$ is of codimension 2, defined by

$$a_{12}z - a_{13}y + a_{23}x = 0 \quad \text{and} \quad a_{12}u - a_{13}v + a_{23}w = 0.$$

By Bézout's theorem, the degree of $D_2(\varphi)$ is $(a+b+c+d)(a+b+c+e) = s_2(F) + s_{1,1}(E)$.

1.21.3 Let $\varphi : E^* \to F$ be skew-symmetric, $n = e - f = 1$ and r is an arbitrary even nonnegative number less than f. Assume that φ is given locally by a matrix

$$\begin{pmatrix} 0 & a_{12} & \dots & a_{1f} & b_1 \\ -a_{12} & 0 & \dots & a_{2f} & b_2 \\ \vdots & \vdots & \ddots & \vdots & \vdots \\ -a_{1f} & -a_{2f} & \dots & 0 & b_f \end{pmatrix}.$$

Then $D_r(\varphi)$ is locally defined by the $(r+2)$-Pfaffians of the extended $e \times e$ skew-symmetric matrix

$$\begin{pmatrix} 0 & a_{12} & \dots & a_{1f} & b_1 \\ -a_{12} & 0 & \dots & a_{2f} & b_2 \\ \vdots & \vdots & \ddots & \vdots & \vdots \\ -a_{1f} & -a_{2f} & \dots & 0 & b_f \\ -b_1 & -b_2 & \dots & -b_f & 0 \end{pmatrix}.$$

This is seen by the following "universal local study". Assume that the a_{ij}'s and the b_p's are variables over k. Let $\mathcal{J}$ be the ideal generated by $(r+1)$-minors of the former matrix, and let $\mathcal{I}$ (resp. $\mathcal{P}$) be the ideal generated by $(r+1)$-minors (resp. $(r+2)$-Pfaffians) of the latter matrix. Of course, $\mathcal{J} \subset \mathcal{I}$; moreover, $\mathcal{I} \subset \mathcal{P}$ by [B-E, p. 462]. Since $\mathcal{P}$ is a prime ideal of height equal to the expected codimension $c(r)$ (see [K-L]), the ideal $\mathcal{P}$ defines (locally) the scheme structure on $D_r(\varphi)$. Hence using [G], ..., we see that $D_r(\varphi)$ is represented by $s_{\rho_{e-r-1}}(E)$. For example, for $e = 5, f = 4, r = 2$, the representing polynomial equals $s_{2,1}(E)$.

In (1.21.4–5) we assume that the ambient space X is a projective space.

1.21.4 Suppose that $E = \mathcal{O}(a) \oplus \mathcal{O}(b) \oplus \mathcal{O}(c)$, $F = \mathcal{O}(a) \oplus \mathcal{O}(b)$, and $\varphi : E^* \to F$ is given by a matrix of forms

$$\begin{pmatrix} 0 & a_{12} & x \\ -a_{12} & 0 & y \end{pmatrix}.$$

Then $D_1(\varphi)$ is of codimension 1, defined by the equation $a_{12} = 0$ whose degree is $a + b = s_1(F)$.

More generally, assume that f is even, $E = \mathcal{O}(p_1) \oplus \ldots \oplus \mathcal{O}(p_{f+1})$, $F = \mathcal{O}(p_1) \oplus \ldots \oplus \mathcal{O}(p_f)$, and $\varphi : E^* \to F$ is given by a matrix of forms

$$\begin{pmatrix} 0 & a_{12} & \ldots & a_{1f} & b_1 \\ -a_{12} & 0 & \ldots & a_{2f} & b_2 \\ \vdots & \vdots & \ddots & \vdots & \vdots \\ -a_{1f} & -a_{2f} & \ldots & 0 & b_f \end{pmatrix}.$$

Then $D_{f-1}(\varphi)$ is of codimension 1, defined by the Pfaffian of the $f \times f$ matrix $A = (a_{ij})$. Indeed, every $(f-1)$-minor of A occuring in the Laplace expansion along the last column of an f-minor that contains the $(f+1)$th column, is a multiple of $\mathrm{Pf}(A)$ by the already-quoted result from [B-E]. The degree of $\mathrm{Pf}(A)$ is $p_1 + \ldots + p_f = s_1(F)$.

1.21.5 Assume that $E = \mathcal{O}(a) \oplus \mathcal{O}(b) \oplus \mathcal{O}(c) \oplus \mathcal{O}(d)$, $F = \mathcal{O}(a) \oplus O(b)$, and $\varphi : E^* \to F$ is given by a matrix of forms

$$\begin{pmatrix} 0 & a_{12} & x & z \\ -a_{12} & 0 & y & w \end{pmatrix}.$$

Then $D_1(\varphi)$ is of codimension 2, defined by

$$a_{12} = 0 \qquad \text{and} \qquad xw - zy = 0.$$

By Bézout's theorem the degree of $D_1(\varphi)$ is $(a+b)(a+b+c+d) = s_1(F)s_1(E)$.

2 Formulas for the top Chern classes of $E \vee F$ and $E \wedge F$

We keep the notation from the Introduction and Section 1. In this section, we give two alternative expressions for the top Chern classes of $E \vee F$ and $E \wedge F$. In particular, these expressions give the polynomials representing $D_r(\varphi)$ for $r = 0$.

In the following, we set $K = \mathrm{Ker}(\alpha : E \twoheadrightarrow F)$.

Proposition 2.1. *We have*

$$c_{\mathrm{top}}(E \vee F) = \sum_I Q_{\rho_f + I}(F) \cdot s_{\mathrm{C}\widetilde{I}}(E - F),$$

where the sum is over all partitions in $(n)^f$ and for the conjugate partition $\widetilde{I} \subset (f)^n$ we write $\mathrm{C}\widetilde{I} = (f - \widetilde{i}_n, \ldots, f - \widetilde{i}_1)$. Similarly,

$$c_{\mathrm{top}}(E \wedge F) = \sum_I P_{\rho_{f-1} + I}(F) \cdot s_{\mathrm{C}\widetilde{I}}(E - F),$$

the same sum as above.

Proof. We give here the proof of the proposition for the bundle $E \vee F$, the case $E \wedge F$ being similar. We have in the Grothendieck group $K(X)$,

$$[E \vee F] = [E \otimes F] - [\wedge^2 F] = [S^2 F] + [K \otimes F].$$

Invoking the formulas

$$c_{\text{top}}(S^2 F) = 2^f s_{\rho_f}(F) = Q_{\rho_f}(F),$$

and

$$c_{\text{top}}(K \otimes F) = \sum_I s_I(F) \cdot s_{\complement \tilde{I}}(K),$$

the same sum as above (see (1.12) and (1.13)), the assertion follows by virtue of the factorization formula for Q-polynomials (1.10) and the equality $[K] = [E] - [F]$ in $K(X)$. will see in the next section that the expressions given in the previous proposition generalize, in a natural way, for higher r. The expressions given in the next proposition do not admit such a generalization.

Proposition 2.2. *We have*

$$c_{\text{top}}(E \vee F) = 2^f \sum_I s_{(e,e-1,\dots,n+2,n+1)/I}(F) \cdot s_{\tilde{I}}(E - F),$$

where the sum runs over all partitions $I \subset (e, e-1, \dots, n+2, n+1)$. Similarly,

$$c_{\text{top}}(E \wedge F) = \textstyle\sum_I s_{(e-1,e-2,\dots,n+1,n)/I}(F) \cdot s_{\tilde{I}}(E - F),$$

where the sum runs over all partitions $I \subset (e-1, e-2, \dots, n+1, n)$.

Proof. We give here the proof of the proposition for the bundle $E \vee F$, the case $E \wedge F$ being similar. By the splitting principle, a (more general) question concerning the total Chern class of $E \vee F$ leads to the calculation of the product

$$\prod_{i \leqslant j} (1 + a_i + a_j) \prod_{i,j} (1 + a_i + b_j), \tag{2.3}$$

where, formally,

$$c(F) = \prod_{i=1}^{f} (1 + a_i) \quad \text{and} \quad c(K) = \prod_{j=1}^{n} (1 + b_j).$$

Write $A = (a_1, \dots, a_f)$ and $B = (b_1, \dots, b_n)$. Let $a_i^+ = 1 + 2a_i$, $1 \leqslant i \leqslant f$ and $b_j^+ = 1 + 2b_j$, $1 \leqslant j \leqslant n$. Set $A^+ = (a_1^+, \dots, a_f^+)$ and $B^+ =$

$(b_1^+, \ldots, b_n^+)$. Then

$$\begin{aligned}\prod_{i,j}(1+a_i+b_j) &= 2^{-fn}\prod_{i,j}[(1+2a_i)+(1+2b_j)]\\ &= 2^{-fn}\prod_{i,j}(a_i^+ - zb_j^+)|_{z=-1}\\ &= 2^{-fn}s_{(n)^f}(A^+ - zB^+)|_{z=-1}\end{aligned}$$

where z is a formal λ-ring element of rank 1. Here, we have used the formula (1.2) for the resultant. Moreover, by (1.13),

$$\begin{aligned}\prod_{i\leqslant j}(1+a_i+a_j) = 2^{-\binom{f+1}{2}}\prod_{i\leqslant j}[(1+2a_i)+(1+2a_j)]\\ = 2^{-\binom{f}{2}}\, s_{\rho_f}(A^+).\end{aligned}$$

Thus (2.3) can be rewritten as

$$2^{-N+f}s_{\rho_f}(A^+)\cdot s_{(n)^f}(A^+ - zB^+)|_{z=-1},$$

where $N = \text{rank } E \vee F$. By the factorization formula (1.9) for elements A^+ and zB^+, we can rewrite this last expression as

$$2^{-N+f}s_{(n+f,n+f-1,\ldots,n+1)}(A^+ - zB^+)|_{z=-1}.$$

The top-degree component of this polynomial is

$$2^f s_{(e,e-1,\ldots,n+1)}(A - zB)|_{z=-1} = 2^f\sum_I s_{(e,e-1,\ldots,n+1)/I}(A)\cdot s_{\tilde{I}}(B),$$

the sum over partitions $I \subset (e, e-1, \ldots, n+1)$, as desired. □

3 The main formulas

The most popular method to compute the fundamental class of a subscheme $D \subset X$ tries to find a scheme G mapping properly to X, on which one has a locus Z that maps birationally onto D and for which one can compute its class $[Z]$. Usually this is because $[Z]$ is the zero locus of a section of some bundle whose rank is equal to $\text{codim}_G Z$, so the class $[Z]$ is evaluated to be the top Chern class of the bundle. For example, this pattern was used in [J-L-P] and many other papers (compare [F1] and [F-P]). We will also follow this pattern in the present section.

We follow the notation from the Introduction and Section 1. Recall that $n = e - f$ and $q = f - r$. Let $\pi : G = G^q(F) \to X$ be the Grassmann bundle parametrizing rank q quotients of the bundle F. On G there exists a tautological sequence

$$0 \to R \to F_G \to Q \to 0,$$

where $\operatorname{rank} R = r$ and $\operatorname{rank} Q = q$. The composite morphism

$$E_G^* \xrightarrow{\varphi_G} F_G \twoheadrightarrow Q$$

gives a section of

$$H = \operatorname{Ker}(E_G \otimes Q \to \wedge^2 Q)$$

when φ is symmetric (and respectively a section of

$$H = \operatorname{Ker}(E_G \otimes Q \to S^2 Q)$$

when φ is skew-symmetric). Let $Z \subset G$ denote the subscheme of zeros of this section, in both respective cases. Observe that π maps Z onto $D_r(\varphi)$. The next proposition will contain the main calculation of the paper.

Proposition 3.1. *(i) Suppose that, for symmetric φ,* $\operatorname{codim}_G Z = \operatorname{rank} H$ *and π restricted to Z establishes a birational isomorphism of Z and $D_r(\varphi)$. Then, the following equality in $A_*(X)$ holds:*

$$[D_r(\varphi)] = \sum_I Q_{\rho_q + I}(F) \cdot s_{\mathrm{C}\widetilde{I}}(E - F) \cap [X], \tag{3.2}$$

where the sum is over all partitions I in $(n)^q$ and for the conjugate partition $\widetilde{I} \subset (q)^n$ we write $\mathrm{C}\widetilde{I} = (q - \widetilde{i}_n, \dots, q - \widetilde{i}_1)$.

(ii) Assume that r is even. Suppose that, for skew-symmetric φ, $\operatorname{codim}_G Z =$ $\operatorname{rank} H$ *and π restricted to Z establishes a birational isomorphism of Z and $D_r(\varphi)$. Then, the following equality in $A_*(X)$ holds:*

$$[D_r(\varphi)] = \sum_I P_{\rho_{q-1} + I}(F) \cdot s_{\mathrm{C}\widetilde{I}}(E - F) \cap [X], \tag{3.3}$$

where the sum is the same as in (3.2).

(iii) Let $n \geqslant 1$ and assume that r is odd. Suppose that, for skew-symmetric φ, $\operatorname{codim}_G Z = \operatorname{rank} H$ *and π restricted to Z establishes a birational isomorphism of Z and $D_r(\varphi)$. Then, the following equality in $A_*(X)$ holds:*

$$[D_r(\varphi)] = \sum_J P_{\rho_q + J}(F) \cdot s_{\mathrm{C}\widetilde{J}}(E - F) \cap [X], \tag{3.4}$$

where the sum is over all partitions J in $(n-1)^q$ and for the conjugate $\widetilde{J} \subset (q)^{n-1}$ we write $C\widetilde{J} = (q - \widetilde{j}_{n-1}, \dots, q - \widetilde{j}_1)$.

Proof. We give a detailed proof in the symmetric case and make some necessary comments about the skew-symmetric cases. Let $K = \mathrm{Ker}(\alpha : E \twoheadrightarrow F)$. In the Grothendieck group $K(G)$, the following equality holds:

$$[E \otimes Q] - [\wedge^2 Q] = [K_G \otimes Q] + [R \otimes Q] + [S^2 Q].$$

Hence we get

$$[D_r(\varphi)] = \pi_*\Big(c_{\mathrm{top}}(K_G \otimes Q) \cdot c_{\mathrm{top}}(R \otimes Q) \cdot c_{\mathrm{top}}(S^2 Q) \cap [G]\Big). \tag{3.5}$$

We have, by (1.12),

$$c_{\mathrm{top}}(K_G \otimes Q) = \sum_I s_I(Q) \cdot s_{\mathrm{C}\,\widetilde{I}}(K_G) = \sum_I s_I(Q) \cdot s_{\mathrm{C}\,\widetilde{I}}(E_G - F_G),$$

where the sum is over partitions $I \subset (n)^q$ and for the conjugate partition $\widetilde{I} \subset (q)^n$ we write $\mathrm{C}\,\widetilde{I} = (q - \widetilde{i}_n, \dots, q - \widetilde{i}_1)$.

We compute the RHS of (3.5). First, using the above expansion of $c_{\mathrm{top}}(K_G \otimes Q)$ and (1.13), we get

$$\begin{aligned}
&\pi_*\big(c_{\mathrm{top}}(K_G \otimes Q) \cdot c_{\mathrm{top}}(R \otimes Q) \cdot c_{\mathrm{top}}(S^2 Q) \cap [G]\big)\\
&= \pi_*\big(\sum_I c_{\mathrm{top}}(R \otimes Q) \cdot Q_{\rho_q}(Q) \cdot s_I(Q) \cdot s_{\mathrm{C}\,\widetilde{I}}(E_G - F_G) \cap [G]\big).
\end{aligned}$$

Secondly, using the factorization formula (1.10), we infer that this last expression equals

$$\begin{aligned}
&\pi_*\big(\sum_I c_{\mathrm{top}}(R \otimes Q) \cdot Q_{\rho_q + I}(Q) \cdot s_{\mathrm{C}\,\widetilde{I}}(E_G - F_G) \cap [G]\big)\\
&= \sum_I Q_{\rho_q + I}(F) \cdot s_{\mathrm{C}\,\widetilde{I}}(E - F) \cap [X],
\end{aligned}$$

where the sum is over $I \subset (n)^q$, and in the last equality we have used the push-forward formula (1.16).

The proof in the skew-symmetric case, when r is even, is analogous; in addition, we must use the push-forward formula (1.17).

The proof in the skew-symmetric case, when $n \geqslant 1$ and r is odd, goes the same way. We use the push-forward formulas (1.18) and (1.16).

The proposition has been proved. □

As usual the formulas for the fundamental classes of degeneracy loci hold under more general assumptions than in the previous proposition.

Theorem 3.6. *(i) If X is a pure-dimensional Cohen–Macaulay scheme and $D_r(\varphi)$ is of expected pure codimension $c(r)$ or empty, then, in the symmetric case, the fundamental class of $D_r(\varphi)$ is evaluated by (3.2).*

(ii) In the skew-symmetric case, under the analogous assumptions, the fundamental class of $D_r(\varphi)$ is evaluated by (3.3) if r is even, and by (3.4) if $n \geqslant 1$ and r is odd.

Proof. We pass to a "universal case". (Our method is similar to the technique explained in [F-P, App. A.2].) For a given morphism $\varphi : E^* \to F$ of one of the two considered types, we define $\tilde{X} = \text{Spec } S^\bullet(E \vee F)^*$ (respectively $\tilde{X} = \text{Spec } S^\bullet(E \wedge F)^*$). Observe that φ induces a section $s : X \to \tilde{X}$. On the other hand, there exists the tautological bundle homomorphism $\tilde{\varphi} : \tilde{E}^* \to \tilde{F}$ where $\tilde{E} = E_{\tilde{X}}$, $\tilde{F} = F_{\tilde{X}}$ such that $s^*(\tilde{\varphi}) = \varphi$. If X is Cohen–Macaulay, then so is $D_r(\tilde{\varphi})$ by virtue of a result from [DC-L], and because an algebraic fibre bundle with Cohen–Macaulay base and fibre is Cohen–Macaulay. Hence, if $D_r(\varphi)$ is of pure codimension $c(r)$ in X, then by [F1, Lemma A.7.1] we get

$$[D_r(\varphi)] = s^*[D_r(\tilde{\varphi})]. \tag{3.7}$$

One checks in a fairly standard way that $\tilde{\varphi}$ satisfies the assumptions of Proposition 3.1. Applying Proposition 3.1 to $\tilde{\varphi}$, we infer that $[D_r(\tilde{\varphi})]$ is evaluated by (3.2) (resp. by (3.3) or (3.4)) with $\tilde{E}$ playing the role of E and $\tilde{F}$ playing the role of F. Consequently, the wanted assertion follows from (3.7) and the pull-back property of Chern classes. that in the case where $E = F$, Giambelli [G] obtained the expression $2^q s_{\rho_q}(E)$ for the class of the degeneracy locus considered in part (i) of the theorem. However, it a classical result of combinatorics that the following equality of symmetric functions in the variables $A = (a_1, a_2, \dots)$ holds:

$$P_{\rho_q}(A) = s_{\rho_q}(A).$$

Example 3.8. (i) Assume first that $\varphi : E^* \to F$ is symmetric. Let $f = 3$ and $r = 2$. If $e = 4$, then the polynomial representing $D_2(\varphi)$ is $Q_2(F) + Q_1(F)s_1(E - F)$; for $e = 5$, the polynomial is $Q_3(F) + Q_2(F)s_1(E - F) + Q_1(F)s_{1,1}(E - F)$.

For the rest of this example, we assume that $\varphi : E^* \to F$ is skew-symmetric.

(ii) Let $f = 3$ and $r = 2$. If $e = 4$ then the polynomial representing $D_2(\varphi)$ is $P_1(F) + s_1(E - F)$; for $e = 5$, the polynomial is $P_2(F) + P_1(F)s_1(E - F) + s_{1,1}(E - F)$.

(iii) If $e = 5$, $f = 4$, and $r = 2$, then the representing polynomial is $P_{2,1}(F) + P_2(F)s_1(E - F) + P_1(F)s_2(E - F)$.

(iv) If f is even and $n = q = 1$, then the polynomial representing $D_{f-1}(\varphi)$ equals $P_1(F)$.

(v) If $e = 4, f = 2$, then $D_1(\varphi)$ is represented by $P_1(F)s_1(E - F) + P_2(F)$.

We leave it to the reader to check that the formulas in this example are consistent with the formulas in (1.21).

Perhaps, the easiest way to remember the formula associated with (e, f, r) in the symmetric case, is to put $T = (e-r, e-r-1, \ldots, n+1)$, and note that for $I \subset (e-f)^{f-r}$ the coefficient of $s_{\tilde{I}}(E-F)$ is $Q_{T-I}(F)$, I being written *increasing* in last index, and subtraction of the sequences being performed componentwise. In the skew-symmetric case, with r even, we respectively put $T = (e-r-1, e-r-2, \ldots, n)$, and note that the coefficient of $s_{\tilde{I}}(E-F)$ is $P_{T-I}(F)$ with the same conventions as above. A similar interpretation can be given in the skew-symmetric case when $n \geqslant 1$ and r is odd.

Example 3.9. $e = 8, f = 4, r = 2$; the symmetric case:

$$\begin{aligned}
&Q_{6,5}(F) + Q_{6,4}(F)s_1(E-F) + Q_{5,4}(F)s_2(E-F) + Q_{6,3}(F)s_{1,1}(E-F)\\
&+ Q_{5,3}(F)s_{2,1}(E-F) + Q_{6,2}(F)s_{1,1,1}(E-F) + Q_{5,2}(F)s_{2,1,1}(E-F)\\
&+ Q_{4,3}(F)s_{2,2}(E-F) + Q_{6,1}(F)s_{1,1,1,1}(E-F) + Q_{4,2}(F)s_{2,2,1}(E-F)\\
&+ Q_{5,1}(F)s_{2,1,1,1}(E-F) + Q_{4,1}(F)s_{2,2,1,1}(E-F) + Q_{3,2}(F)s_{2,2,2}(E-F)\\
&+ Q_{3,1}(F)s_{2,2,2,1}(E-F) + Q_{2,1}(F)s_{2,2,2,2}(E-F)
\end{aligned}$$

So, in this example, $T = (6,5)$ and, for instance, the coefficient of $s_{2,2,1,1}(E-F) = s_{\widetilde{(4,2)}}(E-F)$ is $Q_{(6,5)-(2,4)}(F) = Q_{4,1}(F)$.

A decomposition into a sum of the $s_I(F) \cdot s_J(E)$'s (computed with the help of the library SFA of ACE (see [V])) is four times

$$\begin{aligned}
&s_{6,1}(F)\big(s_{1,1,1,1}(E) + s_{2,2}(E) - s_{2,1,1}(E)\big) + s_{6,5}(F)\\
&+ s_{6,3}(F)\big(-s_2(E) + s_{1,1}(E)\big) + s_{2,1}(F)s_{2,2,2,2}(E) + s_{4,3}(F)s_{2,2}(E)\\
&+ s_{4,1}(F)\big(s_{2,2,1,1}(E) - s_{2,2,2}(E)\big) + s_{8,3}(F) + s_{10,1}(F)\\
&+ s_{8,1}(F)\big(-s_2(E) + s_{1,1}(E)\big).
\end{aligned}$$

4 Some variations

To compute the fundamental classes of subvarieties, one can also use appropriate geometric constructions with a *nontrivial* generic fibre. This method was invented in [P1] in order to give a short proof of the formulas from [J-L-P] and [H-T], and is summarized (and somewhat straightened-out) in the following proposition. In this proposition, we may assume that the Chow groups have rational coefficients.

Proposition 4.1. *Let D be an irreducible (closed) subscheme of a scheme X. Let $\pi : G \to X$ be a proper morphism of schemes and W be a (closed) subscheme of G such that $\pi(W) = D$. We have the following two instances:*
(i) Suppose that G is smooth. Assume that there exists

$$g \in A_{\dim G + \dim D - \dim W}(G)$$

and a point x in the smooth locus of D such that in $A_*(G_x)$, where G_x is the fibre of π over x, one has:

$$i_x^*(g)\cdot[W_x]=[\text{point}].$$

Here, W_x is the fibre of W over x and $i_x: G_x \hookrightarrow G$ is the inclusion. Then the following equality holds in $A_*(X)$:

$$[D]=\pi_*(g\cdot[W]).$$

(ii) Here G is possibly singular. Suppose that there exists a family of vector bundles $\{E^{(\alpha)}\}$ on G and $g=P(\{c.(E^{(\alpha)})\})$ a homogeneous polynomial of degree $\dim W-\dim D$ in the Chern classes of $\{E^{(\alpha)}\}$ ($\deg\ c_i(E^{(\alpha)})=i$) with rational coefficients, such that in $A_*(G_x)$,

$$P(\{c.(i_x^*E^{(\alpha)})\})\cap[W_x]=[\text{point}],$$

where x, G_x, W_x and i_x are as above. Then the following equality holds in $A_*(X)$:

$$[D]=\pi_*(g\cap[W]).$$

Proof. (i) Using a standard dimension argument, we can replace, in the assertion, D by its smooth part, i.e., we can assume D is smooth. Write $G_D=G\times_X D$, $W_D=W\times_X D$, $\eta: G_D\to D$ the projection induced by π, and $k: G_D\to G$ the inclusion. Then, the assertion is a consequence of the following identity in $A_*(D)$:

$$\eta_*(k^*(g)\cdot[W_D])=[D].$$

To prove this last equation, we first remark that the assumptions imply

$$\eta_*(k^*(g)\cdot[W_D])=m[D],$$

where $m\in\mathbb{Z}$. Let x be a point in D and consider the fibre square

$$\begin{array}{ccc} G_x & \stackrel{j}{\hookrightarrow} & G_D \\ {\scriptstyle p}\downarrow & & \downarrow{\scriptstyle \eta} \\ \{x\} & \stackrel{i}{\hookrightarrow} & D. \end{array}$$

Using the assumptions on g and [F1, Th. 6.2], we have

$$\begin{aligned} i^*\eta_*(k^*(g)\cdot[W_D]) &= p_*\Big(j^*(k^*(g)\cdot[W_D])\Big)\\ &= p_*(i_x^*(g)\cdot[W_x]) = p_*([\text{point}])=[\text{point}]. \end{aligned}$$

This implies $m=1$ and assertion (i) is proved.

The proof of (ii) is essentially the same. $F \subset E$ be two vector bundles of ranks f and e on a variety X. We now describe a certain geometric construction associated with a bundle morphism $\varphi : F \to E^*$ induced by a section of $E^* \vee F^*$ (resp. $E^* \wedge F^*$). (In this section, we use a slightly different setup than in the Introduction, Section 1, and Section 3.) This construction generalizes in a natural way the construction used in [P1] and is based on the following characterization of the rank of the above morphisms. Assume that p is a natural number such that $2p < f$. Let $V \subset U$ be two vector spaces of dimensions f and e. Let $\phi : V \to U^*$ be a linear map induced by a section of $U^* \vee V^*$ (resp. $U^* \wedge V^*$). Then $\operatorname{rank}\phi \leqslant 2p$ iff there exists a pair of vector spaces (A, B) such that A is a subspace of U of dimension $e - p$, B is a subspace of V of dimension $f - p$, $B \subset A$, and the composite map

$$B \hookrightarrow V \xrightarrow{\phi} U^* \twoheadrightarrow A^*$$

is zero. Indeed, suppose that such a pair (A, B) exists. Then the matrix of ϕ in some basis has $e - p$ rows whose initial segments of length $f - p$ consist of zeros. Then every $(2p + 1)$-minor of such a matrix vanishes (use the Laplace expansion of this minor w.r.t. the first $p + 1$ from these $e - p$ rows). The opposite implication can be showed using fairly standard linear algebra (by reducing a matrix to its "standard form" via the elementary row and column operations).

Let

$$\pi : G = Fl_{f-p,e-p}(F, E) \to X$$

be the flag bundle parametrizing pairs (A, B) where A is a rank $e - p$ subbundle of E, B is a rank $f - p$ subbundle of F, and $B \subset A$. Let $S \subset R$ be the tautological (sub)bundles of respective ranks $f - p$ and $e - p$ on G.

With $\varphi : F \to E^*$ as above, we associate a locus $W \subset G$ to be the subscheme of zeros of the composite morphism

$$S \hookrightarrow F_G \xrightarrow{\varphi_G} E_G^* \twoheadrightarrow R^*.$$

Let $D = D_{2p}(\varphi)$. By the above discussion, we have $\pi(W) = D$.

We want now to work with some "universal" φ (like that in the proof of Theorem 3.6). Moreover, in this situation, we want now to get information about the generic fibre $W_x =: \mathcal{F}$ of $\pi|_W$ in order to apply Proposition 4.1. To this end, it suffices to make the following "universal local study". Let $V \subset U$ be vector spaces of dimensions f and e, respectively. Let X denote the affine space $U^* \vee V^*$ (resp. $U^* \wedge V^*$). In this situation, there exists a tautological morphism $\varphi : (F = V_X) \to (E = U_X)^*$ and the corresponding subvariety W. Let $\phi : V \to U^*$ be a linear map coming from a section of $U^* \vee V^*$ (resp. $U^* \wedge V^*$). Then the fibre W_ϕ over ϕ is identified with

$$W_\phi = \{(L, M) \in Fl \ : \ i_L^* \circ \phi \circ j_M = 0\},$$

where $Fl = Fl_{f-p,e-p}(V,U)$ and $i_L : L \hookrightarrow U$, $j_M : M \hookrightarrow V$ are the inclusions. By calculation in local coordinates, one checks that $W \subset G = Fl$ is a complete intersection of codimension equal to $\operatorname{rank}(R^* \vee S^*)$ (resp. $\operatorname{rank}(R^* \wedge S^*)$). Hence one easily computes that $\dim \mathcal{F} = \dim W - \dim D = p(p-1)/2$ (resp. $\dim \mathcal{F} = p(p+1)/2$). Consequently, the dimension of the generic fibre $\mathcal{F}$ depends only on p and not on e and f.

The following very simple fact is helpful to find the class g satisfying the requirements of Proposition 4.1.

Lemma 4.2. *Let $i : Y' \hookrightarrow Y$ be a closed embedding of smooth varieties, let $X \subset Y$ and $X' \subset Y'$ be two subvarieties such that $i(X') \subset X$ and* $\dim X' = \dim X$. *Assume that an element $z \in A^*(X)$ satisfies $[X'] \cdot i^*(z) = [\text{point}]$ in $A^*(Y')$. Then, $[X] \cdot z = [\text{point}]$ in $A^*(Y)$.*

Indeed, we have $i_*[X'] = [X]$, and by the projection formula we infer

$$[\text{point}] = i_*\big([X'] \cdot i^*(z)\big) = i_*[X'] \cdot z = [X] \cdot z\,,$$

as claimed.

Proposition 4.3. *The class $g = 2^{-p} s_{\rho_{p-1}}(S^*)$ (resp. $g = s_{\rho_p}(S^*)$) satisfies the assumption of Proposition 4.1(ii).*

Proof. We follow the notation from the discussion before the lemma. Here, the role of "x" from Proposition 4.1 is played by ϕ such that $\operatorname{rank}(V \xrightarrow{\phi} U^* \twoheadrightarrow V^*) = 2p$. Let S' denote the tautological rank $f-p$ bundle on the Grassmannian $G_{f-p}(V)$ parametrizing $(f-p)$-dimensional subspaces of V. 1) Assume first that $e = f = 2p$ so $V = U$ and the corresponding bilinear form is nondegenerate. Then $[\mathcal{F}]$ is evaluated as the top Chern class of the bundle $S^2(S'^*)$ (resp. $\wedge^2(S'^*)$). We get by Section 1.11,

$$[\mathcal{F}] = 2^p s_{\rho_p}(S'^*) \qquad (\text{resp. } [\mathcal{F}] = s_{\rho_{p-1}}(S'^*)\ ).$$

The assertion now follows by taking the dual Schubert cycles in the Grassmannian $G^p(V^*)$ (see, e.g., [F1, Chap. 14]).

2) Let now $2p < e = f$ (so again $V = U$), and let $V' \subset V$ be an inclusion of vector spaces of dimensions $2p$ and f, respectively. Assume that V is endowed with a symmetric (resp. skew-symmetric) form ϕ of rank $2p$ such that the form $\phi|_{V'}$ is nondegenerate. We now use the lemma with the following data: $Y' = G_p(V')$ and $Y = G_{f-p}(V)$; $i : G_p(V') \hookrightarrow G_{f-p}(V)$ being defined by $L \mapsto L \oplus A$, where $V = V' \oplus A$. Moreover, X and X' are the generic fibres under consideration and $z = 2^{-p} s_{\rho_{p-1}}(S'^*)$ (resp. $z = s_{\rho_p}(S'^*)$). Then part 1) and the lemma yield the desired assertion.

3) Finally, suppose that $f < e$ and let $U = V \oplus B$, where $\dim B = n = e - f$. We now apply the lemma to the following embedding:

$$i : \big(Y' = G_{f-p}(V)\big) \hookrightarrow \big(Y = G\big)\,,$$

where $i(L) = (L, L \oplus B)$. Moreover, X and X' are the generic fibres under consideration and $z = 2^{-p} s_{\rho_{p-1}}(S^*)$ (resp. $z = s_{\rho_p}(S^*)$). Then part 2) and the lemma yield the desired result. □

We will be now interested in formal identities of some polynomials in Chern classes and their push-forwards. Therefore we assume that X is smooth and treat the Chern classes as elements of the appropriate Chow rings. Using Propositions 4.1(ii) and (4.3), we infer that in the symmetric case, the degeneracy locus D is represented by

$$\pi_*\big(c_{\text{top}}(R^* \vee S^*) \cdot 2^{-p} s_{\rho_{p-1}}(S^*)\big).$$

Similarly, in the skew-symmetric case, the degeneracy locus D is represented by

$$\pi_*\big(c_{\text{top}}(R^* \wedge S^*) \cdot s_{\rho_p}(S^*)\big).$$

Of course $D = D_{2p}(\varphi^*)$. Combining Propositions 2.1 and 2.2 with Theorem 3.6 applied to $\varphi^* : E \to F^*$, we get the following algebraic equalities. In the symmetric case, writing $T = (e-p, e-p-1, \ldots, n+1)$,

$$\begin{aligned} &\pi_*\big(2^{-p} \sum_I Q_{\rho_{f-p}+I}(S^*) \cdot s_{\complement \tilde{I}}(R^* - S^*) \cdot s_{\rho_{p-1}}(S^*)\big) \\ = \; &\pi_*\big(2^{f-2p} \sum_J s_{T/J}(S^*) \cdot s_{\tilde{J}}(R^* - S^*) \cdot s_{\rho_{p-1}}(S^*)\big) \\ = \; &\sum_L Q_{\rho_{f-2p}+L}(F^*) \cdot s_{\complement \tilde{L}}(E^* - F^*), \end{aligned} \tag{4.4}$$

where I runs over partitions in $(n)^{f-p}$, J runs over partitions in T and L runs over partitions in $(n)^{f-2p}$. In the skew-symmetric case, writing T for the partition $(e-p-1, e-p-2, \ldots, n)$,

$$\begin{aligned} &\pi_*\big(\sum_I P_{\rho_{f-p-1}+I}(S^*) \cdot s_{\complement \tilde{I}}(R^* - S^*) \cdot s_{\rho_p}(S^*)\big) \\ &= \pi_*\big(\sum_J s_{T/J}(S^*) \cdot s_{\tilde{J}}(R^* - S^*) \cdot s_{\rho_p}(S^*)\big) \\ = \; &\sum_L P_{\rho_{f-2p-1}+L}(F^*) \cdot s_{\complement \tilde{L}}(E^* - F^*), \end{aligned} \tag{4.5}$$

where I and L run over the same sets of partitions as above and J runs over partitions in the present T.

These equalities were originally obtained with the help of the library SFA of ACE (see [V]) for small values of f, p, and n. We know no algebraic proofs of (4.4) and (4.5), not invoking geometry.

Let $G' = G_{f-p}(F) \to X$ and set

$$C = \operatorname{Coker}(S' \hookrightarrow F_{G'} \hookrightarrow E_{G'}),$$

where S' is the tautological subbundle on G'. Then using the presentation

$$G = G_n(C) \to G_{f-p}(F) \to X$$

and (1.9), one can rewrite the LHS of (4.4) and (4.5) as purely algebraic expressions involving symmetrizing operators. Or, equivalently, one can embed G in $G_{f-p}(F) \times_X G_{e-p}(E)$ and use the product of the symmetrizing operators corresponding to the factors, at the cost of multiplying the expressions in (4.4) and (4.5) to be push-forwarded by $c_{\mathrm{top}}(S'^* \otimes (E/R'))$, where R' is the tautological subbundle on $G_{e-p}(E)$ (and we omit some pull-back indices).

Finally, the second author takes this opportunity to make the following

Remark 4.6. (Revisions and corrigenda to [P1], [P4], and [P-R]) [P1]: Revisions: The assumptions on g in Proposition 4.1 (in the present paper) straighten out an unprecise expression "the Poincaré dual of" from 5^0 in [P1, Proposition 2.1]. The assumption that the ground field is algebraically closed of characteristic different from 2 was mistakenly omitted in Section 3. The reference "Lemma 9 in [10]" on p. 196 should be replaced by "[2, Lemma A.7.1]". [P4]: Revisions – insert after "... following changes:"

p. 162_5: p.154 l. -10 should read: " Write $a_i = c_1(L_i^E)$ $i = 1, \dots, n.$ ..."
p.154 l. -5 should read: " ... for ω, the fact that ω_* is induced by $\partial^{k-1}_{(a_2,\dots,a_n)}$,"
p.155 l. -14 should read: " ... $= dP_{I,J}(E) \cap \alpha$ ..."
p.155 l. -13 should read: " ... Let $I' = (i_2, \dots, i_k)$."

Misprints – should be: p. 130_3 "$(-1)^{p-1}$" // p. 130_1 "$(-1)^p$" // p. 131^9 "ρ_{n-r-1}" // p. 148^{11} "$-a_{q+1}, \dots$" // p. 162_3 " ...$\cdot\xi^{i_1}\cap$ ... " // p. 164^6 "$\partial_r \circ \dots$" // p .174^{14} " boxes"

[P-R]: Revisions – should read: p. 19_{18} "Whenever, in this paper we speak about Schubert subschemes in OG_nV, we assume that there exists a completely filtered rank n isotropic subbundle of V." // p. 56_{14} and p. 60^3 " ... , so we can apply the induction assumption to M_1 defined below. The partitions ..."

Misprints – should be: p. 39_5 " $\widetilde{P}_J(X_m)$" // p. 86^{12} "Fulton W." // p. 87_2 "Gieseker-Petri"

Acknowledgments. We thank Bill Fulton for his helpful suggestion concerning Proposition 4.1. We are also grateful to Romek Dąbrowski and Witek Kraśkiewicz for some useful discussions. The second author thanks the organizers of the Intersection Theory Conference in Bologna for invitation to attend this interesting meeting, as well as for the encouragement to write up the present paper.

References

[Ba] W. Barth, Counting singularities of quadratic forms on vector bundles, in: *Vector bundles and Differential Equations*; Proc. of the Nice Conf. Birkhäuser, Boston, 1–19, 1979

[B-R] A. Berele, A. Regev, Hook Young diagrams with applications to combinatorics and to representations of Lie superalgebras, *Adv. in Math.*, **64** (1987), 118–175

[Bo] M. Bottasso, Sur une classe de variétés engendrées par des systemes linéaires projectives d'hypersurfaces, in: *Annaes Scientificas de Academia Polytechnica de Porto*, **4** (1909), 193–205

[B-F] A. Buch, W. Fulton, Chern class formulas for quiver varieties, *Invent. Math.* **135** (1999), 665–687

[B-E] D. Buchsbaum, D. Eisenbud, Algebra structures for finite free resolutions and some structure theorems for ideals of codimension 3, *Amer. J. Math.*, **99** (1997), 447–485

[DC-L] C. De Concini, V. Lakshmibai, Arithmetic Cohen-Macaulayness and arithmetic normality of Schubert varieties, *Amer. J. Math.*, **103** (1981), 835–850

[F1] W. Fulton, *Intersection Theory*, Springer-Verlag, New York, 1984

[F2] W. Fulton, Schubert varieties in flag bundles for classical groups, in: *Proceedings of the Hirzebruch 65 Conference on Algebraic Geometry*, IMCP 9 (1996), 241–262

[F3] W. Fulton, Determinantal formulas for orthogonal and symplectic degeneracy loci, *J. Diff. Geom.*, **43** (1996), 276–290

[F-P] W. Fulton, P. Pragacz, *Schubert Varieties and Degeneracy Loci*, Lecture Notes in Mathematics, **1689**, Springer-Verlag, New York, 1998

[G] G. Z. Giambelli, Sulle varieta rappresentate coll'annullare determinanti minori contenuti in un determinante simmetrico od emisimmetrico generico di forme, *Atti della R. Acc. delle Scienze di Torino*, **44** (1905/1906), 102–125

[H-T] J. Harris, L. Tu, On symmetric and skew-symmetric determinantal varieties, *Topology*, **23** (1984), 71–84

[J-L-P] T. Józefiak, A. Lascoux, P. Pragacz, Classes of determinantal varieties associated with symmetric and antisymmetric matrices (in Russian), *Izwiestja AN SSSR*, **45** (1981), 662–673

[K-L] H. Kleppe, D. Laksov, The algebra structure and deformation of pfaffian schemes, *J. of Algebra*, **64** (1980), 167–189

[L] A. Lascoux, Classes de Chern d'un produit tensoriel, *C. R. Acad. Sci. Paris* **286** (1978), 385–387

[L-L-T] A. Lascoux, B. Leclerc, J.-Y. Thibon, Une nouvelle expression des functions P de Schur, *C. R. Acad. Sci. Paris*, **316** (1993), 221–224

[L-P] A. Lascoux, P. Pragacz, Operator calculus for $\widetilde{Q}$-polynomials and Schubert polynomials, *Adv. in Math.*, **140** (1998), 1–43

[L-S] A. Lascoux, M.-P. Schützenberger, *Formulaire raisonné de fonctions symétriques*, Prepublication L.I.T.P., Université Paris 7, 1985

[M] I.G. Macdonald, *Symmetric Functions and Hall Polynomials, Second Edition*, Oxford University Press, 1995

[P1] P. Pragacz, Cycles of isotropic subspaces and formulas for symmetric degeneracy loci, in: *Topics of Algebra*, Banach Center Publications, **26**:2 (1990), 189–199

[P2] P. Pragacz, Enumerative geometry of degeneracy loci, *Ann. Sci. École Norm. Sup.*, **21** (1988), 413–454

[P3] P. Pragacz, Algebro-geometric applications of Schur S- and Q-polynomials, in: *Séminaire d'Algèbre Dubreil-Malliavin 1989–1990*, Lecture Notes in Mathematics, Springer-Verlag, New York, vol. 1478, 1991, 130–191

[P4] P. Pragacz, Symmetric polynomials and divided differences in formulas of intersection theory, in: *Parameter Spaces*, Banach Center Publications, **36** (1996), 125–177

[P-R] P. Pragacz, J. Ratajski, Formulas for Lagrangian and orthogonal degeneracy loci; $\widetilde{Q}$-polynomial approach, *Compositio Math.*, **107** (1997), 11–87

[Sch] I. Schur, Über die Darstellung der symmetrischen und der alternierenden Gruppe durch gebrochene lineare Substitutionen, *J. Reine Angew. Math.*, **139** (1911), 155–250

[St] R. Stanley, Problem 4 in "Problem Session", in: *Combinatorics and Algebra*, Contemporary Mathematics A.M.S., **34**, 1984

[V] S. Veigneau, ACE, an Algebraic Combinatorics Enviroment for the computer algebra system MAPLE, Version 3.0, Université de Marne-la-Vallée, 1998, http://phalanstere. univ-mlv.fr/~ ace/

Alain Lascoux
C.N.R.S.
Institut Gaspard Monge
Université de Marne-la-Vallée
5 Bd Descartes, Champs sur Marne
77454 Marne La Vallée Cedex 2
France
Alain.Lascoux@univ-mlv.fr

Piotr Pragacz
Mathematical Institute
of Polish Academy of Sciences
Chopina 12
87-100 Toruń
Poland
pragacz@mat.uni.torun.pl

Algebraic Cycles and Motives: An Introduction

Robert Laterveer

0 Introduction

This is the written-up version of the lectures I gave in Bologna. It is a pleasure to thank the organizers for doing such a fine job, in such a lovely location.

The aim of study of this paper is algebraic cycles. For X a complex algebraic variety, let $z_i X$ denote the free abelian group generated by i-dimensional closed subvarieties of X (elements of $z_i X$ are called *algebraic cycles*). Since this group is too enormous to handle, it is convenient to impose some equivalence relations. The finest adequate equivalence relation is *rational equivalence*, defined as follows: A cycle in $z_i X$ is rationally trivial if it is the divisor of a rational function on an $(i+1)$-dimensional subvariety. The *Chow group* is defined by setting

$$A_i X := z_i X/\sim_{\text{rat}} \; ,$$

and for X smooth of pure dimension n the codimension type notation is often used: $A^i X = A_{n-i} X$. A well-known Chow group is $A^1 X$ which (again for X smooth) coincides with the Picard group $\text{Pic}X$.

The intersection product makes $A^*X = \oplus A^i X$ (for smooth X) into a ring; moreover, the assignment $X \mapsto A^*X$ is a contravariant functor on smooth varieties (for all this, and much more, cf. [Fu]). This suggests that A^* is close to singular cohomology $H^*(-,\mathbb{Z})$. In fact, not only are the two theories formally very similar, they can also be related via the cycle class map

$$\text{cl}^{\,i}\colon\; A^i X \longrightarrow H^{2i}(X,\mathbb{Z})$$

(sending a subvariety to its fundamental class). The image of this map is conjecturally described (up to torsion) by the famous Hodge conjecture; here we will be concerned instead with the kernel $A^i_{\text{hom}} := \text{Ker}\,\text{cl}^{\,i}$.

On this kernel, one can define a secondary cycle class map, called the Abel–Jacobi map

$$AJ^i\colon\; A^i_{\text{hom}} X \longrightarrow \mathcal{J}^i X := H^{2i-1}(X,\mathbb{C})/\big(F^i + H^{2i-1}(X,\mathbb{Z})\big) \; ,$$

where F^* denotes the Hodge filtration on $H^{2i-1}(X,\mathbb{C})$ [Gri].

In codimension one the story ends here: it follows from the exponential sequence

$$0 \to H^1(X,\mathcal{O})/H^1(X,\mathbb{Z}) \to A^1X \to H^{1,1}(X,\mathbb{Z}) \to 0$$

that AJ^1 is an isomorphism. Otherwise, for $i > 1$ the kernel $A^i_{\mathrm{AJ}} :=$ Ker (AJ^i) is in general very large, as is shown in Part 1 of these notes.

In Part 1, Mumford's theorem is proven, saying that for a general surface, the Chow group of 0-cycles is very large. Another way of phrasing this result is that if Chow groups are small, the cohomology groups are of a very special type, i.e. Chow groups A^* influence cohomology $H^*(-,\mathbb{Q})$. Some more applications in the spirit of Mumford's theorem are given, based on work of Bloch–Srinivas and Jannsen.

Part 2 of these notes centers around a conjectural converse to the influence treated in Part 1: Bloch's conjecture (stating the converse to Mumford's theorem) and the philosophy of mixed motives (embedding Bloch's conjecture into a vast and aesthetically pleasing conjectural theory). The object here is to try and understand the structure of Chow groups in codimension larger than one, more particularly to see whether cohomology H^* somehow influences Chow groups A^*. This section does not pretend to any originality or novelty.

In Part 3, the framework of mixed motives is applied to make some new predictions in codimension 2, and a brief report is given on some special cases that can be proven unconditionally.

Conventions 0. A *variety* will be a reduced, separated equidimensional scheme of finite type over a field k. (Note that a variety is not necessarily irreducible.)

Unless explicitly stated otherwise, the field k will be the field of complex numbers $\mathbb{C}$.

1 From A^* to H^* — Mumford's Theorem

In this part, we will see how Chow groups influence cohomology groups. One way of putting restrictions on cohomology groups is by using Grothendieck's notion of coniveau filtration [Gr]:

Definition 1.0. For any variety X, we say that $H^i(X,\mathbb{Q})$ is supported in codimension c if there exists a closed reduced subscheme $Y \subset X$ of codimension $\geq c$ such that

$$H^i_Y(X,\mathbb{Q}) \longrightarrow H^i(X,\mathbb{Q})$$

is surjective (i.e. such that $H^i(X \setminus Y,\mathbb{Q}) = 0$).

If $H^i(X,\mathbb{Q})$ is supported in codimension c, we will write

$$H^i(X,\mathbb{Q}) = N^c H^i(X,\mathbb{Q}) \ .$$

1.1 Suppose X is a smooth and complete variety. Let $h^{p,q}(X)$ denote the Hodge numbers. Then it is easily seen that

$$H^i(X,\mathbb{Q}) = N^c H^i(X,\mathbb{Q}) \ \Rightarrow \ h^{p,q}(X) = 0 \text{ if } p+q = i \text{ and } |p-q| > i-2c \ .$$

It is expected that the converse to this statement holds; this would follow from Grothendieck's generalized Hodge conjecture [Gr].

One way of trying to measure the Chow group of 0-cycles is the following definition:

Definition 1.2. For any variety X, we say that the group $A_0^{\text{hom}}\, X$ is *finite-dimensional* if there exists an integer $m \in \mathbb{N}$ such that the map

$$\begin{array}{ccc} S^m X \times S^m X & \longrightarrow & A^{\text{hom}}{}_0 X, \\ (a,b) & \mapsto & a-b \end{array}$$

is surjective, where S^m denotes the m-th symmetric power.

This definition is related to cycle class maps:

Lemma 1.3. *Let X be a smooth complete variety of dimension n. The following are equivalent:*

(i) $A_0^{\text{hom}}\, X$ *is finite-dimensional;*

(ii) $A_0^{\text{AJ}}\, X = 0$, *i.e. the cycle class map*

$$A_O X \longrightarrow H^{2n}_{\mathcal{D}}(X,\mathbb{Z}(n))$$

is injective;

(iii) *there exists a one-dimensional closed reduced subscheme $Y \subset X$ such that*

$$A_O(X \setminus Y)_{\mathbb{Q}} = 0$$

(i.e. such that the push-forward map

$$A_0(Y)_{\mathbb{Q}} \longrightarrow A_0(X)_{\mathbb{Q}}$$

is surjective).

(iv) *The cycle class map*

$$A_0 X \longrightarrow H^{2n}_{\mathcal{D}}(X,\mathbb{Z}(n))$$

is an isomorphism.

Proof.

(i)⇒(ii) This is a theorem of Roitman [Ro 1, theorem 4].

(ii)⇒(iii) Let $C \subset X$ be a smooth complete intersection of dimension 1. Consider the commutative diagram

$$\begin{array}{ccc} A_0 C & \xrightarrow{\sim} & H^2_{\mathcal{D}}(C, \mathbb{Z}(1)) \\ \downarrow & & \downarrow \\ A_0 X & \longrightarrow & H^{2n}_{\mathcal{D}}(X, \mathbb{Z}(n)) \end{array}$$

where the top horizontal map is an isomorphism by Remark (0.). The right vertical map is surjective by the Lefschetz hyperplane theorem. Under the hypothesis that the lower horizontal map is an injection, we find that the left vertical map is surjective.

(iii)⇒(i) Normalizing Y, we can suppose there is a proper morphism $C \to X$ from a smooth curve C such that $A_0^{\hom} C \to A_0^{\hom} X$ is surjective. Consider the commutative diagram

$$\begin{array}{ccc} S^m C \times S^m C & \longrightarrow & A_0^{\hom} C \\ \downarrow & & \downarrow \\ S^m X \times S^m X & \longrightarrow & A_0 \hom X\,. \end{array}$$

From the Riemann–Roch theorem for curves, one sees that every degree O divisor on C is rationally equivalent to a difference $\alpha - \beta$, where α and β are effective divisors of degree $g =$ the genus of C. The upper horizontal map is thus surjective for $m \geq g$, and under hypothesis (iii), the same holds for the lower horizontal map after tensoring by $\mathbb{Q}$.

All of this shows that the $\otimes\mathbb{Q}$–statements of (i), (ii), (iii) are equivalent, and the integral statements of (i), (ii) and (iii) are equivalent. However, it follows from a result of Roitman [Ro2] that the cycle class map

$$A^{\hom}{}_0 X \longrightarrow \mathcal{J}^n X$$

is an isomorphism on torsion points, so statement (ii) is equivalent to its $\mathbb{Q}$-version. Hence (iii) actually implies that $A_0(X \setminus Y) = 0$, which implies (i).

(i)⇒(iv) This is immediate from the diagram appearing above, in the proof of the implication (i)⇒(iii).

One of the cornerstones of the study of algebraic cycles is the following famous result of Mumford [Mum] [Bl, Lecture 1]:

Theorem 1.4. [Mumford's theorem] *Suppose X is a surface such that $A_0^{\hom} X$ is finite-dimensional. Then $H^2 X$ is algebraic.*

In this section, we will extend this result from 0-cycles to cycles of arbitrary dimension. The method is an immediate extension of Bloch's idea of "decomposing the diagonal" [Bl, Lecture 1] [B–S].

The next definition is motivated by Grothendieck's coniveau filtration on cohomology 1.0 and by the work of Bloch–Srinivas [B–S]:

Definition 1.5. Let X be a variety of dimension n. For two given integers i, r, we will say that

$$\text{Niveau}\big(A^i(X)_{\mathbb{Q}}\big) \leq r$$

if there exists some closed reduced subscheme $Y \subset X$ of codimension $\geq i-r$ such that one has

$$A^i(X \setminus Y)_{\mathbb{Q}} = 0$$

(equivalently, such that the push-forward

$$A_{n-i}(Y)_{\mathbb{Q}} \longrightarrow A_{n-i}(X)_{\mathbb{Q}}$$

is surjective).

This definition is again related to cycle class maps, extending Lemma 1.3:

Lemma 1.6. *Let X be a smooth complete variety.*

(i) $Niveau\big(A_i(X)_{\mathbb{Q}}\big) \leq 0 \ \forall i \leq s \ \Leftrightarrow \ A_i^{\text{hom}}(X)_{\mathbb{Q}} = 0 \ \forall i \leq s.$

(ii) $Niveau\big(A_i(X)_{\mathbb{Q}}\big) \leq 1 \ \forall i \leq s \ \Leftrightarrow \ A_i^{\text{AJ}}(X)_{\mathbb{Q}} = 0 \ \forall i \leq s.$

Proof. [La 1, 1.5].

Theorem 1.7. [A Mumford-type theorem] *Let X be a smooth complete variety of dimension n. Suppose r, s are two integers such that*

$$\text{Niveau}\big(A_i(X)_{\mathbb{Q}}\big) \leq r \quad \textit{for all } i \leq s.$$

Then one has that: $H^i(X,\mathbb{Q})$ is supported in codimension $\lceil \frac{i-2r}{2} \rceil$ for all $i > 2(n-s-1)$.

Note that the case $(r,s) = (0,1)$ corresponds to Mumford's original theorem, by virtue of Lemma 1.3. For the proof, we will need the concepts of correspondences and of pure motives, introduced by Grothendieck:

1.8 Let $\mathcal{VAR}$ denote the category of smooth complete varieties with arbitrary morphisms as arrows.

Let $\sim$ be an adequate equivalence relation on algebraic cycles [We]; the examples we have in mind are rational, homological and numerical equivalence.

Let $X, Y, Z \in \mathcal{VAR}$. A *correspondence from X to Y* is an element

$$\alpha \in \text{Corr}_{\sim}(X,Y) := \frac{A^{\dim X}(X \times Y)_{\mathbb{Q}}}{\sim}\ .$$

Two correspondences can be composed by the following rule: Let $\alpha \in \mathrm{Corr}_{\sim}(X,Y)$, $\beta \in \mathrm{Corr}_{\sim}(Y,Z)$, then the composition $\beta \circ \alpha$ is by definition

$$\beta\circ\alpha := (p_{XZ})_*\Big((p_{XY})^*\alpha \cdot (p_{YZ})^*\beta\Big) \in \frac{A^{\dim X}(X\times Z)_{\mathbb{Q}}}{\sim} =: \mathrm{Corr}_{\sim}(X,Z)\ ,$$

where p_{XZ}, p_{XY} and p_{YZ} are the projection maps from $X \times Y \times Z$ to $X \times Z$, $X \times Y$ and $Y \times Z$.

Correspondences act on cohomology: Let $\alpha \in \mathrm{Corr}_{\mathrm{rat}}(X,Y)$ be a correspondence with respect to rational equivalence. Then α induces a map

$$\alpha_* : H^i(X,\mathbb{Q}) \longrightarrow H^i(Y,\mathbb{Q})$$

defined as

$$\alpha_*(a) := (p_Y)_*\Big(\mathrm{cl}\,(\alpha)\cdot(p_X)^*(a)\Big) \quad \in H^i(Y,\mathbb{Q})$$

(here p_X, p_Y are the projection maps from $X \times Y$ to X resp. Y).

1.9 Pure motives. For any adequate equivalence relation $\sim$, one forms the *correspondence category* $\mathcal{VAR}_{\sim}$ as follows: The objects of $\mathcal{VAR}_{\sim}$ are the same as the objects of $\mathcal{VAR}$, but the arrows are defined by

$$\mathrm{Mor}_{\mathcal{VAR}_{\sim}}(X,Y) := \mathrm{Corr}_{\sim}(X,Y)\ ,$$

composition being defined by the rule of composition of correspondences.

The category of pure, effective motives $\mathcal{M}_{\sim}$ is then defined as follows: Objects are couples (X,p), with $X \in \mathcal{VAR}$ and $p \in \mathrm{Corr}_{\sim}(X,X)$ a correspondence satisfying $p \circ p = p$ (such a p is called a *projector*). Morphisms are given by

$$\mathrm{Mor}_{\mathcal{M}_{\sim}}\Big((X,p),(Y,q)\Big) := q\circ \mathrm{Corr}_{\sim}(X,Y)\circ p \ \subset\ \mathrm{Corr}_{\sim}(X,Y)\ ,$$

composition being induced by composition of correspondences.

The category $\mathcal{M}_{\sim}$ is an additive tensor-category, which is pseudo-abelian (this means that projectors have images); for more on this cf. [Schol].

Varieties have motives: There is a contravariant functor $h\colon \mathcal{VAR} \to \mathcal{M}_{\sim}$, given by

$$\begin{aligned} h(X) &:= (X,[\Delta]),\\ h(f) &:= [{}^t\Gamma_f], \end{aligned}$$

where $\Delta \subset X\times X$ denotes the diagonal, ${}^t\Gamma_f \subset Y\times X$ denotes the transpose of the graph of the morphism f, and [ ] denotes the class of a cycle modulo the equivalence relation $\sim$.

It is an illuminating exercise to check that the (contravariant) functor H^i from $\mathcal{VAR}$ to abelian groups sending a variety X to $H^i(X,\mathbb{Q})$, factors over h — cohomology of a variety depends only on the associated motive.

Proof of Theorem 1.7. The theorem can be deduced from the following claim:

Claim 1.10. *Let X be a smooth complete n-dimensional variety. Suppose*

$$\text{Niveau}\Big(A_i(X)_{\mathbb{Q}}\Big) \leq r \ \forall i \leq s \ .$$

Then there exist closed subschemes $V_0, \ldots, V_s$ and $W_0, \ldots, W_{s+1}$ of X such that $\dim V_j \leq j + r$ $(j = 0, \ldots, s)$, $\dim W_j \leq n - j$ $(j = 0, \ldots, s+1)$, and such that the diagonal Δ has a decomposition

$$\Delta = \Delta_0 + \Delta_1 + \cdots + \Delta_s + \Delta^{s+1} \quad \in A^n(X \times X)_{\mathbb{Q}} \ ,$$

with Δ_j in the image of

$$A_n(V_j \times W_j)_{\mathbb{Q}} \longrightarrow A^n(X \times X)_{\mathbb{Q}}$$

$(j = 0, \ldots, s)$ and Δ^{s+1} in the image of

$$A_n(X \times W_{s+1})_{\mathbb{Q}} \longrightarrow A^n(X \times X)_{\mathbb{Q}} \ .$$

Assuming this claim for a moment, let us prove Theorem 1.7: The hypothesis of the theorem gives us a decomposition of the diagonal as in the claim. Take some cohomology group $H^i(X,\mathbb{Q})$ with $i > 2(n-s-1)$.

Consider the action of the correspondence Δ on $H^i(X,\mathbb{Q})$. Let $\widetilde{V_j}$, $\widetilde{W_j}$ denote desingularisations of V_j resp. W_j, and let $\widetilde{\Delta_j}$, $\widetilde{\Delta}^{s+1}$ be classes in $A_n(\widetilde{V_j} \times \widetilde{W_j})_{\mathbb{Q}}$, resp. $A_n(X \times \widetilde{W}_{s+1})_{\mathbb{Q}}$ mapping to Δ_j, resp. Δ^{s+1}. One can check [La 1] the action of ${}^t\Delta^{s+1}$ factors according to the commutative diagram

$$\begin{array}{ccc}
H^i(\widetilde{W}_{s+1} \times X, \mathbb{Q}) & \overset{\cdot[{}^t\widetilde{\Delta}^{s+1}]}{\longrightarrow} & H^{i+2(n-s-1)}(\widetilde{W}_{s+1} \times X, \mathbb{Q}) \\
\uparrow & & \\
H^i(\widetilde{W}_{s+1}, \mathbb{Q}) & & \downarrow \\
\uparrow & & \\
H^i(X,\mathbb{Q}) & \overset{({}^t\Delta^{s+1})_*}{\longrightarrow} & H^i(X,\mathbb{Q}) \ ,
\end{array}$$

where the left resp. right vertical maps are the obvious pull-back resp. push-forward maps.

However, the group $H^i(\widetilde{W}_{s+1}, \mathbb{Q})$ vanishes since $\dim W_{s+1} \leq n - s - 1$, so that

$$({}^t\Delta^{s+1})_* H^i(X,\mathbb{Q}) = 0 \ .$$

Likewise, the action of the correspondence ${}^t\Delta_j$ $(j = 0, \ldots, s)$ factors as

$$\begin{array}{ccc} H^i(\widetilde{W_j} \times \widetilde{V_j}, \mathbb{Q}) & \stackrel{\cdot[{}^t\widetilde{\Delta}_j]}{\longrightarrow} & H^{i+2r}(\widetilde{W_j} \times \widetilde{V_j}, \mathbb{Q}) \\ \uparrow & & \downarrow \\ H^i(\widetilde{W_j}, \mathbb{Q}) & & H^{i-2(n-j-r)}(\widetilde{V_j}, \mathbb{Q}) \\ \uparrow & & \downarrow \\ H^i(X, \mathbb{Q}) & \stackrel{({}^t\Delta_j)_*}{\longrightarrow} & H^i(X, \mathbb{Q}) \ . \end{array}$$

Now again for dimension reasons, the group $H^i(\widetilde{W_j}, \mathbb{Q})$ vanishes as soon as $i > 2(n-j)$. So we have $(\Delta_j)_* H^i(X, \mathbb{Q}) = 0$ unless $i \leq 2(n-j)$. But $({}^t\Delta_j)_* H^i(X, \mathbb{Q})$ is supported on V_j, as the above diagram shows. And for $i \leq 2(n-j)$ one has

$$({}^t\Delta_j)_* H^i(X, \mathbb{Q}) \subset N^{\lceil \frac{i-2r}{2} \rceil} H^i(X, \mathbb{Q}) \ .$$

Putting all this together and noting that the diagonal acts as the identity, we find that

$$\begin{aligned} H^i(X, \mathbb{Q}) &= \Delta_* H^i(X, \mathbb{Q}) = ({}^t\Delta)_* H^i(X, \mathbb{Q}) \\ &= ({}^t(\Delta_0 + \cdots + \Delta_s))_* H^i(X, \mathbb{Q}) \subset N^{\lceil \frac{i-2r}{2} \rceil} H^i(X, \mathbb{Q}) \ , \end{aligned}$$

proving the theorem.

It is left to prove the Claim 1.10: The hypothesis

$$\text{Niveau}\Big(A^n(X)_{\mathbb{Q}}\Big) \leq r$$

means that there exists $Y \subset X$ such that

$$A^n(X \setminus Y)_{\mathbb{Q}} = 0 \ .$$

Taking the smallest field of definition of X and Y, we may suppose X and Y are defined over a field k that is finitely generated over its prime subfield. Now a fundamental property of Chow groups is that for a field extension $K \subset L$ the application

$$A^*(M_K)_{\mathbb{Q}} \longrightarrow A^*(M_L)_{\mathbb{Q}}$$

is injective [Bl, Appendix to Lecture 1]. This implies that we actually have

$$A^n\big((X \setminus Y)_K\big)_{\mathbb{Q}} = 0$$

for any finitely generated $K \supset k$ (since such a K is contained in $\mathbb{C}$).

Consider now the restriction map

$$A^n(X_k \times X_k)_{\mathbb{Q}} \longrightarrow A^n(X_{k(X)})_{\mathbb{Q}}$$

($k(X)$ being the function field of X). By assumption, there exists $V_0 \subset X_{k(X)}$ such that the map

$$A_0(V_0)_{\mathbb{Q}} \longrightarrow A^n(X_{k(X)})_{\mathbb{Q}}$$

is surjective. In particular, the restriction of the diagonal to $X_{k(X)}$ comes from a cycle on V_0. Let $\Delta_0 \in A^n(X_k \times X_k)_{\mathbb{Q}}$ denote the closure of this cycle.

By construction, the difference

$$\Delta^1 := \Delta - \Delta_0 \ \in A^n(X_k \times X_k)_{\mathbb{Q}}$$

maps to 0 in

$$A^n(X_{k(X)})_{\mathbb{Q}} = \varinjlim_{U \subset X_k \text{ open}} A^n(X_k \times U)_{\mathbb{Q}} ,$$

so Δ^1 maps to 0 when restricted to some sufficiently small U. Denoting by $W_1 \subset X$ the complement of such a U, we find that the cycle Δ^1 comes from $A_n(X \times W_1)_{\mathbb{Q}}$.

If $s = 0$, we have found a decomposition $\Delta = \Delta_0 + \Delta^1$ satisfying the claim, and we are done. If $s > 0$, we apply the same argument to

$$\Delta^1 \in A_n(X \times W_1)_{\mathbb{Q}}$$

and the restriction

$$A_n(X \times W_1) \longrightarrow A^{n-1}(X_{k(W_1)})_{\mathbb{Q}} .$$

After $s+1$ steps we arrive at a decomposition as in the claim. □

Corollary 1.11. *Let X be a smooth complete variety of dimension n.*

(i) *Suppose $A_i^{\text{hom}}(X)_{\mathbb{Q}} = 0$ for all $i < n/2$. Then the cycle class map induces an isomorphism of graded rings*

$$A^*(X)_{\mathbb{Q}} \xrightarrow{\sim} H^*(X, \mathbb{Q}) ;$$

in particular Hodge numbers $h^{p,q}(X)$ vanish if $p \neq q$;

(ii) *Suppose $A_i^{\text{AJ}}(X)_{\mathbb{Q}} = 0$ for all $i < n/2 - 1$. Then the cycle class map induces an isomorphism of graded rings*

$$A^*(X)_{\mathbb{Q}} \xrightarrow{\sim} H^*_{\mathcal{D}}(X, \mathbb{Q}(*)) ;$$

in particular Hodge numbers $h^{p,q}(X)$ vanish if $|p-q| > 1$.

Proof. [La 1].

Theorem 1.7 shows how the Chow groups $A_0, \ldots, A_s$ influence cohomology. A variant result is possible, where we start not with 0-cycles, but with divisors. For this result, we need to make one extra assumption:

1.12 Let X be a smooth complete variety. One says that *the Künneth components of the diagonal are algebraic* if the classes

$$\pi_i \in H^{2n-i}(X,\mathbb{Q}) \otimes H^i(X,\mathbb{Q}) \subset H^{2n}(X \times X,\mathbb{Q})$$

appearing in the Künnneth decomposition of the diagonal, are in the image of the cycle class map $A^n(X \times X)_{\mathbb{Q}} \to H^{2n}(X \times X,\mathbb{Q})$.

Note that if the Künneth components are algebraic, one can decompose the motive $h(X) \in \mathcal{M}_{\text{hom}}$ as

$$h(X) = h^0(X) \oplus h^1(X) \oplus \cdots \oplus h^{2n}(X) \quad \in \mathcal{M}_{\text{hom}} \ ,$$

where the motive $h^i(X) := (X,\pi_i)$ is responsible for the cohomology group $H^i(X,\mathbb{Q})$.

For general X, this hypothesis is not known to hold; however it would follow from Grothendieck's standard conjectures [Gr] [Kl], or from the Hodge conjecture. The Künneth components are known to be algebraic in case X can be defined over a finite field [K–M].

Theorem 1.13. [Variant Mumford-type theorem] *Let X be a smooth complete variety, and suppose the Künneth components of the diagonal of X are algebraic. If*

$$\text{Niveau}\big(A^i(X)_{\mathbb{Q}}\big) \le r \quad \forall i \le s \ ,$$

then

$$H^i(X,\mathbb{Q}) = N^{\lceil\frac{j-2r}{2}\rceil} H^i(X,\mathbb{Q})$$

for $i \le s$, and for $i \ge 2n-s$.

Proof. Consider the Künneth component

$$\pi_{2n-i} \in H^i(X,\mathbb{Q}) \otimes H^{2n-i}(X,\mathbb{Q}) \ ,$$

for $i \le s$.

It follows from the Lefschetz hyperplane theorem that π_{2n-i} is in the image of

$$H^i(X,\mathbb{Q}) \otimes H^i(H^{n-i},\mathbb{Q}) \ \to \ H^{2n}(X \times X,\mathbb{Q}) \ ,$$

where H^k denotes k-fold general hyperplane section.

On the other hand, we suppose π_{2n-i} is algebraic. This means there is a cycle underlying π_{2n-i}, and this cycle is homologically trivial when restricted to $X \times (X \setminus H^{n-i})$. We can thus suppose this cycle is supported on $X \times H^{n-i}$, i.e., we have an element

$$\alpha \in A^i(X \times H^{n-i})_{\mathbb{Q}}$$

mapping to π_{2n-i}.

Since $i \leq s$, we have by hypothesis $\mathrm{Niveau}(A^i(X)_\mathbb{Q}) \leq r$. Therefore, we can apply the procedure of the proof of (1.) to the cycle α. The result is a decomposition

$$\alpha = \alpha_0 + \cdots + \alpha_l \quad \in A^n(X \times X)_\mathbb{Q} ,$$

where each α_j has support on $V_j \times W_j$ for some closed subvarieties $V_j, W_j \subset X$ of dimension $n-j+r+i$ resp. $j-i$. A fortiori $\pi_{2n-i} \in H^{2n}(X \times X, \mathbb{Q})$ has such a decomposition. Applying the correspondence π_{2n-i} to $H^{2n-i}(X, \mathbb{Q})$ as in (1.) yields the result for $i \geq 2n-s$. For $i \leq s$, just use that $\pi_i = {}^t\pi_{2n-i}$. □

Remark 1.14. (i) The method of proof in this section is far from being original. Indeed, Bloch proves Mumford's theorem along these lines [Bl, Appendix to Lecture 1], and Bloch–Srinivas give further applications to 0-cycles [B–S].

Results similar to 1.11 (i) resp. (ii) have been proven by Jannsen [Ja 4] resp. Esnault–Levine [E–L].

(ii) Stronger Mumford type results have been proven by Lewis [Le] and by Schoen [Sc], both assuming some standard conjecture.

(iii) For a sample of some further applications, cf. [La 1] or [Pe].

2 From H^* to A^*? — Conjectures

In his book on algebraic cycles [Bl], Bloch conjectured that the converse to Mumford's theorem should be true:

Bloch's conjecture 2.0. Let S be a smooth complete surface such that $H^O(S, \Omega^2) = 0$. Then $A_0 S$ is finite-dimensional.

This conjecture has been verified for surfaces S which are not of general type [B–K–S], but in general it is still wide open.

Bloch's conjecture forms part of a vast conjectural framework, which is sometimes referred to as the "philosophy of mixed motives", developed mainly by Beilinson and Bloch (a good introduction to this theory is [Ja 4]). This framework admits several independent equivalent formulations.

Beilinson's conjectures 2.1. For every smooth complete variety X, there exists a descending filtration F^* on $A^i(X)_\mathbb{Q}$, for all $i \geq 0$, satisfying

(i)

$$\begin{aligned} F^0 A^i(X)_\mathbb{Q} &= A^i(X)_\mathbb{Q} , \\ F^1 A^i(X)_\mathbb{Q} &= A^i_{\mathrm{hom}}(X)_\mathbb{Q} , \\ F^2 A^i(X)_\mathbb{Q} &= A^i_{\mathrm{AJ}}(X)_\mathbb{Q} , \\ F^N A^i(X)_\mathbb{Q} &= 0 \ \text{for}\ N >> 0 ; \end{aligned}$$

(ii) F^* is preserved by f_* and f^* for arbitrary morphisms f, and by exterior products;

(iii) the Künneth components of the diagonal are algebraic, and the graded

$$\mathrm{Gr}\,^r_F A^i(X)_{\mathbb{Q}}$$

depends only on the motive $h^{2i-r}(X) \in \mathcal{M}_{\mathrm{hom}}$.

Remarks 2.2. (i) The meaning of 2.1 (ii) is that for any morphism $f: X \to Y$, $f^*: A^*(Y)_{\mathbb{Q}} \to A^*(X)_{\mathbb{Q}}$ and $f_*: A^*(X)_{\mathbb{Q}} \to A^*(Y)_{\mathbb{Q}}$ send F^r into F^r, and that for any two X, Y, the exterior product

$$A^i(X)_{\mathbb{Q}} \otimes A^j(Y)_{\mathbb{Q}} \longrightarrow A^{i+j}(X \times Y)_{\mathbb{Q}}$$

sends $F^r \otimes F^s$ into F^{r+s}.

As a consequence, the filtration F^* is compatible with the intersection product:

$$A^i(X)_{\mathbb{Q}} \otimes A^j(X)_{\mathbb{Q}} \longrightarrow A^{i+j}(X)_{\mathbb{Q}}$$

sends $F^r \otimes F^s$ to F^{r+s}.

(ii) The meaning of 2.1 (iii) is as follows: As a consequence of 2.1 (ii), correspondences for rational equivalence preserve F^*, and hence act on $\mathrm{Gr}\,^r_F A^i(X)_{\mathbb{Q}}$. However, since by (i) one has

$$A^n_{\mathrm{hom}}(X \times X)_{\mathbb{Q}} = F^1 A^n(X \times X)_{\mathbb{Q}},$$

homologically trivial correspondences act trivially on $\mathrm{Gr}\,^r_F A^i(X)_{\mathbb{Q}}$. Hence correspondences for homological equivalence act on the graded $\mathrm{Gr}\,^r_F A^i(X)_{\mathbb{Q}}$.

Now 2.1 (iii) means that the action of the correspondence π_j on $\mathrm{Gr}\,^r_F A^i(X)_{\mathbb{Q}}$ is 0 for $j \neq 2i - r$, and is the identity otherwise.

(iii) A trivial case of 2.2 is the divisor case: on $A^1(X)_{\mathbb{Q}}$ there is a one-step filtration $F^1 A^1(X)_{\mathbb{Q}} = A^1_{\mathrm{hom}}(X)_{\mathbb{Q}}$, $F^2 A^1(X)_{\mathbb{Q}} = 0$, such that $\mathrm{Gr}\,^1_F A^1(X)_{\mathbb{Q}} = A^1_{\mathrm{hom}}(X)_{\mathbb{Q}}$, being isomorphic to

$$\frac{H^1(X, \mathbb{C})}{F^1 + H^1(X, \mathbb{Z})}$$

(cf. (0.)) only depends on $h^1(X)$.

(iv) There is a more fancy formulation of Beilinson's conjectures, where the dependence 2.1 (iii) is explained by a concise formula. This formulation conjectures the existence of a category $\mathcal{M}M$ ("mixed motives over $\mathbb{C}$") containing $\mathcal{M}_{\mathrm{hom}}$ as a full subcategory, and such that the gradeds for the filtration F^* appear as extension groups:

$$\mathrm{Gr}\,^r_F A^i(X)_{\mathbb{Q}} = \mathrm{Ext}^r_{\mathcal{M}M}\big(h(^{\mathbb{C}}), h^{2i-r}(X)(i)\big) .$$

Examples 2.3. The conjecture 2.1 implies Bloch's conjecture (2.0). Indeed, suppose S is a surface such that $H^0(S,\Omega^2)=0$ i.e. $H^2(S,\mathbb{Q})$ is algebraic. Then the Künneth component $\pi_2 \in H^2(S,\mathbb{Q}) \otimes H^2(S,\mathbb{Q})$ consists of a sum $\sum a_{ij} C_i \times C_j$, where the $C_l \subset S$ are curves.

Conjecture 2.1 (iii) states that $F^2A^2(S)_{\mathbb{Q}} = A^2_{\mathrm{AJ}}(S)_{\mathbb{Q}}$ only depends on π_2. But since a 0-cycle on S can be moved away from a finite union of curves, π_2 acts trivially on $A^2(S)_{\mathbb{Q}}$, hence $F^2A^2(S)_{\mathbb{Q}}=0$.

Alternatives 2.4. An independent proposal for a filtration F^* on Chow groups has been given by Murre [Mu], based on a (conjectural) decomposition of the motive $h(X) \in \mathcal{M}_{\mathrm{rat}}$ refining the Künneth decomposition in $\mathcal{M}_{\mathrm{hom}}$. Jannsen then proved the surprising fact that Murre's conjectures are actually equivalent to Beilinson's, and that if they are true, the two filtrations coincide [Ja 4, Section 5].

Another interesting construction of a filtration on Chow groups has been given by S. Saito [Sa].

We elaborate on yet another approach to the filtration F^*, which was developed by Jannsen [Ja 4] and Raskind [Ra]. In this approach, we need the notion of continuous étale cohomology, defined by Jannsen [Ja 2].

2.5 Let X now be a scheme over an arbitrary field k. One problem with the l-adic cohomology groups

$$H^i(X,\mathbb{Z}_l(j)) := \lim_{\substack{\longleftarrow \\ m}} H^i(X,\mathbb{Z}/l^m(j))$$

of X, is that they are not defined as derived functors, but as inverse limits of derived functors.

To remedy this defect, Jannsen [Ja 2] introduced a new cohomology theory, by deriving the left exact functor

$$\begin{array}{rcl}\{\text{inverse systems of étale sheaves on } X\} & \longrightarrow & \{\text{abelian groups}\} \\ (F_m) & \mapsto & \lim\limits_{\substack{\longleftarrow \\ m}} H^0(X,F_m)\ .\end{array}$$

Defining continuous l-adic cohomology as

$$H^i_{\mathrm{cont}}(X,\mathbb{Z}_l(j)) := H^i(X,(\mathbb{Z}/l^m(j)))\ ,$$

one gets the usual l-adic cohomology in the case k is algebraically closed, but a different theory in general.

In fact, one can construct a Hochschild–Serre spectral sequence

$$E_2^{p,q} = H^p_{\mathrm{cont}}(G_k, H^q(X_{\overline{k}},\mathbb{Z}_l(j))) \ \Rightarrow\ H^{p+q}_{\mathrm{cont}}(X,\mathbb{Z}_l(j))\ ,$$

which in general does not exist for the usual l-adic cohomology.

One also defines $\mathbb{Q}_l$-adic continuous étale cohomology, setting

$$H^i_{\text{cont}}(X, \mathbb{Q}_l(j)) := H^i_{\text{cont}}(X, \mathbb{Z}_l(j)) \otimes \mathbb{Q}_l \ .$$

For a smooth and complete variety X, the Hochschild–Serre spectral sequence with $\mathbb{Q}_l$-coefficients degenerates at the E_2-term [Ra].

2.6 Let X be a smooth complete variety over a field k. As is to be expected, there exist cycle class maps

$$A^i(X) \longrightarrow H^{2i}_{\text{cont}}(X, \mathbb{Z}_l(i))$$

with the usual functorial properties [Ja 2, Section 3]. So one can try to define a filtration on Chow groups using the filtration on $H^{2i}_{\text{cont}}(X, \mathbb{Z}_l(i))$ coming from the Hochschild–Serre spectral sequence.

Certainly, this can only give a finite filtration on $A^i(X)$ if the cycle class map is injective, which cannot be true in general (for instance, for k algebraically closed). However, Jannsen and Raskind have conjectured the following:

(*Injectivity conjecture*) Let X be defined over a field k which is finitely generated over its prime subfield, and let l be a prime different from the characteristic of k. Then the cycle class map

$$A^i(X)_{\mathbb{Q}} \longrightarrow H^{2i}_{\text{cont}}(X, \mathbb{Q}_l(i))$$

is injective.

Assuming this conjecture, the Hochschild–Serre spectral sequence defines a finite filtration F^* on $A^i(X)_{\mathbb{Q}}$. For a general field k, a filtration F^* is then defined by

$$F^r A^i(X)_{\mathbb{Q}} := \varinjlim_{L} F^r A^i(X_L)_{\mathbb{Q}} \ ,$$

the limit over all finitely generated fields $L \subset k$ such that X has a model over L.

This conjectural filtration has wonderful properties. In particular, it shows how Chow groups should be governed by cohomology groups, just as in Beilinson's conjectures:

2.7 Let $\mathcal{INJ}$ be the class of smooth complete varieties over $\mathbb{C}$, which satisfy the injectivity conjecture over every finitely generated field for which there is a model.

Proposition 2.8. *Let $f: Y \to X$ be a morphism between two varieties in $\mathcal{INJ}$. Suppose $f^*: H^j(X, \mathbb{Q}) \to H^j(Y, \mathbb{Q})$ is an isomorphism for $i \le j < 2i$. Then also*

$$f^*: \ A^i_{\text{hom}}(X)_{\mathbb{Q}} \ \to \ A^i_{\text{hom}}(Y)_{\mathbb{Q}}$$

is an isomorphism.

Proof. Immediate from the construction of the filtration F^*. □

Remarks 2.9. (i) Similar statements can also be made about the push-forward f_* instead of the pull-back.

(ii) It would be very interesting to try and prove injectivity for, say, Fermat hypersurfaces. This would yield that

$$A^i(X)_{\mathbb{Q}} \cong \mathbb{Q} \ \ \forall i < \dim X/2 \ ,$$

for X a Fermat hypersurface.

It was first conjectured by Hartshorne [Ha] that this should be true for any smooth complete intersection. Of course, such a result would follow from Beilinson's conjectures 2.1.

3 On Codimension 2 Cycles

In this last section, we further elaborate on the consequences of the Beilinson type conjecture treated in Section 2. For simplicity, we will restrict to codimension 2 cycles.

3.0 Suppose Beilinson's conjectures hold in the strong version mentioned in 2.2. Consider a morphism $f: Y \to X$ between two smooth, complete varieties, and suppose that the pull-back map

$$f^*:\ H^0(X, \Omega^2) \longrightarrow H^0(Y, \Omega^2)$$

is an injection. By the truth of the Hodge conjecture for divisors, this means that the kernel of

$$f^*:\ H^2(X, \mathbb{Q}) \longrightarrow H^2(Y, \mathbb{Q})$$

is generated by divisors, and using Manin's identity principle [Schol] for pure motives, it follows that the kernel of the map of motives in $\mathcal{M}_{\mathrm{hom}}$

$$f^*:\ h^2(X) \longrightarrow h^2(Y)$$

is equal to a sum of trivial motives $h(^{\mathbb{C}})(-1)$. If the category $\mathcal{M}_{\mathrm{hom}}$ is semi-simple, as is expected (this is true if and only if homological and numerical equivalence coincide, one of the standard conjectures, cf. [Ja 3]), this would imply that the map

$$\mathrm{Ext}^2_{\mathcal{MM}}(h(^{\mathbb{C}}), h^2(X)(2)) \longrightarrow \mathrm{Ext}^2_{\mathcal{MM}}(h(^{\mathbb{C}}), h^2(Y)(2))$$

is injective (since $h(^{\mathbb{C}})(1)$ does not have any Ext^2). Using Beilinson's formula representing A^2_{AJ} as an Ext^2-group, we thus get the following conjecture:

Conjecture 3.1. *Let $f: Y \to X$ be a morphism between smooth complete varieties, such that*

$$f^*:\ H^0(X,\Omega^2) \longrightarrow H^0(Y,\Omega^2)$$

is an injection (resp. a surjection). Then also

$$f^*:\ A^2_{\mathrm{AJ}}(X)_{\mathbb{Q}} \longrightarrow A^2_{\mathrm{AJ}}(Y)_{\mathbb{Q}}$$

is an injection (resp. a surjection).

Remarks 3.2. (i) We get back Bloch's conjecture 2.0 by taking one of the two varieties a surface, and the other a point. In fact, 3.1 is a relative version of Bloch's conjecture.

(ii) An interesting open question is whether the converse to 3.1 is true; this would be a relative version of Mumford's theorem 1.0, so it should be easier to prove than 3.1.

3.3 One special case of conjecture 3.1 is that $A^2_{\mathrm{AJ}}()_{\mathbb{Q}}$ should be a birational invariant. This can actually be proven [La 2].

3.4 Another special case of conjecture 3.1 is obtained by taking $Y \subset X$ to be a smooth hyperplane section. One would then get that

$$A^2_{\mathrm{AJ}}(X)_{\mathbb{Q}} \longrightarrow A^2_{\mathrm{AJ}}(Y)_{\mathbb{Q}}$$

is an injection for $\dim Y \geq 2$, and an isomorphism for $\dim Y \geq 3$. Under somewhat stronger restrictions, this can again actually be proven [La 2].

References

[Ba] L. Barbieri–Viale, Des invariants birationnels associés aux théories cohomologiques, *C. R. Acad. Sci. Paris* **315** (1992), 1259–1262

[Bl] S. Bloch, *Lectures on algebraic cycles*, Duke Univ. Math. Ser. 4, Durham 1980

[B–K–L] S. Bloch, A. Kas and D. Lieberman, Zero-cycles on surfaces with $p_g = 0$, *Comp. Math.* **33** (1976), 135–145

[B–O] S. Bloch and A. Ogus, Gersten's conjecture and the homology of schemes, *Ann. Scient. de l'E.N.S.* 4ᵉ série 7 (1974), 181–202

[B–S] S. Bloch and V. Srinivas, Remarks on correspondences and algebraic cycles, *Amer. J. Math.* **105** (1983), 1235–1253

[C–H] A. Corti and M. Hanamura, Motivic decomposition and intersection Chow groups I, preprint 1998

[Co] J.–L. Colliot–Thélène, Birational invariants, purity and the Gersten conjecture, in: *Proceedings of Symposia in Pure Mathematics* Vol. 58 No. 1 (1995)

[C–H–K] J.-L. Colliot-Thélène, R. Hoobler and B. Kahn, The Bloch–Ogus–Gabber theorem, in: *The Fields Institute Communications* Vol. 16 (1997)

[E–V] H. Esnault and E. Viehweg, Deligne–Beilinson cohomology, in: *Beilinson's conjectures on special values of L-functions* (M. Rapoport, N. Schappacher and P. Schneider, eds.), Academic Press 1988

[Fu] W. Fulton, *Intersection theory*, Springer–Verlag, 1984

[Gi] H. Gillet, Deligne homology and Abel–Jacobi maps, *Bulletin of the AMS* Vol. 10 No. 2 (1984), 285–288

[Gri] P. Griffiths, ed., *Topics in transcendental algebraic geometry*, Annals of Mathematics Study 106, Princeton University Press, 1984

[Gr] A. Grothendieck, Hodge's general conjecture is false for trivial reasons, *Topology* **8** (1969), 299–303

[Gr 2] A. Grothendieck, Standard conjectures on algebraic cycles, in: *Algebraic geometry*, Bombay 1968, Oxford, 1969

[Ha] R. Hartshorne, Equivalence relations on algebraic cycles and subvarieties of small codimension, in: *Algebraic geometry* (Arcata 1974), Proceedings of Symposia in Pure Mathematics 29 (1975), 129–164

[Ja 1] U. Jannsen, Deligne cohomology, Hodge $\mathcal{D}$-conjecture and motives, in: *Beilinson's conjectures on special values of L-functions* (M. Rapoport, N. Schappacher and P. Schneider, eds.), Academic Press, 1988

[Ja 2] U. Jannsen, Continuous étale cohomology, *Math. Ann.* **280** (1988), 207–245

[Ja 3] U. Jannsen, Motives, numerical equivalence, and semi-simplicity, *Invent. Math.* **107** (1992), 447–452

[Ja 4] U. Jannsen, Mixed motivic sheaves and filtrations on Chow groups, in: *Motives*, Proceedings of Symposia in Pure Mathematics 55 (1994) Part 1

[Ki] S. Kimura, Fractional intersection and bivariant theory, *Comm. Algebra* **20** (1992), 285–302

[K–M] N. Katz and W. Messing, Some consequences of the Riemann hypothesisd for varieties over finite fields, *Inv. Math.* **23** (1974), 73–77

[Kl] S. Kleiman, The standard conjectures, in: *Motives*, Proceedings of Symposia in Pure Mathematics 55 (1994) Part 1

[K–S] T. Katsura and T. Shioda, On Fermat varieties, *Tôhoku Math. J.* **31** (1979), 97–115

[La 1] R. Laterveer, Algebraic varieties with small Chow groups, to appear in *J. Math. Kyoto Univ.*

[La 2] R. Laterveer, Codimension 2 algebraic cycles, in preparation

[Le] J. Lewis, A generalization of Mumford's theorem II, *Illinois Journal of Mathematics* **39** (1995), 288–304

[Mu] J.P. Murre, On a conjectural filtration on the Chow groups of an algebraic variety, I and II, *Indag. Math.* **4** (1993), 177–201

[No] M. Nori, Algebraic cycles and Hodge theoretic connectivity, *Inv. Math.* **111** (1993), 349–373

[Pe] C. Peters, Lawson homology for varieties with small Chow groups, preprint 1998

[Ro] J. Roberts, Chow's moving lemma, in: *Algebraic geometry* Oslo 1970 (F. Oort, ed.), Wolters–Noordhoff Publ. Groningen 1972

[Ro 1] A. Roitman, Rational equivalence of zero-cycles, *Math. USSR-Sb.* **18** (1974), 571–588

[Ro 2] A. Roitman, The torsion in the group of zero-cycles modulo rational equivalence, *Ann. of Math.* **111** (1980), 553–569

[Sa] S. Saito, Motives and filtrations on Chow groups, *Invent. Math.* **125** (1996), 149–196

[Sc] C. Schoen, On Hodge structures and non-representability of Chow groups, *Comp. Math.* **88** (1993), 285–316

[Schol] A.J. Scholl, Classical motives, in: *Motives*, Proceedings of Symposia in Pure Mathematics 55 (1994) Part 1

[Sh] T. Shioda, The Hodge conjecture for Fermat varieties, *Math. Ann.* **245** (1979), 175–184

[We] A. Weil, Sur les critères d'équivalence en géométrie algébrique

CNRS, IRMA,
7 rue R. Descartes,
67084 Strasbourg Cedex,
France
`laterv@math.u-strasbg.fr`

Zero Estimates on Abelian Varieties

Michael Nakamaye

ABSTRACT We give two proofs that a principally polarized abelian variety with maximally singular theta divisor must split into a product of elliptic curves. Both proofs are based on a systematic use of intersection theory, applied to derivatives of the theta divisor. The second of the two proofs, inspired by zero estimate techniques of transcendental number theory, is closely related to the study of Seshadri contants.

0 Introduction

Let (A, Θ) be a principally polarized abelian variety of dimension g. Whereas on a general principally polarized abelian variety the theta divisor is smooth, it is well known that when A is a Jacobian

$$\dim(\Theta_{\text{sing}}) \geq g - 4.$$

This observation forms the basis for one of the classical approaches to the Schottky problem, developed by Andreotti and Mayer. In light of this theory, it is not surprising that information about Θ_{sing} can yield interesting global information about the abelian variety A.

The n-fold singular locus of Θ, denoted S_n, is defined for all non-negative integers n by

$$S_n = \{P \in A : \text{mult}_P(\Theta) \geq n\};$$

here $\text{mult}_P(\Theta)$ is the multiplicity at P of a local equation for Θ in a neighborhood of P. Kollár ([K] Theorem 17.13) has recently shown that the pair (A, Θ) is log-canonical which implies

Theorem 0.1. *Let (A, Θ) be a principally polarized abelian variety of dimension g. Then*

$$\text{codim}(S_n, A) \geq n.$$

Ein and Lazarsfeld [EL2] have established that extremal situations where $\text{codim}(S_n, A) = n$ occur only when (A, Θ) splits into a product:

Theorem 0.2. *Let (A, Θ) be a principally polarized abelian variety of dimension g and suppose* $\operatorname{codim}(S_n, A) = n$ *for some $n \geq 2$. Then*

$$(A, \Theta) = \prod_{i=1}^{n} (B_i, \Theta_i),$$

where (B_i, Θ_i) are principally polarized abelian varieties for $1 \leq i \leq n$.

Theorem 0.2 has several interesting corollaries. At one extreme, when $n = 2$, Theorem 0.2 solves the following famous problem:

Corollary 0.3. *Suppose (A, Θ) is a principally polarized abelian variety and that Θ is singular in codimension 1. Then there exist principally polarized abelian varieties (A_1, Θ_1) and (A_2, Θ_2) such that*

$$(A, \Theta) = (A_1, \Theta_1) \times (A_2, \Theta_2).$$

At the other extreme, when $n = g$, one finds the following concise criterion for deciding whether or not a principally polarized abelian variety splits into a product of g elliptic curves:

Corollary 0.4. *Let (A, Θ) be a principally polarized abelian variety of dimension g and suppose there is a point $P \in A$ such that* $\operatorname{mult}_P(\Theta) = g$. *Then there exist elliptic curves $E_1, \ldots, E_g$ such that*

$$A = E_1 \times \ldots \times E_g \ \ \textit{and} \ \ \Theta = \sum_{i=1}^{g} \pi_i^{-1}(\pi_i(P));$$

here $\pi_i : E_1 \times \ldots \times E_g \to E_i$ denotes the projection to the ith factor.

Corollary 0.4 was established prior to the work of Ein and Lazarsfeld, by Smith and Varley [SV] and by the present author, independently. The goal of this note is to present two proofs of Corollary 0.4. The first proof, using intersection theory and the Matsusaka–Ran criterion for Jacobians ([BL] Chapter 11, 8.1), is similar in flavor to that of Smith and Varley although the language is different as we use the techniques developed in [ELN]. In the second proof, we use techniques developed in transcendence theory [B, N1] to produce explicitly the elliptic curves $E_1, \ldots, E_g$ and the isomorphism in Corollary 0.4, independent of the Matsusaka–Ran criterion. This proof is motivated by a result on Seshadri constants of abelian varieties which is sharpened in [N2] and which is presented at the end of this paper.

The fundamental observation behind this work, due to R. Lazarsfeld, is that if Θ has a singularity of multiplicity g at a point P then a general i^{th} order derivative $D_i(\Theta)$ will have multiplicity $g - i$ and consequently the intersection multiplicity at P satisfies

$$i(P, \Theta \cdot D_1(\Theta) \cdot \ldots \cdot D_{g-1}(\Theta); A) \geq g!\,;$$

the fact that the intersection is proper follows from Kollár's Theorem 0.1. On the other hand, assuming that $D_i(\Theta)$ is numerically equivalent to Θ,

$$\Theta \cdot D_1(\Theta) \cdot \ldots \cdot D_{g-1}(\Theta) = c_1(\Theta)^g = g!.$$

Thus the entire intersection product is supported at P and one expects a simple geometric explanation for this phenomenon. It remains to define rigorously the derivatives $D_i(\Theta)$ and this will be accomplished in §1.

F. Angelini, expanding on Lazarsfeld's idea, used Demailly's self-intersection inequality ([D1] 10.7) to show that if Θ has a singularity of multiplicity g, then $\deg_\Theta(S_{g-1}) \leq g$ (see [ELN] Remark 2.7). If S_{g-1} were non-degenerate, this would imply that $\deg_\Theta(S_{g-1}) = g$ and one could then apply the Matsusaka–Ran criterion to conclude that A is a Jacobian in which case Corollary 0.4 is clear. We will show below by intersection theoretic techniques that S_{g-1} is non-degenerate and this will give the first of our two proofs of Corollary 0.4.

As a warm-up for our second proof of Corollary 0.4, consider a singular conic $C \subset \mathbf{P}^2$. Of course, C is either a union of two distinct lines or it is a double line. Geometrically, this can be seen as follows: suppose $P \in C$ is a singular point. By Bézout's theorem, if L is any line through P meeting C properly, then $C \cap L$ is supported entirely at the point P. Thus if $Q \neq P$ is any other point on C then the line $\overline{PQ}$ must be contained in C; hence C is either a union of two lines or a double line.

Our second method of proof for Corollary 0.4 mimics closely this argument for singular conics. In particular, we will construct curves $C_i \subset A$ (the irreducible components of S_{g-1}) and a linear series of divisors (obtained by differentiating Θ) whose intersection with the C_i is entirely supported at a point P. Using the method of differentiating along a one-parameter subgroup (see [B, N1]), fundamental in many transcendence theory arguments, we then show that this implies that the C_i are elliptic curves and that A splits into a product of the C_i.

This paper is written in the same spirit as [ELN], namely as an attempt to illustrate how ideas from diophantine approximation and transcendence theory can be applied in certain purely algebro-geometric situations. The arguments in this paper, revolving around intersection theory and Bézout's theorem, occur in nearly identical form in the zero estimates of transcendence theory [B, N1]. In diophantine approximation as well, one encounters a virtually identical set-up: instead of a highly singular principal polarization Θ one has a so-called 'auxilliary polynomial' $P \in \mathbf{Z}[X_1, \ldots, X_m]$ with a very complicated singularity at a fixed point $(x_1, \ldots, x_m)$. Just as in the proof of Corollary 0.4, one takes derivatives of P to construct an intersection product, most of which will be supported at $(x_1, \ldots, x_m)$ because of the singularity of P at this point. The conclusion is that the polynomial could not have large order of vanishing elsewhere (or else Bézout's theorem would be violated) and this allows one to establish such important results as Roth's theorem. The same argument has been further refined and developed by Faltings and Vojta to prove the Mordell conjecture as well as higher dimensional analogues (for an account of this work written in the same spirit as this paper, see [N3]).

Notation and Conventions

- If L is a line bundle on a variety X and $\sigma \in H^0(X, \mathcal{O}_X(L))$ then we will write $Z(\sigma)$ for the zero scheme of σ.
- Suppose X is a smooth projective variety. Let $V_1, \ldots, V_k$ be pure dimensional subschemes of X, and Z a proper component of the intersection $V_1 \cap \ldots \cap V_k$. Then the intersection multiplicity $i(Z, V_1 \cdot \ldots \cdot V_k; X)$ is defined as in [Fu] §8.2. The intersection multiplicity can be defined for cycles using the definition for subvarieties and extending by linearity.

Acknowldegments. As noted above, the original idea of differentiating Θ and using intersection theory to obtain information about its singularities is due to R. Lazarsfeld while the idea of examining S_{g-1} and using the Matsusaka–Ran criterion is due to F. Angelini; these ideas were independently worked out by Smith and Varley. This work would never have been undertaken, let alone completed, without several stimulating conversations with Lazarsfeld and Angelini; the argument of §1 is a direct outgrowth of these conversations.

1 Proof by Matsusaka–Ran

We begin with some remarks about differentiating sections of line bundles on an abelian variety A. A more complete, rigorous account can be found in [D1, ELN]. The tangent bundle T_A is trivial and so in particular it is *nef.* This implies that $T_A(\epsilon L)$ is ample (in the sense of [ELN] §2) for any $\epsilon > 0$. In what follows we will always assume

$$\epsilon \in \mathbf{Q}. \tag{1.1}$$

For any non-negative integer a, we let $\mathcal{D}^a_L$ be the sheaf of differential operators of order $\leq a$ on L (cf. [ELN] §2). By 1.1 and [ELN] Lemma 2.5 there exists $q > 0$ such that $q\epsilon \in \mathbf{Z}$ and such that $\mathrm{Sym}^q(\mathcal{D}^a_L) \otimes \mathcal{O}_A(q\epsilon L)$ is generated by global sections for $0 \leq a \leq g-1$. Consider the natural map of sheaves induced by the section Θ

$$\mathrm{Sym}^q(\mathcal{D}^a_L) \otimes \mathcal{O}_A(q\epsilon L) \xrightarrow{\psi_a} \mathcal{O}_A(qL + q\epsilon L).$$

Choose an open subset U over which both $\mathcal{D}^a_L$ and L are trivial. Let $D_1, \ldots, D_{r_a}$ be a basis of differential operators of weight $\leq a$ on U and let

f be a local equation for Θ on U. Suppose $\sigma \in H^0(\mathrm{Sym}^q(\mathcal{D}_L^a) \otimes \mathcal{O}_A(q\epsilon L))$ and let $\psi_a(\sigma)|U$ denote the restriction of $\psi_a(\sigma)$ to U. Then there is a homogeneous polynomial $Q \in \mathcal{O}_U[X_1, \ldots, X_{r_a}]$ of degree q such that

$$\psi_a(\sigma)|U = Q(D_1(f), \ldots, D_{r_a}(f)). \tag{1.2}$$

We will call $\psi_a(\sigma)$ a *differential section* of order a.

The drawback to differentiating as in [D1, ELN] is that one is forced to twist T_A by ϵL and then take symmetric powers. In certain cases, however, it is possible to avoid both the perturbation term and the symmetric powers. Given any line bundle L on A and a global section $\sigma \in H^0(A, L)$, it is always possible to take derivatives of σ locally by choosing a local trivialization of L. Unfortunately this local data will not in general patch together to give a global section of a line bundle. There are two obstructions to globalizing these derivatives; firstly, the derivations on the open subsets must patch together to give a vector field on A and secondly the transition functions for the local derivatives must satisfy the 1-cocycle condition. The first obstruction is not a problem since T_A is trivial and consequently there are many non-trivial vector fields on A. The transition functions are not in general well-behaved, but if one restricts to the zero locus $Z(\sigma)$ it is easy to verify that they are exactly the transition functions of L. In other words, given a vector field F, one can differentiate along F to produce a section $D_F(\sigma) \in H^0(Z(\sigma), \mathcal{O}_{Z(\sigma)}(L))$.

To formalize the above discussion, let P_L^k denote the principal parts bundle of L whose fibre at a point $x \in A$ is given by $L \otimes \mathcal{O}_A/\mathcal{I}_x^{k+1}$. Recall the exact sequence used in [ELN] §2

$$0 \to \mathrm{Sym}^k(\Omega_A^1) \otimes L \to P_L^k \xrightarrow{\pi_k} P_L^{k-1} \to 0.$$

The divisor Θ determines a section $\sigma \in H^0(A, \mathcal{O}_A(L))$ which lifts to a section $j_k(\sigma) \in H^0(A, P_L^k)$. Recall that S_k denotes the locus of k-fold singular points of Θ. Suppose $V \subset S_k$ is an irreducible subvariety. Since Θ vanishes to order at least k along V, the $k-1$-jets of Θ are identically zero along V; in particular $\pi_k(j_k(\sigma))|V = 0$. Hence

$$j_k(\sigma)|V \in H^0\left(V, \mathcal{O}_V\left(\mathrm{Sym}^k(\Omega_A^1) \otimes L\right)\right).$$

Choose tangent vectors $v_1, \ldots, v_k \in T_0(A)$ and consider the associated translation invariant sections $F_1, \ldots, F_k \in H^0(A, \mathcal{O}_A(T_A))$. Let

$$F = \prod_{i=1}^{k} F_i \in H^0\left(\mathrm{Sym}^k \mathcal{O}_A(T_A)\right).$$

Contracting $j_k(\sigma)$ against F determines a section $D_F(\sigma) \in H^0(V, \mathcal{O}_V(L))$.

Let $\pi : Y \to A$ be the blow up of A at P with exceptional divisor E and let $\eta : E \to A$ be the restriction of π to E. Thus $E \simeq \mathbf{P}^{g-1}$ and for a divisor $D \subset A$ passing through P, the tangent cone of D at P is defined by

$$C_P(D) = \tilde{D} \cap E \subset \mathbf{P}^{g-1}.$$

Lemma 1.3. *Suppose $U \subset A$ is a small affine open subset containing P on which L is trivial and let $f \in \Gamma(U, \mathcal{O}_U)$ be a local equation for Θ. Let D_i be a general differential operator of order i for $0 \leq i \leq g-1$ and let $Z_i = Z(D_i(f))$. Then*

$$\bigcap_{i=0}^{g-1} C_P(Z_i) = \emptyset.$$

Proof of Lemma 1.3 Choose general sections

$$\sigma_a \in H^0\left(\operatorname{Sym}^q\left(\mathcal{D}_L^a\right) \otimes \mathcal{O}_A(q\epsilon L)\right),\ 0 \leq a \leq g-1$$

and let $E_a = Z(\psi_a(\sigma_a))$. The intersection multiplicity $i(P, E_0 \cdot \ldots \cdot E_{g-1}; A)$ is determined locally and so, shrinking A if necessary, we can assume that P is the only point of intersection of $E_0, \ldots, E_{g-1}$. Let $\tilde{E}_a \subset Y$ denote the strict transform of E_a. By [Fu; Theorem 12.4]

$$\begin{aligned} i(P, E_0 \cdot \ldots \cdot E_{g-1}; A) \quad &= \prod_{a=0}^{g-1} \operatorname{mult}_P(E_a) \\ &+ \eta_*\left(\tilde{E}_0 \cdot \ldots \cdot \tilde{E}_{g-1}\right)_P, \end{aligned} \tag{1.4}$$

where $\eta_*\left(\tilde{E}_0 \cdot \ldots \cdot \tilde{E}_{g-1}\right)_P$ denotes the coefficient of $[P]$ in $\eta_*\left(\tilde{E}_0 \cdot \ldots \cdot \tilde{E}_{g-1}\right)$. Since the sections σ_a are general, 1.2 implies that $\operatorname{mult}_P(E_a) = q(g-a)$ and consequently

$$\prod_{a=0}^{g-1} \operatorname{mult}_P(E_a) = q^g g!.$$

On the other hand,

$$i(P, E_0 \cdot \ldots \cdot E_{g-1}; A) \leq \deg(E_0 \cdot \ldots \cdot E_{g-1}) = (1+\epsilon)^g q^g g!,$$

so 1.4 implies

$$\eta_*\left(\tilde{E}_0 \cdot \ldots \cdot \tilde{E}_{g-1}\right)_P \ \leq \ \left((1+\epsilon)^g - 1\right) q^g g!. \tag{1.5}$$

Suppose that $\bigcap_{i=0}^{g-1} C_P(Z_i)$ is not empty and let $V \subset \bigcap_{i=0}^{g-1} C_P(Z_i)$ be an irreducible component. Since the derivatives $D_i(f)$ are general, $V \subset C_P(Z(D(f)))$ for any differential operator D of degree $\leq g-1$. By 1.2, this implies that $\mathrm{mult}_V(\tilde{E}_a) \geq q$ for all a and since the sections σ_a are general 1.2 also implies that V is an irreducible component of $\bigcap_{a=0}^{g-1} \tilde{E}_a$. Applying [ELN] Lemma 1.4 gives

$$\eta_* \left(\tilde{E}_0 \cdot \ldots \cdot \tilde{E}_{g-1} \right)_P \geq q^g$$

and for ϵ sufficiently small this contradicts 1.5.

Lemma 1.3 will be used in connection with [Fu; Theorem 12.4] to compute intersection multiplicities. We will also use the following refinement:

$$C_P(\tilde{Z}_0) \cap \ldots C_P(\tilde{Z}_r) \cap C_P(\tilde{E}_{r+1}) \cap \ldots \cap C_P(\tilde{E}_{g-1}) = \emptyset,\ 0 \leq r \leq g-1. \quad (1.6)$$

This follows immediately from Lemma 1.3 by 1.2.

We now state the main result from which both proofs of Corollary 0.4 will follow.

Proposition 1.6. *Let (A, Θ) be a principally polarized abelian variety of dimension g. Suppose $P \in A$ with $\mathrm{mult}_P(\Theta) = g$ and let S_{g-1} denote the $g-1$-fold singular locus of Θ. Then S_{g-1}, considered as a cycle on A, is rationally equivalent to $c_1(\Theta)^{g-1}/(g-1)!$.*

Proof of Proposition 1.6. As mentioned in the introduction, the main idea behind the proof of Proposition 1.6, as in the proof of Lemma 1.3, is to use the fact that $\mathrm{mult}_P(\Theta) = g$ together with our knowldege of the intersection number $c_1(\Theta)^g = g!$. As stated earlier, this means that if one could simply take derivatives of Θ and remain in the same linear equivalence class, then the intersection class $\Theta \cap D_1(\Theta) \cap \ldots \cap D_{g-1}(\Theta)$ would necessarily be supported entirely at the point $P = S_{g-1}$. Similarly,

$$\Theta \cap D_1(\Theta) \cap \ldots \cap D_{g-2}(\Theta)$$

would be supported on S_{g-1} and a refinement will yield Proposition 1.6. Since these derivatives require twisting and taking symmetric powers, slightly more subtle computations are necessary but the idea remains the same.

Choose general tangent vectors $v_1, \ldots, v_g \in T_0(A)$. If $V \subset S_n$ is an irreducible component, $2 \leq n \leq g-1$, then the tangent vectors $v_1, \ldots, v_n$ define a differential section $D_{n,V}(\sigma) \in H^0(V, \mathcal{O}_V(L))$. Letting $[Z(D_{n,V}(\sigma))]$ denote the cycle associated to the subscheme $Z(D_{n,V}(\sigma)) \subset V$, we have

$$[Z(D_{n,V}(\sigma))] = F_{n,V} + G_{n,V} \quad (1.7)$$

where $F_{n,V}$ is supported on S_{n+1} and no component of $G_{n,V}$ is supported on S_{n+1}. Observe that $F_{n,V} + G_{n,V}$ represents $c_1(\Theta) \cap V$. To prove Proposition 1.6, we will first show that $G_{n,V} = 0$ for all n and all V. Suppose to the contrary that $G_{n,V} \neq 0$ for some $2 \leq n \leq g-1$ and for some $V \subset S_n$. Choose V and r with $G_{r,V} \neq 0$ and r minimal. Thus for all $n < r$ and for all $V \subset S_n$, $G_{n,V} = 0$ and consequently, applying 1.8 inductively gives a cycle supported on S_{n+1} representing $c_1(\Theta)^{n+1}$; call this cycle W_{n+1}. Since W_{n+1} is supported on S_{n+1}, applying 1.7 again gives a cycle W_{n+2}, not necessarily supported on S_{n+2}, representing $c_1(\Theta)^{n+2}$. When $n = r-1$ this gives (or, if $r = 2$ then 1.7 gives)

$$c_1(\Theta)^{r+1} \sim F_r + G_r \tag{1.8}$$

where F_r is supported on S_{r+1} and no component of G_r is supported on S_{r+1}.

Choose general differential sections $\psi_r(\sigma), \ldots, \psi_{g-1}(\sigma)$ as in Lemma 1.3 and let $E_a = Z(\psi_a(\sigma))$ for $r \le a \le g-1$. Let f be a local equation defining Θ in a neighborhood of P and let $Z_i = Z(D_i(f))$ for $0 \le i \le g-1$ (here D_i is the same differential operator used to obtain $D_{i,V}(\sigma)$). The cycle $F_r + G_r$ was obtained using 1.7; locally about P, this amounts to successively intersecting $Z_0, \ldots, Z_r$. By [Fu; Example 7.1.8]

$$\begin{aligned} i(P, F_r \cdot E_{r+1} \cdot \ldots \cdot E_{g-1}; A) + i(P, G_r \cdot E_{r+1} \cdot \ldots \cdot E_{g-1}; A) \\ = i(P, Z_0 \cdot \ldots \cdot Z_r \cdot E_{r+1} \cdot \ldots \cdot E_{g-1}; A). \end{aligned} \tag{1.9}$$

We can compute the intersection multiplicity in 1.9 by 1.6 and [Fu; Theorem 12.4]:

$$i(P, Z_0 \cdot \ldots \cdot Z_r \cdot E_{r+1} \cdot \ldots \cdot E_{g-1}; A) = q^{g-r-1} g!. \tag{1.10}$$

On the other hand by 1.8

$$\deg((F_r + G_r) \cdot E_{r+1} \cdot \ldots \cdot E_{g-1}) = (1+\epsilon)^{g-r-1} q^{g-r-1} g!\,. \tag{1.11}$$

Thus 1.10 and 1.11 imply that most of the intersection class $(F_r + G_r) \cdot E_{r+1} \cdot \ldots \cdot E_{g-1}$ is concentrated at P; we obtain a contradiction by intersecting G_r with different differential sections, increasing the intersection multiplicity at P so that it becomes larger than the total intersection degree.

Observe that $P \in Y$ for any irreducible component $Y \subset G_r$: if not, then the contribution of Y to $G_r \cdot E_{r+1} \cdot \ldots \cdot \ldots E_{g-1}$ would have degree at least q^{g-r-1}, none of which would be supported at P, violating 1.9, 1.10, and 1.11. Since no component of G_r is supported on S_{r+1}, $E_r \cap Y_r$ is proper. But $\mathrm{mult}_P E_r = \left(\frac{g-r}{g-r-1}\right) \mathrm{mult}_P(E_{r+1})$. Using Lemma 1.3 and estimating the intersection multiplicity by [Fu; Theorem 12.4] gives

$$i(P, G_r \cdot E_r \cdot E_{r+2} \cdot \ldots \cdot E_{g-1}; A) \ge \left(\frac{g-r}{g-r-1}\right) i(P, G_r \cdot E_{r+1} \cdot \ldots \cdot E_{g-1}; A).$$

Since

$$i(P, F_r \cdot E_{r+1} \cdot \ldots \cdot E_{g-1}; A) + i(P, G_r \cdot E_r \cdot E_{r+2} \cdot \ldots \cdot E_{g-1}; A) \le q^g g! + O(\epsilon) q^g,$$

1.9, 1.10, 1.11, and 1.12 imply that

$$i(P, G_r \cdot E_{r+1} \cdot \ldots \cdot E_{g-1}; A) \le O(\epsilon) q^{g-r-1}. \tag{1.12}$$

But since $P \in Y$ for all irreducible components $Y \subset G_r$, [Fu; Theorem 12.4] gives

$$i(P, G_r \cdot E_{r+1} \cdot \ldots \cdot E_{g-1}) \geq q^{g-r-1}$$

and this contradicts 1.12 for ϵ sufficiently small. Hence $G_{n,V} = 0$ for all $2 \leq n \leq g-1$ and for all irreducible components $V \subset S_n$.

Since $G_{n,V} = 0$ for all n and V, we can use 1.7 to inductively define a class W_{g-1} supported on S_{g-1} which represents $c_1(\Theta)^{g-1}$. We claim that this implies that S_{g-1} is non-degenerate. Suppose to the contrary that S_{g-1} is degenerate so there exists a point $a \in A$ and a proper abelian subvariety $B \subset A$ such that $S_{g-1} \subset a + B$. Let $\pi : A \to A/B$ be the projection map and choose an ample line bundle L on A/B. Since S_{g-1} is contracted by π it follows that $\pi^* L \cdot C = 0$ for all $C \subset S_{g-1}$. But since $c_1(\Theta)^{g-1}$ is rationally equivalent to a cycle supported on S_{g-1} this would imply that $c_1(\Theta)^{g-1} \cap c_1(\pi^* L) = 0$ which is a contradiction since Θ is ample. Thus S_{g-1} is non-degenerate, giving

$$c_1(\Theta) \cap S_{g-1} \geq g. \tag{1.13}$$

Let $C \subset S_{g-1}$ be an irreducible component. Using [Fu; Example 7.1.8] the coefficient of C in W_{g-1} is $i(C, Z_0 \cdot \ldots \cdot Z_{g-2}; A)$. By [Fu; Example 12.4.4] we have

$$i(C, Z_0 \cdot \ldots \cdot Z_{g-2}; A) \geq \prod_{i=0}^{g-2} \operatorname{mult}_C(Z_i) = (g-1)!. \tag{1.14}$$

Thus $W_{g-1} - (g-1)! S_{g-1}$ is represented by an effective cycle. But by 1.13,

$$c_1(\Theta) \cap (W_{g-1} - (g-1)! S_{g-1}) \leq c_1(\Theta)^g - g! = 0.$$

Thus $W_{g-1} = (g-1)! S_{g-1}$ concluding the proof of Proposition 1.6.

Corollary 0.4 follows easily from Proposition 1.6: if $\operatorname{mult}_P(\Theta) = g$ for some point $P \in \Theta$ then Proposition 1.6 and the Matsusaka–Ran criterion ([BL] Chapter 11, 8.1) imply that A is a Jacobian and the fact that (A, Θ) decomposes as a product of elliptic curves then follows from the Riemann singularity theorem.

We conclude this section with two results which will be used in the second proof of Corollary 0.4 and which follow readily from the proof of Proposition 1.6. Let $C \subset S_{g-1}$ be an irreducible component. Since $G_{g-1,C} = 0$, 1.7 implies that $D_{g-1,C}(\sigma) \in H^0(C, \Theta)$ vanishes only at $P = S_g$ and hence $\Theta \cdot C = i(P, Z_{g-1} \cdot C; A)$. Moreover by Lemma 1.3 and [Fu; Theorem 12.4]

$$i(P, Z_{g-1} \cdot C; A) = \operatorname{mult}_P(Z_{g-1}) \cdot \operatorname{mult}_P(C) = \operatorname{mult}_P(C).$$

Hence we have

Lemma 1.15. *Let $C \subset S_{g-1}$ be an irreducible curve. Then*

$$\Theta \cdot C = \text{mult}_P(C). \tag{1.16}$$

Finally, let $\text{Sing}_{g-1}(\Theta)$ denote the subscheme of A locally defined by all derivatives of Θ of order at most $g-2$. The support of $\text{Sing}_{g-1}(\Theta)$ is, by definition, S_{g-1}, the locus of $g-1$–fold singular points of Θ. Let $[\text{Sing}_{g-1}(\Theta)]$ denote the cycle associated to the subscheme $\text{Sing}_{g-1}(\Theta)$. We claim that

$$[\text{Sing}_{g-1}(\Theta)] = S_{g-1}. \tag{1.17}$$

Let $C \subset S_{g-1}$ be an irreducible component and let f be a local equation for Θ in a neighborhood of C with $Z_i = D_i(f)$ for $0 \leq i \leq g-2$. Let $\pi : Y \to A$ be the blow-up of A along C (in the small open set on which f represents Θ) and let $\tilde{Z}_i$ be the strict transform of Z_i. Since $W_{g-1} \sim c_1(\Theta)^{g-1}$ and since $c_1(\Theta)^g = g!$, 1.13 and 1.14 imply that equality holds in 1.14. By [Fu; Theorem 12.4], for a general point $\eta \in C$,

$$\pi^{-1}(\eta) \cap \tilde{Z}_0 \cap \ldots \cap \tilde{Z}_{g-2} = \emptyset. \tag{1.18}$$

Let $h_1, \ldots, h_{g-1}$ be general derivatives of f of order $g-2$ and let $\tilde{Y}_i$ be the strict transform of $Z(h_i)$. Then 1.18 implies that

$$\pi^{-1}(\eta) \cap \tilde{Y}_1 \cap \ldots \cap \tilde{Y}_{g-1} = \emptyset$$

and consequently $i(C, Y_1 \cdot \ldots \cdot Y_{g-1}; A) = 1$. By [Fu; Example 7.1.10], this means that $\text{Sing}_{g-1}(\Theta)$ is generically reduced along C and this implies 1.17.

2 Proof by Zero Estimates

In this section we present an alternative proof of Corollary 0.4 using zero estimates. The key idea in the second proof is that the equality $\text{mult}_P(C) = \Theta \cdot C$ shown in Lemma 1.15 is enough to show that C must be an elliptic curve, independent of the Matsusaka–Ran criterion. In [N2], it is shown that 1.16 implies that C is an elliptic curve and that (A, Θ) decomposes with $(C, \Theta \cdot C)$ as a direct summand (see Proposition 2.7 below). Here we give an alternative, slightly quicker argument using the fact that $C \subset S_{g-1}$.

Proposition 2.1. *Suppose $C \subset S_{g-1}$. Then C is an elliptic curve and $\deg_\Theta(C) = 1$.*

Proof of Proposition 2.1. Let Q and Q' be two distinct general points of C. Suppose $v \in T_Q(A)$ generates $T_Q(C)$. The tangent vector v can be translated to any point $P \in A$ and we will continue to denote the new tangent vector by v. Denote the constant vector field with value v at each point by F_v and let ∂_v be the associated derivation of $\mathcal{O}_A$. Suppose $v \notin T_{Q'}(C)$. If f is a local equation for Θ about Q and Q', then 1.17 implies that there is a derivative $D(f)$ or order $\leq g-2$ such that $\partial_v(D(f))(Q') \neq 0$. Thus there exist $r \leq g-2$ translation invariant vector fields $F_1, \ldots, F_r$ such that if $F = F_1 \cdot \ldots \cdot F_r \cdot F_v$ then

$$D_F(\Theta) \in H^0(C, \mathcal{O}_C(\Theta)) \text{ is not identically zero.}$$

On the other hand, a local computation shows that

$$\text{mult}_P(C)[P] + [Q] \leq [Z(D_F(\Theta))]$$

where $[P]$, $[Q]$, and $[Z(D_F(\Theta))]$ are the zero cycles associated to the points P and Q and to the section $D_F(\Theta)$ respectively. Since $c_1(\Theta) \cap C = \deg[Z(D_F(\Theta))]$ this contradicts 1.16 above and hence we conclude that v generates $T_{Q'}(C)$ for a general point $Q' \in C$. This implies that C is an elliptic curve and by 1.16 again $c_1(\Theta) \cap C = 1$.

We now derive Corollary 0.4 from Proposition 2.1. Lemma 1.15 and Proposition 2.1 imply that S_{g-1} consists of elliptic curves. Since $\deg_\Theta(S_{g-1}) = g$, there are exactly g distinct elliptic curves in S_{g-1} which we write as $E_1, \ldots, E_g$. Consider the product $E = E_1 \times \ldots \times E_g$ and let $\pi_i : E \to E_i$ denote the projection to the ith factor. For each i choose a point $P_i \in E_i$. Let D be a divisor on $E_1 \times \ldots \times E_g$ and define its degree along the ith factor by

$$d_i = D \cdot \left(\bigcap_{j \neq i} \pi_i^{-1}(P_i) \right).$$

The degree of D is defined to be $d(D) = (d_1, \ldots, d_g)$. Consider the addition map

$$\phi : E_1 \times \ldots \times E_g \to A.$$

Since $\deg_\Theta(E_i) = 1$ for all i it follows that $d(\phi^*\Theta) = (1, \ldots, 1)$. Since S_{g-1} is non-degenerate ϕ is dominant and consequently $\phi^*(\Theta)$ is ample. Theorem 0.3 follows from

Lemma 2.2. *Let D be an ample effective divisor on a product of elliptic curves*

$$E = E_1 \times \ldots \times E_g.$$

If $d(D) = (1, \ldots, 1)$ then there exist points $P_i \in E_i$ such that

$$D = \sum_{i=1}^{g} \pi_i^{-1}(P_i).$$

Proof of Lemma 2.2. For a subset $I = \{i_1, \ldots, i_s\} \subset \{1, \ldots, g\}$ let

$$\pi_I : E_1 \times \ldots \times E_g \to E_{i_1} \times \ldots E_{i_s}$$

denote the natural projection. Since D is effective of degree $(1, \ldots, 1)$ there exists a partition $I_1, \ldots, I_r$ of $\{1, \ldots, g\}$ such that

$$D = \sum_{j=1}^{r} \pi_{I_j}^* D_j, \quad D_j \text{ irreducible and } d(D_j) = (1, \ldots, 1). \tag{2.3}$$

It is easy to verify that

$$c_1(D)^g = \frac{g!}{\prod_{j=1}^{r} |I_j|!} \left(\prod_{j=1}^{r} c_1(D_j)^{|I_j|} \right). \tag{2.4}$$

We will show that $c_1(D_j)^{|I_j|} = 0$ unless $|I_j| = 1$. Since $c_1(D)^g > 0$, 2.4 implies Lemma 2.2. So suppose $|I_j| \neq 1$ for some j. Let $I_j = \{i_1, \ldots, i_s\}$ for $s \geq 2$ and let

$$p : E_{i_1} \times \ldots \times E_{i_s} \to E_{i_2} \times \ldots \times E_{i_s}$$

denote the natural projection. Since D_j is irreducible and $d(D_j) = (1, \ldots, 1)$ the map $p : D_j \to E_{i_2} \times \ldots \times E_{i_s}$ is birational. This is only possible if $D_j \subset E_{i_1} \times \ldots \times E_{i_s}$ is a translate of an abelian subvariety in which case $c_1(D_j)^s = 0$.

We conclude this note with the result which motivated the entire argument. The reason why 1.16 gives information about the curve C is because it means that C is a *Seshadri exceptional curve*. We recall here the definition of Seshadri constants. Let X be a smooth projective variety and let L be a numerically effective line bundle on X. The *Seshadri constant* of L at a point $x \in X$ is defined [D2, EL1] by

$$\epsilon(L, x) = \inf_{C \ni x} \frac{L \cdot C}{\mathrm{mult}_x(C)};$$

the infimum is taken over all irreducible curves C passing through x. The global Seshadri constant is defined by

$$\epsilon(L) = \inf_{x \in X} \epsilon(L, x).$$

On an abelian variety, the global Seshadri constant is uniformly bounded from below:

Lemma 2.5. *Let L be an ample line bundle on an abelian variety A. Then* $\epsilon(L) \geq 1$.

Proof of Lemma 2.5. Suppose $x \in A$ and C is an irreducible curve containing x. Since L is ample $|L|$ is non-empty. Choose a divisor $D \in |L|$. We claim that there exists $a \in A$ such that

$$(a+D) \cap C \text{ is proper and } x \in a+D. \tag{2.6}$$

Lemma 2.5 follows from 2.6: since $a+D$ is numerically equivalent to D, [Fu; Theorem 12.4] implies that

$$\begin{aligned} L \cdot C &= (a+D) \cdot C \\ &= \sum_{P \in (a+D) \cap C} i(P, (a+D) \cdot C; A) \\ &\geq \operatorname{mult}_x(a+D) \cdot \operatorname{mult}_x(C) \geq \operatorname{mult}_x(C). \end{aligned}$$

To prove 2.6 let

$$V = \{x - y : y \in D\} = \{a \in A : x \in a + D\}.$$

It suffices to show that there exists $a \in V$ such that $C \not\subset a+D$. But if $C \subset a+D$ for all $a \in V$ it follows that $C - a \subset D$ for all $a \in V$. By definition of V this means that $y-x+C \subset D$ for all $y \in D$ and consequently $C - x \subset \operatorname{Stab}(D)$. In particular, $\operatorname{Stab}(D)$ is positive dimensional and hence D must be degenerate. But this is impossible since D is ample and this concludes the proof of Lemma 2.5.

We conclude by giving a statement of the main result of [N2] which estabishes Proposition 2.1 *without* the assumption that $C \subset S_{g-1}$.

Proposition 2.7. *Let A be an abelian variety of dimension $g \geq 2$ and L an ample line bundle on A. Then $\epsilon(L) = 1$ if and only if $A \simeq E \times B$ where E is an elliptic curve and B is an abelian variety of dimension $g-1$.*

References

[B] D. Bertrand, Lemmes de zéros et nombres transcendants, Séminaire Bourbaki, 38ème année, 1985–86, exposé n° 652, S.M.F. *Astérisque*, **145–6**, 1987, 21–44

[BL] C. Birkenhake and H. Lange, *Complex Abelian Varieties*, Springer, 1993

[D1] J.-P. Demailly, A numerical criterion for very ample line bundles, *Journal of Differential Geometry*, **37**, 1993, 323–374

[D2] J.-P. Demailly, Singular Hermitian metrics on positive line bundles, in Hulek, Peternell, Schneider and Schreyer (eds), *Complex Algebraic Varieties*, Proceedings, Bayreuth 1990, Lecture Notes in Math, **1507**, 1992, 87–104

[EL1] L. Ein and R. Lazarsfeld, Seshadri Constants on Smooth Surfaces, *Astérisque*, **218**, 1993, 177–186

[EL2] L. Ein and R. Lazarsfeld, Singularities of theta divisors and the birational geometry of irregular varieties, *JAMS*, **10**, Number 1, 1997, 243–258

[ELN] L. Ein, R. Lazarsfeld, and M. Nakamaye, Zero Estimates, Intersection Theory, and a Theorem of Demailly, in Andreatta and Peternell (eds.), *Higher-Dimensional Complex Varieties*, de Gruyter, 1996, 183–208

[Fu] W. Fulton, *Intersection Theory*, Springer Ergebnisse, 1984

[K] J. Kollár, *Shafarevich maps and automorphic forms*, Princeton University Press, 1995

[N1] M. Nakamaye, Multiplicty Estimates and the Product Theorem, *Bulletin de la Société Mathématique de France*, **123**, 1995, 155–188

[N2] M. Nakamaye, Seshadri Constants on Abelian Varieties, *American Journal of Math*, **118**, Number 3, 1996, 621–635

[N3] M. Nakamaye, *Diophantine Approximation on Algebraic Varieties*, Journal de Théorie de Nombres de Bordeau

[SV] R. Smith and R. Varley, Multiplicity g points on theta divisors, *Duke Mathematical Journal*, **82**, Number 2, 1996, 319–326

Department of Mathematics
Harvard University
Cambridge, MA 02138
nakamaye@math.harvard.edu

Generalized Plücker Formulas

Anders Thorup

ABSTRACT The classical Plücker formula for a plane curve was generalized by Teissier to the case of a hypersurface with isolated singularities and further by Kleiman to the case of an arbitrary n-dimensional projective variety V with isolated singularities. The formula relates the zeroth rank of V (the degree of the dual variety) to the Segre numbers of the conormal module and certain Buchsbaum–Rim multiplicities associated to the singular points of V. A second generalization was obtained by Pohl. It relates the $(n-1)$th rank of V to the first Chern class of a desingularization of V and the degree of the cuspidal divisor. We describe, for a projective variety V with arbitrary singularities, a natural class in the Chow group of the singular locus whose top dimensional part is given by Buchsbaum–Rim multiplicities, and we obtain generalizations of both formulas. The formulas are equations in the Chow group of V. They imply numerical formulas for all the ranks of V.

1 Introduction

Fix a reduced variety V of pure dimension n in projective N-space $\mathbb{P} = \mathbb{P}^N$, say $\mathbb{P} = \mathbb{P}(E)$, where E is vector space of rank $N+1$ over a perfect field. Let d be the degree of V, and r_0 the *class* (or the zeroth *rank*) of V, that is, the number of hyperplanes through a fixed general codimension-2 plane in $\mathbb{P}$ and tangent to V. If the dual variety $V' \subseteq \check{\mathbb{P}}(E)$ is a hypersurface, then r_0 is its degree; otherwise $r_0 = 0$.

Assume that V is a plane curve. It is an old result that if V is smooth, then $r_0 = d(d-1)$. Plücker justified in 1834 the proper correction to this equation when V has double points. It is the first of the following two formulas:

$$r_0 = d(d-1) - 2\,(\#\text{nodes}) - 3\,(\#\text{cusps}), \tag{1.1}$$

$$r_0 = 2d + (2g-2) - (\#\text{cusps}). \tag{1.2}$$

The second formula is equivalent to the first, given the well-known formula for the genus g of a curve. Of course, the second form is later, since the

---

The author was supported in part by the Danish Natural Science Research Council, grant 9502145.

Mathematics Subject Classification: 14C17.

genus was only introduced in 1857.

The two formulas have been generalized by many geometers. Teissier [T, App. II], in 1975, generalized the first formula to the case when V is a hypersurface in $\mathbb{P}^N$ with isolated singularities:

$$r_0 = d(d-1)^n - \sum_{w \in V_{\text{Sing}}} e_w, \tag{1.3}$$

where e_w is the Samuel multiplicity of the Jacobian ideal of $\mathcal{O}_{V,w}$.

In turn, Teissier's formula was generalized by Kleiman [K3] in 1992: If V is of arbitrary dimension in $\mathbb{P}^N$, but still with isolated singularities, then

$$r_0 = \check{s}_0 - \sum_{w \in V_{\text{Sing}}} e_w, \tag{1.4}$$

where the numbers on the right side are defined as follows: Let $\mathcal{N} := \mathcal{I}_V/\mathcal{I}_V^2$ be the conormal module. Assume first that V is lci. Then $\mathcal{N}$ is locally free, and the dual $\mathcal{N}(1)^*$ is the twisted normal module. The number $\check{s}_0$ is (the degree of) the zero-dimensional part of the Segre class of $\mathcal{N}(1)^*$. Moreover, the number e_w is the Buchsbaum–Rim multiplicity of the Jacobian map $E^*_{V,w} \to \mathcal{N}(1)^*_w$. Assume for instance that V is a global intersection of $q = N - n$ hypersurfaces $F_k = 0$ of degrees $d_1, \ldots, d_q$. Then $\mathcal{N}(1)^* = \mathcal{O}_V(d_1 - 1) \oplus \cdots \oplus \mathcal{O}_V(d_q - 1)$. The degree of the i-dimensional part of the Segre class of $\mathcal{N}(1)^*$ is given by the formula,

$$\check{s}_i = ds^{n-i}(d_1 - 1, \ldots, d_q - 1),$$

where s^j is the complete symmetric polynomial of degree j. Moreover, the Jacobian map is essentially the matrix of partial derivatives of the F_k.

In general, if V is not lci, the number $\check{s}_0$ is a certain natural generalization of the Segre number of the dual of $\mathcal{N}(1)$, and e_w is a generalized Buchsbaum–Rim multiplicity (which is only strictly positive when w is a distinguished singular point).

The second formula (1.2) was generalized by Pohl [Po] in 1964: Let $f : X \to V$ be a proper birational map from smooth variety X. Let r_{n-1} be the $(n-1)$th *rank* of V, that is, the number of simple points v of V lying on a given general codimension-$(n-1)$ linear subspace and such that the tangent space of V at v passes through a given general point. Then

$$r_{n-1} = (n+1)d + c^1 - \kappa, \tag{1.5}$$

where c^1 and κ are the degrees, with respect to $\mathcal{O}_V(1)$, of the first Chern class of Ω^1_X and the cuspidal divisor on X.

1.1 As a consequence of the main results proved in this paper, we obtain generalizations of both formulas (1.4) and (1.5). In the two generalized formulas, V is allowed to have arbitrary singularities, and each formula implies equations for all the ranks of V. The formulas are equations in the Chow group $\mathrm{A}(V)$. In this group there is a natural class $\mathfrak{r} = \mathfrak{r}(V)$ whose homogeneous components are represented by the effective cycles of the polar subschemes $V(A)$ of V; in particular, the degree r_i of the i-dimensional part of $\mathfrak{r}$ is the ith *rank* of V, see [K2, p. 338]. The class $\mathfrak{r}$ may be defined as the pushforward to $\mathrm{A}(V)$ of the class $s(\mathcal{L}')[CV]$, where CV is the conormal variety of V and $\mathcal{L}' = \mathcal{O}_{\check{\mathbb{P}}(E)}(1)$. It is well known, see for instance [S, Lemma 1.2.1, p. 164] or the proof of (2.3) below, that the class $\mathfrak{r}$ is the Mather Chern class of the module of principal parts of $\mathcal{O}_V(1)$,

$$\mathfrak{r} = \mathfrak{c}_{\mathrm{M}}\mathcal{P}^1(\mathcal{O}_V(1)). \tag{1.6}$$

The two generalizations have the following form, where the right-hand sides are described below, and in detail in Section 5:

The generalized Kleiman–Teissier–Plücker formula

$$\mathfrak{r} = \check{\mathfrak{s}} - \sum\nolimits_W \mathfrak{e}_W. \tag{1.7}$$

The generalized Pohl–Plücker Formula. *If $f : X \to V$ is a proper birational map from a smooth variety X, then*

$$\mathfrak{r} = \sum\nolimits_{i=0}^{n} (1+h)^{i+1}\mathfrak{c}^{n-i} - \mathfrak{k}. \tag{1.8}$$

The class $\check{\mathfrak{s}}$ in (1.7) is a natural generalization of the Segre class of the dual of the twisted conormal module. In the lci case, it is simply the total Segre class $s(\mathcal{N}(1)^*)[V]$; it should be emphasized that the class, in general, it is not the Segre class in the sense of Fulton and Johnson [F] of $\mathcal{N}(1)^*$. The sum on the right side of (1.7) is over certain *distinguished singular subsets* W of V, and each term $\mathfrak{e}_W$ is a class naturally supported in the Chow group $\mathrm{A}(W)$. In particular, if V has isolated singularities, then (1.4) follows from (1.7) by taking the zero-dimensional part; moreover, since the terms $\mathfrak{e}_W$ are supported on a zero-dimensional subscheme, we obtain for $i > 0$ the equation for the ith rank,

$$r_i = \check{s}_i.$$

In general, by the same argument, the latter equation holds when i is bigger than the dimension of the singular locus of V.

In Formula (1.8), h is the hyperplane class, i.e., the first Chern class of $\mathcal{O}_V(1)$, the class $\mathfrak{c}^j$ is the pushforward of the codimension-j Chern class of Ω^1_X, and $\mathfrak{k} = \mathfrak{k}_{V,\mathbb{P}}(f)$ is a natural class supported on the discriminant locus of f. The codimension-1 part of (1.8) is the equation,

$$\mathfrak{r}_{n-1} = (n+1)h[V] + \mathfrak{c}^1 - \mathfrak{k}_{n-1},$$

from which Pohl's Plücker Formula (1.5) follows by taking the degree.

1.2 The classes appearing in (1.7) are natural invariants associated to the "homogeneous" Jacobian $\tilde{d}_V$, fitting into the exact sequence,

$$\mathcal{N}_{V/\mathbb{P}}(1) \xrightarrow{\tilde{d}_V} E_V \longrightarrow \mathcal{P}^1(\mathcal{O}_V(1)) \longrightarrow 0. \tag{1.9}$$

The usual Jacobian d_V fits into the exact sequence,

$$\mathcal{N}_{V/\mathbb{P}} \xrightarrow{d_V} \Omega^1_{\mathbb{P}}|V \longrightarrow \Omega^1_V \longrightarrow 0. \tag{1.10}$$

In turn, $\tilde{d}_V$ is obtained from d_V by twisting with $\mathcal{O}(1)$ and composing with the natural inclusion $\Omega^1_{\mathbb{P}}(1) \to E_{\mathbb{P}}$.

The main theme of this work is to associate certain natural invariants with any generically injective homomorphism of (coherent) modules on V,

$$\alpha : \mathcal{N} \to \mathcal{E},$$

with a locally free target $\mathcal{E}$, and to prove formulas relating these invariants. Among the formulas is a general Plücker formula, from which (1.7) follows when α is the homogeneous Jacobian $\tilde{d}_V$. Similarly, (1.8) follows essentially by considering the homomorphism $\Omega^1_V|X \to \Omega^1_X$.

The general setup and the invariants are introduced in Section 2. There are invariants of α corresponding to the classes $\check{\mathfrak{s}}$ and $\mathfrak{r}$ in (1.7). In addition, associated with α, there are certain distinguished singular subsets W contained in the locus where α is not split injective, and for each W a certain class $\mathfrak{e}_W$ in the Chow group $\mathrm{A}(W)$. The top-dimensional part of $\mathfrak{e}_W$ is essentially an integer, the Buchsbaum–Rim multiplicity.

The general Plücker formula is proved in Section 3, and other natural properties of the invariants are described in Section 4. A fundamental property, and maybe the main observation of this work, is that the invariants, after a normalization given by multiplication by the Chern class $c(\mathcal{E})$, are invariants of α viewed as a complex in the derived category $D(V)$.

The case of a projective variety is treated in Section 5. The Plücker formula (1.7) is obtained when α is the homogeneous Jacobian $\tilde{d}_V$ in (1.9). When α is the Jacobian d_V of (1.10), the complex is the cotangent complex of V; hence the normalized invariants of d_V are intrinsic invariants of V, that is, they are independent of the embedding of V into $\mathbb{P}^N$. Other properties described in Section 4 permit us to relate the invariants of $\tilde{d}_V$ with those of d_V. As a consequence we obtain, for the classes $\mathfrak{r}$, $\check{\mathfrak{s}}$, and $\mathfrak{e}_W$ in (1.7), expressions in terms of intrinsic invariants of V. Moreover, from the general properties we obtain adjunction formulas relating the invariants of V with those of a general hypersurface section of V. The end of Section 5 contains an application to the dimension of the dual variety.

The general theory is applied in Section 6 to obtain invariants of a generically smooth map, and in Section 7 to obtain invariants of a hypersurface section of a smooth variety. In Section 8 we describe the generalization (1.8) of Pohl's Plücker Formula. As a special case we recover Piene's [Pi1,

Theorem (2.5), p. 259] Plücker formula in the hypersurface case. Finally, as an example, we relate in Section 9 the invariants of a surface in $\mathbb{P}^3$ to the classical invariants.

1.3 The results in this paper may be proved by standard arguments of intersection theory for (possibly singular) algebraic varieties, as presented in [F]. For a more general version of the results, using intersection theory for arbitrary Noetherian schemes as in [Th1], and with more details, see [Th2]. Our notation for Segre classes differs by a sign from Fulton's: If $\mathcal{E}$ is a locally free module on a variety V, of dimension n, we define the total Segre operator $s(\mathcal{E})$ from the symmetric algebra $\operatorname{Sym}\mathcal{E}$, and the fundamental relation between the Segre and Chern operators is the equation,

$$c(\mathcal{E})s(\mathcal{E}^*) = 1.$$

Whenever possible, a dimension will be indicated by a subscript and a codimension by a superscript. Thus the jth Chern class operator, lowering dimension by j, is denoted $c^j(\mathcal{E})$. The i-dimensional Chern class, in the Chow group $\mathrm{A}(V)$, is obtained by applying the operator $c^{n-i}(\mathcal{E})$ to the fundamental cycle of V,

$$\mathfrak{c}_i(\mathcal{E}) = c^{n-i}(\mathcal{E})[V].$$

If $f : V' \to V$ is a proper map, the pushforward f_*, defined on cycles and on classes on V' will be denoted as the integral $\int_{V'/V}$. If the source of the map is clear from the context, it will often be omitted in the notation.

We define the *Mather Chern class* of a coherent module $\mathcal{N}$, assumed to be generically locally free, essentially as in Fulton [F, p. 79]: Let $f : \widetilde{V} \to V$ be *modification* for $\mathcal{N}$, that is, f is a proper birational map and on $\widetilde{V}$ there is locally free module $\widetilde{\mathcal{N}}$ and a surjection $\mathcal{N}|\widetilde{V} \to \widetilde{\mathcal{N}}$ which is an isomorphism generically. (For instance, $\widetilde{V}$ could be the Nash modification: Assume that $\mathcal{N}$ is locally free of rank t on a dense open subset U of V. Let $G := \operatorname{Grass}^t(\mathcal{N})$ be the Grassmannian of locally free rank-t quotients of $\mathcal{N}$, and take as $\widetilde{V}$ the closure of U under the natural embedding $U \to G$.) Then the Mather Chern class is the pushforward,

$$\mathfrak{c}_{\mathrm{M}}(\mathcal{N}) := \int_{\widetilde{V}/V} c(\widetilde{\mathcal{N}})[\widetilde{V}].$$

Several algebraic operations involving a "twist" by an invertible module $\mathcal{K}$ on V are "mirrored" in the Chow group $\mathrm{A}(V)$ by the following action of $\mathcal{K}$ on classes:

$$\mathcal{K} \star \mathfrak{z} := \sum c(\mathcal{K})^i \mathfrak{z}_i, \tag{1.11}$$

where $c(\mathcal{K}) = 1 + c^1(\mathcal{K})$ is the total Chern class (operator), and $\mathfrak{z}_i$ is the i-dimensional component of the class $\mathfrak{z}$. For instance, the Mather Chern class of $\mathcal{K} \otimes \mathcal{N}$ is given by the formula,

$$\mathfrak{c}_{\mathrm{M}}(\mathcal{K} \otimes \mathcal{N}) = c(\mathcal{K})^{\mathrm{rk}\,\mathcal{N} - \dim V} \mathcal{K} \star \mathfrak{c}_{\mathrm{M}}(\mathcal{N}),$$

where $\mathrm{rk}\,\mathcal{N}$ is the (constant) generic rank of $\mathcal{N}$.

The operation $\mathcal{K} \star \mathfrak{z}$ is closely related to the "tensor product" $\mathfrak{z} \otimes \mathcal{K}$ introduced by Aluffi [A]; in fact, $\mathcal{K} * \mathfrak{z} = c(\mathcal{K})^{\dim V}(\mathfrak{z} \otimes \mathcal{K})$. We prefer the product (1.11) because it extends naturally to the case of a scheme which is not assumed to be equidimensional.

2 The Invariants of a Homomorphism

Work throughout in the following general setup: Let V be a proper variety (reduced) over a perfect base field (it suffices that V is generically smooth over an arbitrary field), of pure dimension n. In addition, fix a homomorphism of (coherent) modules on V,

$$\alpha : \mathcal{N} \to \mathcal{E}.$$

Assume that the target $\mathcal{E}$ is locally free (of constant rank) and that α generically is injective. In addition, assume for simplicity, that $\mathcal{N}$ is of constant and positive rank generically. It is convenient to define the *rank* of α as the difference,

$$\mathrm{rk}\,\alpha := \mathrm{rk}\,\mathcal{E} - \mathrm{rk}\,\mathcal{N}.$$

2.1 The conormal geometry. Associate with α two graded algebras: Form the dual homomorphism $\alpha^* : \mathcal{E}^* \to \mathcal{N}^*$. If $\mathcal{N}$ is locally free, let $\check{\mathrm{S}}\mathrm{ym}\mathcal{N} := \mathrm{Sym}\,\mathcal{N}^*$. In general, $\mathcal{N}$ is locally a quotient of a free module $\mathcal{F}$, and $\check{\mathrm{S}}\mathrm{ym}\mathcal{N}$ is defined (locally) as the subalgebra generated by $\mathcal{N}^*$ in $\mathrm{Sym}\,\mathcal{F}^*$. Alternatively, since V is reduced, the algebra $\check{\mathrm{S}}\mathrm{ym}\mathcal{N}$ is the quotient of the symmetric algebra $\mathrm{Sym}\,\mathcal{N}^*$ modulo $\mathcal{O}_V$-torsion. Let $\mathcal{C} = \mathcal{C}_\alpha$ be the image of the natural map $\mathrm{Sym}\,\mathcal{E}^* \to \check{\mathrm{S}}\mathrm{ym}\mathcal{N}$,

$$\mathrm{Sym}\,\mathcal{E}^* \twoheadrightarrow \mathcal{C}_\alpha \hookrightarrow \check{\mathrm{S}}\mathrm{ym}\mathcal{N}.$$

Associated to the inclusion of graded algebras $\mathcal{C} \subseteq \check{\mathrm{S}}\mathrm{ym}\mathcal{N}$ is a *central projection*, see for instance [KT1]: Let $P = P_\alpha := \check{\mathbb{P}}(\mathcal{N})$ be the 'Proj' of $\check{\mathrm{S}}\mathrm{ym}\mathcal{N}$ and $\mathcal{L} := \mathcal{O}_P(1)$. Let $Z = Z_\alpha$ be the zero scheme of the induced map $\mathcal{E}^*_P \to \mathcal{N}^*_P \to \mathcal{L}$, let $B = B_\alpha$ be the blowup of P with center Z, and $D = D_\alpha$ the exceptional divisor. On the other side, the 'Proj' $C = C_\alpha := \mathrm{Proj}\,\mathcal{C}$ is a closed subscheme of $\check{\mathbb{P}}(\mathcal{E})$, and its $\mathcal{O}(1)$ is the restriction of the tautological $\mathcal{L}' = \mathcal{O}(1)$ on $\check{\mathbb{P}}(\mathcal{E})$. Then there is a commutative diagram of schemes and natural proper maps,

$$\begin{array}{ccc} B_\alpha & \xrightarrow{\ b\ } & P_\alpha = \check{\mathbb{P}}(\mathcal{N}) \\ {\scriptstyle q}\big\downarrow & & \big\downarrow{\scriptstyle p} \\ \check{\mathbb{P}}(\mathcal{E}) \supseteq C_\alpha & \xrightarrow{\ c\ } & V. \end{array}$$

The maps b and q are birational, and the three schemes P, B, and C are of pure dimension $\dim V + \operatorname{rk} \mathcal{N} - 1$. It is a fundamental observation that the exceptional divisor D is the zero scheme of a natural injection $\mathcal{L}'_B \to \mathcal{L}_B$; in particular, $\mathcal{O}(D) = \mathcal{L}_B \otimes \mathcal{L}'^{\otimes -1}_B$.

2.2 The singular subsets. There are several "singular subsets" associated with the homomorphism α: First, the zero scheme Z_α (with its restricted map $p : Z_\alpha \to V$) will be called the *projecting singular subscheme* of α. Next, there are three closed subsets of V with obvious inclusions,

$$\operatorname{Sing}(\alpha) \supseteq \operatorname{Supp} \operatorname{Cok}(\alpha^*) \supseteq p(Z_\alpha),$$

where $\operatorname{Sing}(\alpha)$ is the subset where α is not locally split injective. The projecting singular subscheme Z_α, as a subscheme of P_α, has *distinguished subsets*, defined as the images of the components of D_α. Their images in V are, by definition, the *distinguished singular subsets* of α. We will write $\operatorname{Dsing}(\alpha)$ both for the set (with multiplicities) of distinguished singular subsets of α and for their union $p(Z_\alpha)$. Note that all singular subsets are nowhere dense, because α is assumed to be generically split injective. If $\mathcal{N}$ is locally free, then the three singular subsets agree, that is, every component of $\operatorname{Sing}(\alpha)$ is distinguished; moreover, then $\operatorname{Sing}(\alpha)$ has a natural scheme structure defined by the zeroth Fitting ideal of $\operatorname{Cok} \alpha^*$, or, the ideal of maximal minors (of size equal to $\operatorname{rk}\mathcal{N}$) of α. Note that, also when $\mathcal{N}$ is locally free, there may be distinguished singular subsets of α that are not components of $\operatorname{Sing}(\alpha)$.

2.3 The associated classes. Associate with α the following three Segre classes:

$$\check{\mathfrak{s}} := \check{\mathfrak{s}}(\mathcal{N}) = \mathfrak{s}(\check{\operatorname{S}}\operatorname{ym}\mathcal{N}), \quad \mathfrak{r} := \mathfrak{s}(\mathcal{C}_\alpha), \quad \mathfrak{s}(Z_\alpha, \mathcal{L}_\alpha).$$

The first two classes $\check{\mathfrak{s}}$ and $\mathfrak{r}$ are the Segre classes of (the affine cones of) the algebras, see [F, p. 70]. They belong to the Chow group $\mathrm{A}(V)$. The algebra $\check{\operatorname{S}}\operatorname{ym}\mathcal{N}$ and its subalgebra $\mathcal{C}_\alpha$ are locally contained in a polynomial algebra, and, generically the algebras are equal. It follows that the Segre classes are defined from the 'Proj's, cf. [F, Example 4.1.2, p. 71],

$$\check{\mathfrak{s}} = \int_{/V} s(\mathcal{L})[\check{\mathbb{P}}(\mathcal{N})], \qquad \mathfrak{r} = \int_{/V} s(\mathcal{L}')[C_\alpha].$$

The third class is the *twisted Segre class* of Z in P, $\mathfrak{s}(Z,\mathcal{L}) := s(Z,\mathcal{L})[P]$, see [KT2, p. 547]. It belongs to the Chow group $\mathrm{A}(Z)$. By [KT2, (4.5.4) p. 548], the twisted Segre class is the pushforward to $\mathrm{A}(Z)$ of the class $s(\mathcal{L}')s(\mathcal{L})[D]$ in $\mathrm{A}(D)$,

$$\mathfrak{s}(Z,\mathcal{L}) = \int_{/Z} s(\mathcal{L}')s(\mathcal{L})[D];$$

in particular, the class is a sum of terms supported on the distinguished subsets of Z. It is easy to relate the twisted Segre class $\mathfrak{s}(Z,\mathcal{L})$ to the absolute Segre class $\mathfrak{s}(Z)$, see [KT2, (4.4.3), p. 548]; in terms of the operation defined in (1.11), the relation is the equation,

$$\mathfrak{s}(Z,\mathcal{L}) = s(\mathcal{L})^{\dim V + \mathrm{rk}\,\mathcal{N}}\mathcal{L}^* \star \mathfrak{s}(Z). \tag{2.1}$$

Finally, the pushforward of the twisted Segre class,

$$\mathfrak{e} := \int_{/\,\mathrm{Dsing}\,(\alpha)} \mathfrak{s}(Z,\mathcal{L}),$$

is a class in the Chow group $\mathrm{A}(\mathrm{Dsing}\,(\alpha))$, and naturally a sum of terms,

$$\mathfrak{e} = \sum \mathfrak{e}_W,$$

over the distinguished singular subsets W of α.

2.4 The canonical classes. Define the *canonical classes* of α as those obtained from the associated classes by multiplying by the total Chern class of $\mathcal{E}$:

$$\mathfrak{c} := c(\mathcal{E})\,\check{\mathfrak{s}}, \quad \mathfrak{c}_{\mathrm{M}} := c(\mathcal{E})\,\mathfrak{r}, \quad \Delta := c(\mathcal{E})\,\mathfrak{s}(Z,\mathcal{L}), \quad \mathfrak{d} := c(\mathcal{E})\,\mathfrak{e}.$$

The classes $\mathfrak{c}$ and $\mathfrak{c}_{\mathrm{M}}$ belong to the Chow group $\mathrm{A}(V)$, the class Δ belongs to $\mathrm{A}(Z_\alpha)$, and the class $\mathfrak{d}$ belongs to $\mathrm{A}(\mathrm{Dsing}\,(\alpha))$. The class $\mathfrak{d}$ splits into components, $\mathfrak{d} = \sum \mathfrak{d}_W$, where the sum is over the distinguished singular subsets W of α.

Obviously, if $\mathcal{N}$ is locally free, then $\check{\mathfrak{s}}(\mathcal{N})$ is the total Segre class $\mathfrak{s}(\mathcal{N}^*)$ and

$$\mathfrak{c} = c(\mathcal{E} - \mathcal{N})[V] \tag{2.2}$$

is the usual Chern class of α; in general, the class $\mathfrak{c} = \mathfrak{c}(\alpha)$ will be called the *Chern class* of α. The class $\mathfrak{d} = \mathfrak{d}(\alpha)$ will be called the *singular class* of α. Finally, the class $\mathfrak{c}_{\mathrm{M}}$ is the Mather Chern class of the cokernel of α,

$$\mathfrak{c}_{\mathrm{M}} = \mathfrak{c}_{\mathrm{M}}(\mathrm{Cok}\,\alpha). \tag{2.3}$$

Indeed, the cokernel $\mathrm{Cok}\,\alpha$ is generically locally free of rank $q = \mathrm{rk}\,\alpha$. Take a modification $f : \widetilde{V} \to V$ for $\mathrm{Cok}\,\alpha$, see 1.3. Then, on $\widetilde{V}$ there is a locally free module $\mathcal{Q}$ of rank q and a surjection $\operatorname{Cok}\alpha|\widetilde{V} \to \mathcal{Q}$ which is an isomorphism over a dense open subset U of $\widetilde{V}$.

Let $\mathcal{E}_0$ be the kernel of surjection $\mathcal{E}|\widetilde{V} \to \operatorname{Cok}\alpha|\widetilde{V} \to \mathcal{Q}$. Then $\mathcal{E}_0$ is locally free, and contains the image of $\alpha|\widetilde{V}$. Moreover, the homomorphism $\mathcal{N}|\widetilde{V} \to \mathcal{E}_0$ is surjective on U. So the dual map $\mathcal{E}_0^* \to (\mathcal{N}|\widetilde{V})^*$ is injective on all of $\widetilde{V}$. [Locally, cover $\mathcal{N}|\widetilde{V}$ by a locally free $\mathcal{F}$. Then $\mathcal{F} \to \mathcal{N}|\widetilde{V} \to \mathcal{E}_0$ is surjective (and hence locally split surjective) on U. So the dual $\mathcal{E}_0^* \to \mathcal{F}^*$ is locally split injective on U. It follows that $\mathcal{E}_0^* \to \mathcal{F}^*$ is injective on all of $\widetilde{V}$. Hence, so is $\mathcal{E}_0^* \to (\mathcal{N}|\widetilde{V})^*$.] So there is a canonical surjection $\mathcal{C}_\alpha|\widetilde{V} \to \operatorname{Sym}\mathcal{E}_0^*$, and it is an isomorphism over the open subset where $\operatorname{Cok}\alpha$ is locally free.

It follows that $\check{\mathbb{P}}(\mathcal{E}_0)$ is a closed subscheme of $C \times_V \widetilde{V}$, and we obtain a birational V-map $g : \check{\mathbb{P}}(\mathcal{E}_0) \to C$. Therefore,

$$g_*[\check{\mathbb{P}}(\mathcal{E}_0)] = [C].$$

Hence $\mathfrak{s}(\mathcal{C}) = \int_{/V} s(\mathcal{L}')[C] = \int_{/V} s(\mathcal{O}_{\check{\mathbb{P}}(\mathcal{E}_0)}(1))[\check{\mathbb{P}}(\mathcal{E}_0)]$. Consequently,

$$\mathfrak{c}_{\mathrm{M}} = c(\mathcal{E}) \int_{\widetilde{V}/V} \int_{/\widetilde{V}} s(\mathcal{O}_{\check{\mathbb{P}}(\mathcal{E}_0)}(1))[\check{\mathbb{P}}(\mathcal{E}_0)] = \int_{\widetilde{V}/V} c(\mathcal{E}) s(\mathcal{E}_0^*)[\widetilde{V}] = \int_{\widetilde{V}/V} c(\mathcal{Q})[\widetilde{V}].$$

By definition, the right-hand side is the Mather Chern class of $\operatorname{Cok}\alpha$.

2.5 Buchsbaum–Rim multiplicities. For each distinguished singular subset W of α, the class $\mathfrak{e}_W$ belongs to the Chow group $\mathrm{A}(W)$, and its component of top dimension has the form $e_W[W]$, with a positive integer coefficient e_W. To determine the coefficient, localize at the generic point w of W. Then V_w is the spectrum of the local ring $\mathcal{O}_{V,w}$, and w is the closed point of V_w. Say P_w is of dimension r. Then D_w is of dimension $r-1$. Since w is distinguished, there is some component of D_w lying in the closed fiber. If W is component of $p(Z_\alpha)$, then the entire divisor D_w lies in the closed fiber. In this case, the coefficient is the coefficient of a Hilbert polynomial; it is the classical or generalized Buchsbaum–Rim multiplicity according as $\mathcal{N}_w$ is free or not. Moreover, if $\mathcal{N}$ is locally free, then any component of $\operatorname{Sing}(\alpha)$ is distinguished. In general, for an arbitrary distinguished singular subset W, the coefficient is the leading coefficient of some Hilbert type polynomial, cf. Kleiman–Thorup [KT2, (3.5), p. 543] or Achilles–Manaresi [AM]. As $\mathfrak{d}_W = c(\mathcal{E})\mathfrak{e}_W$, it follows that the class $\mathfrak{d}_W$ has the same top-dimensional component $e_W[W]$. In particular, if $m = \dim \operatorname{Dsing}(\alpha)$, then the top-dimensional part of $\mathfrak{d}$ is given by the sum,

$$\mathfrak{d}_m(\alpha) = \sum e_W[W],$$

where the sum is over all distinguished m-dimensional singular subsets W of α.

2.6 Dimensions of the distinguished components. The local theory gives information on dimension. Assume that W as above is a component of Dsing(α). The scheme P_w over V_w has dimension,

$$r = \dim V_w + \operatorname{rk} \mathcal{N} - 1 = \dim V - \dim W + \operatorname{rk} \mathcal{N} - 1.$$

Hence, by the local theory, see [KT1, (10.2), p.226], we obtain the inequality $\operatorname{rk} \mathcal{E} \geq r$, that is,

$$\dim W \geq \dim V - \operatorname{rk} \alpha - 1; \tag{2.4}$$

moreover, the dimension of the fiber $c^{-1}(w)$ is equal to $r - 1$, that is,

$$\dim c^{-1}(w) = \dim C - \dim W - 1. \tag{2.5}$$

Assume in addition that $\mathcal{N}$ is locally free. Then the homomorphism α is injective (since it is assumed to be generically injective), and its singular subset is the subscheme R defined by the zeroth Fitting ideal of α^*. It is a result of Buchsbaum and Rim [BR, Corollary 4.5, p. 223] that if $\mathcal{O}_{V,w}$ is Cohen–Macaulay and equality holds in (2.4), then the Buchsbaum–Rim multiplicity is given by the length,

$$e_W = \operatorname{length} \mathcal{O}_{R,w} = \operatorname{length} \operatorname{Cok} \alpha_w^*. \tag{2.6}$$

3 The Plücker Formula for Homomorphisms

The invariants associated with $\alpha : \mathcal{N} \to \mathcal{E}$ are related by the simple equation $\check{\mathfrak{s}} - \mathfrak{r} = \mathfrak{e}$ in the Chow group $\mathrm{A}(V)$. Similarly, the canonical invariants are related by the equation $\mathfrak{c} - \mathfrak{c}_{\mathrm{M}} = \mathfrak{d}$. In more detail, we have the following formulas:

The general Plücker formula. *The invariants of α are related by the following equations in the Chow group $\mathrm{A}(V)$:*

$$\check{\mathfrak{s}} = \mathfrak{r} + \sum \mathfrak{e}_W, \qquad \mathfrak{c} = \mathfrak{c}_{\mathrm{M}} + \sum \mathfrak{d}_W, \tag{3.1}$$

where the sums are over the distinguished subsets W of α. In particular,

$$\check{\mathfrak{s}}_i = \mathfrak{r}_i \text{ and } \mathfrak{c}_i = \mathfrak{c}_{\mathrm{M},i} \quad \text{for } i > \dim \operatorname{Dsing}(\alpha), \text{ and} \tag{3.2}$$

$$\mathfrak{d}_i = \mathfrak{c}_i \quad \text{for } i < \dim V - \operatorname{rk} \alpha. \tag{3.3}$$

The first formula in (3.1) is almost a tautology: The map $q : B \to P$ is birational, and the Segre class $\check{\mathfrak{s}}$ is the pushforward to V of the class $s(\mathcal{L})[P]$. Hence $\check{\mathfrak{s}}$ is the pushforward of the class $s(\mathcal{L})[B]$. Similarly, since $q : B \to C$

is birational, $\mathfrak{r}$ is the pushforward of the class $s(\mathcal{L}')[B]$. Therefore, the difference $\check{\mathfrak{s}} - \mathfrak{r}$ is the pushforward to V of the difference,

$$\big(s(\mathcal{L}) - s(\mathcal{L}')\big)[B] = s(\mathcal{L})s(\mathcal{L}')\big(c^1\mathcal{L} - c^1\mathcal{L}'\big)[B]. \tag{$*$}$$

Now, on B, the difference $c^1\mathcal{L} - c^1\mathcal{L}'$ is the first Chern class of $\mathcal{O}(D)$. Hence $(c_1\mathcal{L} - c_1\mathcal{L}')[B]$ is represented by the fundamental cycle $[D]$. So, by definition of the twisted Segre class, the pushforward to P of (*) is equal to $\mathfrak{s}(Z, \mathcal{L})$ (viewed in $\mathrm{A}(P)$). Therefore, the pushforward to V is equal to $\mathfrak{e}$. Thus $\check{\mathfrak{s}} - \mathfrak{r} = \mathfrak{e}$, and the first equation in (3.1) has been verified. To obtain the second, multiply the first by $c(\mathcal{E})$.

The equalities in (3.2) hold, because the classes $\mathfrak{e}$ and $\mathfrak{d}$ have support in the distinguished singular subset $\mathrm{Dsing}\,(\alpha)$.

Finally, to prove (3.3), note that $\mathfrak{c}_{\mathrm{M}}$ is the pushforward of the class $c(\mathcal{E})s(\mathcal{L}')[C]$. On C there is a surjection $\mathcal{E}_C^* \to \mathcal{L}'$. Hence $c(\mathcal{E})s(\mathcal{L}')$ is the Chern class of the dual of the kernel. As the kernel is locally free of rank equal to $e-1$, where $e := \mathrm{rk}\,\mathcal{E}$, it follows that $c(\mathcal{E})s(\mathcal{L}')[C]$ vanishes in dimension i for $i < \dim C - e + 1$. Hence $\mathfrak{c}_{\mathrm{M}}$ vanishes in the same dimensions. As $\dim C = \dim V + \mathrm{rk}\,\mathcal{N} - 1$ and $\mathfrak{c} - \mathfrak{d} = \mathfrak{c}_{\mathrm{M}}$, we obtain the equations of (3.3).

4 Properties of the Invariants

View $\alpha : \mathcal{N} \to \mathcal{E}$ as a complex (with $\mathcal{E}$ in degree 0 and $\mathcal{N}$ in degree -1), and hence as an object in the derived category $D(V)$. The basic invariance states that the singular subsets of α (that is, $\mathrm{Sing}\,(\alpha)$, $\mathrm{Supp}\,\mathrm{Cok}\,(\alpha^*)$, Z_α, and the various distinguished singular subsets W), and the canonical classes (that is, $\mathfrak{c}$, $\mathfrak{c}_{\mathrm{M}}$, Δ, $\mathfrak{d}$, and the components $\mathfrak{d}_W$), depend only on the isomorphism class of α in the derived category $D(V)$. More generally, consider a second homomorphism β, with locally free target and satisfying the same assumptions as α. If α and β are related by a certain distinguished triangle in the derived category, then there is a simple relation between the canonical invariants of α and β. This basic invariance is the main observation of this work. In addition, we describe below other natural properties of the canonical invariants.

4.1 Basic invariance. Assume there is given in the derived category $D(V)$ a distinguished triangle $\alpha \to \beta \to \mathcal{F}[0]$, where $\mathcal{F}[0]$ is the complex consisting of a single locally free module $\mathcal{F}$ in degree 0. Then α and β have the same singular subsets and, for any of the canonical classes $\mathfrak{j}$, we have the equation,

$$\mathfrak{j}(\beta) = c(\mathcal{F})\,\mathfrak{j}(\alpha). \tag{4.1}$$

4.2 Twisting. Clearly, if $\mathcal{K}$ is an invertible module, then the homomorphisms α and $\mathcal{K}\otimes\alpha$ have the same singular subsets. In the notation of (1.11), if $\mathfrak{j}$ is any of the canonical classes, then

$$\mathfrak{j}(\mathcal{K}\otimes\alpha) = c(\mathcal{K})^{\operatorname{rk}\alpha-\dim V}\mathcal{K}\star\mathfrak{j}(\alpha). \tag{4.2}$$

4.3 Flat base change. The formation of the invariants of α commutes in the obvious sense with flat base change: Assume that $f : V' \to V$ is flat with equidimensional fibers. Then the singular subsets of $f^*\alpha$ are obtained from those of α by pullback; if $\mathfrak{j}$ is any of the canonical (or associated) classes, then, in the appropriate Chow group,

$$\mathfrak{j}(f^*\alpha) = f^*\mathfrak{j}(\alpha). \tag{4.3}$$

In general, the formation of the invariants is poor with respect to general base change.

4.4 The Adjunction formula. If K is a *general* Cartier divisor on V, then, for any of the canonical classes $\mathfrak{j}$,

$$\mathfrak{j}(\alpha|K) = K\cdot\mathfrak{j}(\alpha).$$

(In this assertion, a Cartier divisor K is *general*, if it intersects a well-determined finite set of irreducible closed subsets properly.)

4.5 Positivity. Assume that the target $\mathcal{E}$ is trivial, say $\mathcal{E} = E_V$. Then $\check{\mathbb{P}}(\mathcal{E}) = \check{\mathbb{P}}(E)\times V$, and C is a subscheme of the product. Let V' be the image of the projection $C \to \check{\mathbb{P}}(E)$, and set $n' := \dim V'$. By prime avoidance, the i-dimensional part $(c_1\mathcal{L}')^{\dim C - i}[C]$ of $s(\mathcal{L}')[C]$ is non-negative. Hence the pushforward $\mathfrak{r}_i$ is non-negative. Obviously, $\mathfrak{r}_i = 0$ for $\dim C - i > n'$. In fact, it is easy to prove that

$$\mathfrak{r}_i \succ 0 \iff \dim C - n' \le i \le n$$

(that is, the class $\mathfrak{r}_i$ is represented by a strictly positive cycle in the stated range). Since $\mathcal{E}$ is trivial, $\mathfrak{c} = \check{\mathfrak{s}}$, $\mathfrak{c}_{\mathrm{M}} = \mathfrak{r}$, and $\mathfrak{d} = \mathfrak{e}$. Of course, in general we would expect $\check{\mathfrak{s}} \succcurlyeq 0$ only if $\mathcal{N}^*$ is generated by global sections. Similarly, the twisted Segre class $\mathfrak{s}(Z,\mathcal{L})$ and its pushforward $\mathfrak{e}$ are non-negative if $\mathcal{N}^*$ is generated by global sections.

4.6 Question. Assume that $\mathcal{N}$ is locally free of rank equal to the rank of $\mathcal{E}$, that is, the homomorphism $\alpha : \mathcal{N} \to \mathcal{E}$ is an injective homomorphism of locally free modules of the same rank. Then, clearly, the singular locus is the divisor R, of dimension $n-1$, defined as the zero scheme of the determinant $\det \alpha$. Clearly, $\check{\mathbb{P}}(\mathcal{N}) = \mathbb{P}(\mathcal{N}^*)$ and $C = \mathbb{P}(\mathcal{E}^*)$. Hence, the Plücker formula (3.1) is the equation, $\mathfrak{r}(\alpha) = \mathfrak{s}(\mathcal{N}^*) - \mathfrak{s}(\mathcal{E}^*)$. The same consideration applies to the dual homomorphism $\alpha^* : \mathcal{E}^* \to \mathcal{N}^*$, and it has the same singular subscheme R. In $\mathrm{A}(V)$, we have the equations,

$$\mathfrak{r}(\alpha) = \mathfrak{s}(\mathcal{N}^*) - \mathfrak{s}(\mathcal{E}^*), \quad \mathfrak{r}(\alpha^*) = \mathfrak{s}(\mathcal{E}) - \mathfrak{s}(\mathcal{N}).$$

Now the Segre class of the dual is given by the equation $\mathfrak{s}_i(\mathcal{E}^*) = (-1)^{n-i}\mathfrak{s}_i(\mathcal{E})$, that is, the component of codimension j with respect to V is multiplied by $(-1)^j$. Let $\check{\mathfrak{r}}$ be the class obtained from $\mathfrak{r}$ by multiplying the component of codimension j in R by $(-1)^j$. Then we obtain the following equation in the Chow group $\mathrm{A}(V)$:

$$\mathfrak{r}(\alpha^*) = \check{\mathfrak{r}}(\alpha).$$

It is natural to ask if the latter equation holds in $\mathrm{A}(R)$.

5 The Invariants of a Projective Variety

Assume, as in the introduction, that V is a subvariety of $\mathbb{P} = \mathbb{P}^N = \mathbb{P}(E)$. The Jacobian d_V of (1.10), viewed as a complex, is the cotangent complex, a well-known intrinsic invariant of V in the derived category $D(V)$. Hence, from the cotangent complex we obtain singular subsets (denoted $\mathrm{Sing}(V)$, $\mathrm{Dsing}(V)$, Z_V, etc) and canonical (intrinsic) classes $\mathfrak{j}(V) := \mathfrak{j}(d_V)$, (say $\mathfrak{c}(V)$, $\mathfrak{c}_{\mathrm{M}}(V)$, $\Delta(V)$, $\mathfrak{d}(V)$, etc). As noted in (2.2) and (2.3), the Chern class $\mathfrak{c}(V) := \mathfrak{c}(d_V)$ in the lci case and the Mather Chern class $\mathfrak{c}_{\mathrm{M}}(V) := \mathfrak{c}_{\mathrm{M}}(\mathrm{Cok}\, d_V)$ in all cases are given by the formulas,

$$\mathfrak{c}(V) = c(\Omega^1_{\mathbb{P}}|V - \mathcal{N}_{V/\mathbb{P}})\,[V], \qquad \mathfrak{c}_{\mathrm{M}}(V) = \mathfrak{c}_{\mathrm{M}}(\Omega^1_V).$$

It should be emphasized that the classes according to this definition differ by the sign $(-1)^{n-i}$ in dimension i from the classes defined by Fulton [F, p. 77 and 79].

Similarly, from the homogeneous Jacobian $\tilde{d}_V$ of (1.9), we obtain relative classes $\mathfrak{j}(V,\mathbb{P}) := \mathfrak{j}(\tilde{d}_V)$ depending on the embedding of V in $\mathbb{P}$. As the target of the homomorphism is the trivial module E_V, we may use the notation for the associated classes,

$$\check{\mathfrak{s}}(V,\mathbb{P}) = \check{\mathfrak{s}}(\mathcal{N}_{V/\mathbb{P}}(1)), \quad \mathfrak{r}(V,\mathbb{P}) = \mathfrak{c}_{\mathrm{M}}(\mathcal{P}^1(\mathcal{O}_V(1))), \quad \mathfrak{e}(V,\mathbb{P}) = \int_{/V} \mathfrak{s}(Z_V, \mathcal{L}),$$

where $\mathcal{L}$ is the tautological $\mathcal{O}(1)$ on $\mathrm{Proj}\,\check{\mathrm{S}}\mathrm{ym}\mathcal{N}(1)$.

Note that for both homomorphisms d_V and $\tilde{d}_V$, the scheme C is the conormal variety, $C = CV$, of V; however, the two tautological $\mathcal{O}(1)$'s on C are different. For the homogeneous Jacobian $\tilde{d}_V$ we obtain the usual embedding of CV in the product $\mathbb{P}(E_V^*) = \check{\mathbb{P}}(E) \times V$, and the invertible module $\mathcal{L}' = \mathcal{O}_C(1)$ is the pullback of $\mathcal{O}_{\check{\mathbb{P}}(E)}(1)$. As $\mathfrak{r} = \int_{/V} s(\mathcal{L}')[C]$, we obtain, with $\ell' := c^1\mathcal{O}_{\check{\mathbb{P}}(E)}(1)$, the equation,

$$\mathfrak{r}_i = \int_{/V} \ell'^{N-1-i}[CV]. \tag{5.1}$$

In particular, it follows, see Kleiman [K2, pp. 338–39], that the class $\mathfrak{r}_i$ is represented by the effective cycle of a polar variety $V(A)$ of V, and that its degree r_i is the ith *rank* of V.

Clearly, in this notation, the Plücker Formula (1.7) follows by applying (3.1) to the homomorphism $\tilde{d}_V$. By the basic invariance (4.1) and the twisting property (4.2), the intrinsic classes $\mathfrak{j}(V)$ and the relative classes $\mathfrak{j}(V, \mathbb{P})$ are related by the equation,

$$\mathfrak{j}(V, \mathbb{P}) = c(\mathcal{O}_{\mathbb{P}}(1))\, \mathcal{O}_{\mathbb{P}}(1) \star \mathfrak{j}(V). \tag{5.2}$$

In particular, for the Mather Chern classes we obtain the equation, with $h := c^1\mathcal{O}_{\mathbb{P}}(1)$,

$$\mathfrak{r}(V, \mathbb{P}) = c(\mathcal{O}_{\mathbb{P}}(1))\, \mathcal{O}_{\mathbb{P}}(1) \star \mathfrak{c}_{\mathrm{M}}(\Omega_V^1) = \sum (1+h)^{i+1} \mathfrak{c}_{\mathrm{M}}^{n-i}(\Omega_V^1). \tag{5.3}$$

If V is smooth, then the Mather Chern class $\mathfrak{c}_{\mathrm{M}}^j(\Omega_V^1)$ on the right is the usual jth Chern class of Ω_V^1, and (5.3) is essentially the classical formula relating the ranks of V to the Chern classes of V, see Kleiman [K1, (4.15), p. 342].

In the lci case, $\check{\mathfrak{s}} = s(\mathcal{N}_{V/\mathbb{P}}(1)^*)[V]$. In the gci case, say V is an intersection of q hypersurfaces of degrees $d_1, \ldots, d_q$, we obtain the equation,

$$\check{\mathfrak{s}} = \frac{1}{(1-(d_1-1)h)\cdots(1-(d_q-1)h)}[V]; \tag{5.4}$$

in particular, $\check{s}_i = ds^{n-i}(d_1 - 1, \ldots, d_q - 1)$.

The set of equations (3.3), for the homomorphisms d_V and $\tilde{d}_V$, is empty, since $\operatorname{rk} d_V = \dim V$. Consider the equations of (3.2). It follows, by taking degrees, that

$$r_i = \check{s}_i \quad \text{for } i > \dim \operatorname{Dsing}(V). \tag{5.5}$$

Moreover, in dimension $m = \dim \operatorname{Dsing}(V)$, we obtain the equation,

$$r_m = \check{s}_m - \sum{}' e_W \deg W, \tag{5.6}$$

where the sum is over the distinguished singular subsets W of maximal dimension and e_W is a corresponding Buchsbaum–Rim type multiplicity.

In the lci case, $\operatorname{Dsing}(V)$ is simply the singular locus of V, and the e_W are classical Buchsbaum–Rim multiplicities corresponding to components of maximal dimension of $\operatorname{Sing}(V)$; in the general, singular case, it may happen that $\operatorname{Dsing}(V)$ is empty, even for a singular curve in $\mathbb{P}^3$, cf. [K3, p. 258].

Example 5.1. For the Veronese embedding $\mathbb{P}^N \to \mathbb{P}^{N_k}$, where $N_k + 1$ is the rank of $\mathrm{Sym}^k E$, we obtain for the relative invariants,

$$\mathrm{j}(V, \mathbb{P}^{N_k}) = c(\mathcal{O}(k-1))\, \mathcal{O}(k-1) \star \mathrm{j}(V, \mathbb{P}^N). \tag{5.7}$$

Since $r_i(V, \mathbb{P}^N) \geq 0$ and $r_n(V, \mathbb{P}^N) = d > 0$, it follows for any $k \geq 2$, that $r_i(V, \mathbb{P}^{N_k}) > 0$ for $0 \leq i \leq n$. In particular, the dual of V with respect to $\mathbb{P}^{N_k}$ is a hypersurface for $k \geq 2$.

5.2 The dimension of the dual variety. Work over an algebraically closed field, with V integral and such that duality holds for V. Assume that some component W (with generic point w) of $\mathrm{Sing}\,(V)$ is distinguished (this is true for any component of $\mathrm{Sing}\,(V)$ in lci case). Then we have the following inequality for the dimension of the dual variety:

$$\dim V' \geq r - 1, \quad \text{where } r = N - \dim W - 1. \tag{5.8}$$

Indeed, since $\dim C = N - 1$, it follows from (3.2) that the fiber $c^{-1}w$ is of dimension $r - 1$. Hence, for any closed point x of W, the fiber $c^{-1}x$ is at least of that dimension. The fiber $c^{-1}x$ is projected isomorphically onto its image in $\check{\mathbb{P}}(E)$. Hence $\dim V' \geq r - 1$.

Assume equality in (5.8). Then V' is equal to the image in $\check{\mathbb{P}}(E)$ of the fiber $c^{-1}x$, for any closed point x of W. In particular, since the fiber consists of hyperplanes through x, it follows that V' is contained in the hyperplane x' of $\check{\mathbb{P}}(E)$, dual to the point x. Hence we have the inclusion,

$$V' \subseteq L',$$

where L is the linear span of W. The inclusion is strict, since otherwise V would be linear. Hence,

$$r - 1 = \dim V' < \dim L' = N - 1 - \dim L \leq N - 1 - \dim W = r.$$

It follows that $\dim W = \dim L$. Hence $W = L$ is linear and V' is a hypersurface in L'. Since W is assumed to be a component of $\mathrm{Sing}\,(V)$, it follows easily that V is cone with vertex $W = L$ over a smooth base. Therefore, except in the latter case, the inequality in (5.8) is strict, that is,

$$\dim V' \geq N - \dim W - 1. \tag{5.9}$$

Note in particular, that if V has an isolated singularity that is distinguished, and V is not a cone, then V' is a hypersurface.

5.3 Adjunction Formula. Let K be a hypersurface in $\mathbb{P}^N$ of degree not divisible by the characteristic, and let $\mathcal{K} = \mathcal{O}(K)$. Set $V_0 := V \cap K$.

Assume that K is generic. Then, for any of the absolute invariants $\mathfrak{j}(V)$, we have the equation,

$$\mathfrak{j}(V_0) = s(\mathcal{K})\, K \cdot \mathfrak{j}(V) = \frac{K}{1-K}\, \mathfrak{j}(V). \tag{5.10}$$

For the relative invariants, the corresponding equations are the following:

$$\begin{gathered} \mathfrak{r}(V_0) = s(\mathcal{K}(-1))K \cdot \mathfrak{r}(V), \quad \check{\mathfrak{s}}(V_0) = s(\mathcal{K}(-1))K \cdot \check{\mathfrak{s}}(V), \\ \mathfrak{e}(V_0) = s(\mathcal{K}(-1))K \cdot \mathfrak{e}(V). \end{gathered} \tag{5.11}$$

6 The Invariants of a Generically Smooth Map

Clearly, it follows from the basic invariance that if a module $\mathcal{Q}$ on V is a quotient of a locally free module $\mathcal{E}$ and $\mathcal{Q}$ has generic constant rank, then there are invariants of $\mathcal{Q}$ defined from any generation $\mathcal{Q} = \mathcal{E}/\mathcal{N}$.

As a second application of the invariance, consider a generically smooth map of proper varieties $f : V \to T$. Let t be the dimension of T. Assume that f is *smoothable* [SGA6, p. 475], that is, there exists a factorization,

$$\begin{array}{ccc} V & \longrightarrow & M \\ {\scriptstyle f}\downarrow & \swarrow & \\ T, & & \end{array}$$

where $V \to M$ is a closed embedding and M/T is smooth of constant relative dimension. (Assume for instance that V embeds into a smooth variety.) Then there is an exact sequence,

$$\mathcal{N}_{V/M} \xrightarrow{d_f} \Omega^1_{M/T}|V \longrightarrow \Omega^1_{V/T} \longrightarrow 0,$$

where $\mathcal{N}_{V/M} := \mathcal{I}_V/\mathcal{I}_V^2$ is the conormal module, $\Omega^1_{V/T} = \Omega^1_f$ is the module of relative differentials, and d_f is induced by the universal derivation $d_{M/T}$. The homomorphism d_f is generically split injective because f is generically smooth.

It is well known that the complex defined by d_f (the cotangent complex), in the derived category $D(V)$, is independent of the choice of factorization, see [SGA6, VIII.2.2, p. 476]. Therefore, by the basic invariance, we obtain singular subsets Z_f, Sing (f), and Dsing (f) etc, and canonical classes $\mathfrak{c}(f)$, $\mathfrak{c}_{\mathrm{M}}(f)$, $\Delta(f)$, $\mathfrak{d}(f)$ etc. Clearly, the set Sing (f) is the subset of points of V at which f is not smooth. Note that the rank of d_f is equal to $\dim V - \dim T = n - t$. Hence, by (2.4), if W is any component of Dsing (f), then

$$\dim W \geq t - 1. \tag{6.1}$$

From the general formula (3.1), we obtain in the Chow group $\mathrm{A}(V)$ the Plücker formula,

$$\mathfrak{d}(f) = \mathfrak{c}(f) - \mathfrak{c}_{\mathrm{M}}(\Omega^1_{V/T}), \tag{6.2}$$

and from (3.3), we obtain the equations,

$$\mathfrak{d}_i(f) = \mathfrak{c}_i(f) \quad \text{for } i \leq t-1. \tag{6.3}$$

Assume that f is appropriately generic is the sense that $\mathrm{Sing}\,(f)$ is (empty or) of dimension $t-1$. It follows from (6.1) that $\mathrm{Dsing}\,(f)$ is of pure dimension $t-1$. The top-dimensional component of $\mathfrak{d}$ is represented by the cycle,

$$\mathfrak{d}_{t-1} = \sum_W e_W[W], \tag{6.4}$$

where the sum is over the distinguished components W of $\mathrm{Sing}\,(f)$, and the e_W are Buchsbaum–Rim multiplicities.

Assume that f is *regular*, that is, the embedding $V \to M$ is regular. Then the source $\mathcal{N}$ of d_f is locally free, and $\mathrm{Sing}\,(f) = \mathrm{Dsing}\,(f)$ may be viewed as the subscheme defined by the ideal of $t \times t$ minors of d_f. It follows from (6.1) that any component of $\mathrm{Sing}\,(f)$ is of dimension at least $t-1$. Moreover, if f is appropriately generic, then $\mathrm{Sing}\,(f)$ is of pure dimension $t-1$; in this case, the cycle (6.4) is called the *cuspidal cycle*. If, in addition, V is Cohen–Macaulay, then, by (2.6), the cuspidal cycle is the fundamental cycle of the subscheme $\mathrm{Sing}\,(f)$. Hence, when f is regular and appropriately generic, and V is Cohen–Macaulay, we obtain the equation for the cuspidal cycle,

$$\mathfrak{c}_{t-1}(f) = [\mathrm{Sing}\,(f)]. \tag{6.5}$$

Example 6.1. Assume that V and T are smooth. Then d_f fits into the exact sequence,

$$\Omega^1_T|V \xrightarrow{d_f} \Omega^1_V \longrightarrow \Omega^1_{V/T} \longrightarrow 0.$$

and d_f is injective, because it is injective generically. The singular subset $\mathrm{Sing}\,(f)$ is the ramification subscheme defined by the ideal of $t \times t$ minors of d_f, and any component of $\mathrm{Sing}\,(f)$ is of dimension at least $t-1$. The Chern class of f is the class $\mathfrak{c}(f) = c(\Omega^1_V - \Omega^1_T|V)[V]$. If f is appropriately generic, then the Chern class is related to the Mather Chern class of $\Omega^1_{V/T}$ and the singular class $\mathfrak{d}(f)$ by the equations,

$$\mathfrak{c}_i(f) = \begin{cases} \mathfrak{c}_{\mathrm{M},i}(\Omega^1_{V/T}), & \text{for } t \leq i \leq n, \\ \mathfrak{d}_i, & \text{for } 0 \leq i \leq t-1. \end{cases} \tag{6.6}$$

In particular, in dimension $t-1$, equation (6.5) is the classical formula for the class of the cuspidal cycle in terms of the Chern classes of Ω^1_V and Ω^1_T, see [K1, III.8, p. 330].

6.2 Composition of maps. Consider a composition gh of generically smooth maps h and g. Then h and g embed in a commutative diagram,

$$\begin{array}{ccccc} V & \longrightarrow & R & \longrightarrow & M \\ & \overset{h}{\searrow} & \downarrow & & \downarrow \\ & & S & \longrightarrow & Q \\ & & & \overset{g}{\searrow} & \downarrow \\ & & & & T, \end{array}$$

such that the horizontal maps are closed embeddings, and the vertical maps are smooth, and such that the square is cartesian. Since the square is cartesian, we have identifications,

$$\mathcal{N}_{R/M} = \mathcal{N}_{S/Q}|R, \qquad \Omega^1_{R/S} = \Omega^1_{M/Q}|R.$$

Moreover, there is a commutative diagram with exact rows,

$$\begin{array}{ccccccccc} & & \mathcal{N}_{S/Q}|V & \longrightarrow & \mathcal{N}_{V/M} & \longrightarrow & \mathcal{N}_{V/R} & \longrightarrow & 0 \\ & & \downarrow{\scriptstyle d_g|V} & & \downarrow{\scriptstyle d_{gh}} & & \downarrow{\scriptstyle d_h} & & \\ 0 & \longrightarrow & \Omega^1_{Q/T}|V & \longrightarrow & \Omega^1_{M/T}|V & \longrightarrow & \Omega^1_{R/S}|V & \longrightarrow & 0. \end{array}$$

Furthermore, the upper left horizontal map is injective if $V \to R$ is a regular embedding.

Assume that h is smooth. Then $V \to R$ is a regular embedding, and so the top row is exact. Moreover, d_h is injective, and its cokernel is the locally free module Ω^1_h. In other words, there is, in the derived category $D(V)$ an exact triangle $d_g|V \to d_{gh} \to \Omega^1_h[0]$ relating the cotangent complexes. Whence, by the basic invariance of Section 4, we deduce the equality $Z_{gh} = Z_g \times_S V$, and, in the Chow group $\mathrm{A}(Z_{gh})$, the equations,

$$\Delta(gh) = c(\Omega^1_h)\Delta(d_g|V) = c(\Omega^1_h)h^*\Delta(g). \tag{6.7}$$

There are similar equations for the invariants $\mathfrak{c}(gh)$, $\mathfrak{c}_{\mathrm{M}}(\Omega^1_{gh})$ and $\mathfrak{d}(gh)$.

7 The Invariants of a Hypersurface

Assume that V is a hypersurface section of a smooth variety M, say V is the zero scheme of a regular section F of an invertible module $\mathcal{L}$ on M. The cotangent complex of V is defined by the homomorphism d in the exact sequence,

$$0 \longrightarrow \mathcal{L}^*_V \xrightarrow{d} \Omega^1_M|V \longrightarrow \Omega^1_V \longrightarrow 0.$$

Here the conormal module $\mathcal{N} = \mathcal{L}_V^*$ is free of rank 1. In particular, $\check{\mathbb{P}}(\mathcal{N})$ is equal to V, and its $\mathcal{L}$ is simply $\mathcal{L}_V$. The projecting singular scheme Z is the singular locus of V. The classes $\check{\mathfrak{s}}$, $\mathfrak{r}$ and $\mathfrak{e}$ of Section 2 are the following:

$$\mathfrak{e} = \mathfrak{s}(Z, \mathcal{L}), \quad \check{\mathfrak{s}} = s(\mathcal{L})[V], \quad \mathfrak{r} = \int_{/V} s(\mathcal{L}(1))[B],$$

where B is the blowup of V along Z. The Plücker formula is the simple equation in $\mathrm{A}(V)$,

$$\mathfrak{s}(Z, \mathcal{L}) = s(\mathcal{L})[V] - \mathfrak{r}. \tag{7.1}$$

Consider the Schwarz–MacPherson class $\mathfrak{c}_{\mathrm{SM}}(V)$ in the Chow group $\mathrm{A}(V)$. It is related to the Segre classes of Z in V and in M by the formula of Aluffi [A, Theorem 1, p. 456]. If $\check{\mathfrak{c}}_{\mathrm{SM}}$ denotes the "dual" class, obtained from $\mathfrak{c}_{\mathrm{SM}}$ by multiplying in dimension i by $(-1)^{n-i}$, Aluffi's Formula is equivalent to the equation,

$$c(\Omega_M^1)^{-1}\check{\mathfrak{c}}_{\mathrm{SM}}(V) = s(\mathcal{L})[V] - s(Z, \mathcal{L})[M], \tag{7.2}$$

where the last Segre class is the twisted Segre class of Z in M. The intrinsic form is the equation,

$$\check{\mathfrak{c}}_{\mathrm{SM}}(V) = \mathfrak{c}(V) - c(\Omega_M^1)s(Z, \mathcal{L})[M];$$

recall that the Chern class $\mathfrak{c}(V)$ is related to Fulton's Chern class $\mathfrak{c}_{\mathrm{F}}(V)$ by the equation $\mathfrak{c}(V) = \check{\mathfrak{c}}_{\mathrm{F}}(V)$.

The two twisted Segre classes $s(Z, \mathcal{L})[V]$ and $s(Z, \mathcal{L})[M]$ are related as follows: Let $\hat{b} : \widehat{B} \to M$ be the blowup of with center Z, and $\widehat{D}$ the exceptional divisor. Let R be the residual divisor defined by the equation $\hat{b}^{-1}V = \widehat{D} + R$. Its associated invertible module is $\mathcal{O}(R) = \mathcal{L}(1) =: \mathcal{L}'$. Then,

$$s(Z, \mathcal{L})[V] = \int_{/Z} s(\mathcal{L}(1))[R]^Z + s(Z, \mathcal{L})[M], \tag{7.3}$$

where the superscript Z indicates the part over Z. As $\mathfrak{r} = \int_{/V} s(\mathcal{L}')[B]$, we obtain from (7.1) and (7.3) the following expression for the right-hand side of (7.2):

$$s(\mathcal{L})[V] - s(Z, \mathcal{L})[V] + \int_{/Z} s(\mathcal{L}')[R]^Z = \int_{/V} s(\mathcal{L}')\{[R]^Z + [B]\}.$$

Obviously $[R] = [R]^Z + [B]$. In other words, the right-hand side of (7.2) equals $\int_{/Z} s(\mathcal{L}(1)[R]$, that is, Aluffi's Formula is equivalent to the following, see also Parusiński–Pragacz [PP, Theorem 3.1]:

$$\check{\mathfrak{c}}_{\mathrm{SM}}(V) = c(\Omega_M^1)\int_{/V} s(\mathcal{L}')[R] = c(\Omega_M^1)\int_{/V} \frac{[R]}{1-R}. \tag{7.4}$$

An expression for the residual divisor R and the class in (7.4) may be obtained as follows: Consider the diagram of modules on M,

$$\begin{array}{ccccccccc} & & & & \mathcal{O}_M & & & & \\ & & & d^1F & \downarrow & \searrow^{F} & & & \\ 0 & \longrightarrow & \Omega^1_M \otimes \mathcal{L} & \longrightarrow & \mathcal{P}^1_M(\mathcal{L}) & \longrightarrow & \mathcal{L} & \longrightarrow & 0. \end{array}$$

Here V is the zero scheme of F, and the zero scheme of d^1F is equal to Z. Set $\mathcal{Q} := \mathcal{P}^1_M(\mathcal{L})^* \otimes \mathcal{L}$. Take the duals and tensor with $\mathcal{L}$ to obtain the diagram,

$$\begin{array}{ccccccccc} & & & & \mathcal{L} & & & & \\ & & & \partial^1F & \uparrow & \nwarrow^{F} & & & \\ 0 & \longleftarrow & (\Omega^1_M)^* & \longleftarrow & \mathcal{Q} & \longleftarrow & \mathcal{O}_M & \longleftarrow & 0. \end{array}$$

So Z is the zero scheme of ∂^1F. Pull the diagram back to the blowup $\widehat{B}$. On $\widehat{B}$, the homomorphism $\mathcal{Q}|\widehat{B} \to \mathcal{L}|\widehat{B}$ factors as the surjection $\mathcal{Q}|\widehat{B} \to \mathcal{L}'$ followed by the inclusion $\mathcal{L}' \to \mathcal{L}|\widehat{B}$. In particular, the map $\mathcal{O}_{\widehat{B}} \to \mathcal{L}|\widehat{B}$ factors into an inclusion $\mathcal{O}_{\widehat{B}} \to \mathcal{L}'$ followed by the inclusion $\mathcal{L}' \to \mathcal{L}_{\widehat{B}}$. Hence, the residual scheme R is the zero scheme of the map $\mathcal{O}_{\widehat{B}} \to \mathcal{L}'$. Clearly, on the zero scheme R, the surjection $\mathcal{Q}_R \to \mathcal{L}'_R$ factors over the quotient $(\Omega^1_M)^*_R$. Hence there is a canonical V-map $p : R \to \check{\mathbb{P}}(\Omega^1_M|V)$ such that $(\Omega^1_M)^*_R \to \mathcal{L}'_R$ is the pullback of the universal invertible module on $\check{\mathbb{P}}(\Omega^1_M)$. Let $Q := \check{\mathbb{P}}(\Omega^1_M)$. Then there is a well-known exact sequence,

$$0 \to \Omega^1_{Q/M}(1) \to (\Omega^1_M)^*_Q \to \mathcal{O}_Q(1) \to 0.$$

In particular, since $\mathcal{L}'_R = p^*\mathcal{O}_Q(1)$, we obtain on R the equation of operators,

$$c(\Omega^1_M)s(\mathcal{L}'_R) = c\big((\Omega^1_{Q/M})^*\big).$$

Hence, from (7.4) we obtain the equation,

$$\check{c}_{\mathrm{SM}}(V) = \int_{/V} c(\mathcal{T})[R] = \int_{/V} c(\mathcal{T})\, p_*[R], \tag{7.5}$$

where $Q = \check{\mathbb{P}}(\Omega^1_M|V)$ and $\mathcal{T} = \mathcal{T}_{Q/M}(-1)$ is the twisted relative tangential module (the dual of $\Omega^1_{Q/M}(1)$).

8 Pohl's Plücker Formula

Assume in the general setup of Section 2 that the homomorphism $\alpha : \mathcal{N} \to \mathcal{E}$ generically is an isomorphism. Then the singular locus of α is the subset where α is not an isomorphism. The cokernel of α is generically zero, and so the Mather Chern class $\mathfrak{c}_{\mathrm{M}}(\operatorname{Cok}\alpha)$ reduces to its top term $[V]$. Thus the Plücker formula is the equation $\mathfrak{d} = \mathfrak{c} - [V]$. A more interesting equation may be obtained as follows.

Let R be the *ramification locus* of α: As a set R is the support of the cokernel of α, and as a subscheme R is defined by the zeroth Fitting ideal of $\operatorname{Cok}\alpha$, or equivalently, R is the zero scheme of the homomorphism (where $e := \operatorname{rk}\mathcal{E}$),

$$\bigwedge^{e}\mathcal{N} \to \bigwedge^{e}\mathcal{E}.$$

Clearly, as a set R is contained in the singular locus $\operatorname{Sing}(\alpha)$, and R contains $\operatorname{Dsing}(\alpha)$. Let $\widetilde{V} \to V$ be the blowup of V with center R, and let $\widetilde{R}$ be the exceptional divisor. Then, on $\widetilde{V}$, the image of $\bigwedge^{e}\alpha|\widetilde{V}$ is an invertible submodule of $\bigwedge^{e}\mathcal{E}$. As a consequence, the image of $\alpha|\widetilde{V}$ is a locally free submodule $\widetilde{\mathcal{N}}$ of $\mathcal{E}|\widetilde{V}$. In particular, we obtain a factorization of $\alpha|\widetilde{V}$,

$$\mathcal{N}|\widetilde{V} \to \widetilde{\mathcal{N}} \xrightarrow{\tilde{\alpha}} \mathcal{E}|\widetilde{V},$$

as the surjection $\mathcal{N}|\widetilde{V} \to \widetilde{\mathcal{N}}$ followed by the injective homomorphism $\tilde{\alpha}$. The singular locus of $\tilde{\alpha}$ is the exceptional divisor $\widetilde{R}$. In particular, the singular locus of $\tilde{\alpha}$ maps into the singular locus R. Define the *ramification class* of α as the class in the Chow group $\mathrm{A}(R)$,

$$\mathfrak{k}(\alpha) := \int_{\tilde{R}/R} c(\widetilde{\mathcal{N}})\mathfrak{d}(\tilde{\alpha}).$$

The Plücker formula for a generically bijective homomorphism. *The ramification class of α is given by the following equation in the Chow group* $\mathrm{A}(V)$*:*

$$\mathfrak{k}(\alpha) = \mathfrak{c}(\mathcal{E}) - \mathfrak{c}_{\mathrm{M}}(\mathcal{N}). \tag{8.1}$$

Proof. Indeed, the homomorphism $\tilde{\alpha}$ is a homomorphism of locally free modules of the same rank. So the Plücker formula (3.1) for $\tilde{\alpha}$ is the equation $\mathfrak{d}(\tilde{\alpha}) = c(\mathcal{E})s(\widetilde{\mathcal{N}}^{*})[\widetilde{V}] - [\widetilde{V}]$. Multiply the equation by $c(\widetilde{\mathcal{N}})$, and push forward to the Chow group $\mathrm{A}(V)$. The result is the equation (8.1). □

It follows from (2.6), or directly, that the top-dimensional part of $\mathfrak{d}(\tilde{\alpha})$, or of $c(\widetilde{\mathcal{N}})\mathfrak{d}(\tilde{\alpha})$, is the fundamental cycle of the determinant,

$$\mathfrak{d}_{n-1}(\tilde{\alpha}) = [\widetilde{R}].$$

Hence, the codimension-1 part of $\mathfrak{k}(\alpha)$ is the image cycle,

$$\mathfrak{k}_{n-1}(\alpha) = \int_{/V} [\widetilde{R}]; \tag{8.2}$$

it vanishes if $\dim R < n-1$. In turn, the image cycle is the codimension-1 *multiplicity cycle* of R, that is, if w is the generic point of an $(n-1)$-dimensional component of R, then the corresponding coefficient in the cycle is the Samuel multiplicity of the ideal $\mathcal{I}_{R,w}$. If the 1-dimensional local ring $\mathcal{O}_{V,w}$ is regular, it is simply the colength of the ideal. In particular, if V is regular in codimension 1, then

$$\mathfrak{k}_{n-1} = [R]_{n-1}. \tag{8.3}$$

8.1 A special case. Assume that the source $\mathcal{N}$ of α is a quotient of a trivial free module F_V of rank equal to $e+1$. In this case, we may relate the Mather Chern class of $\mathcal{N}$ to the Segre class of R, or better, to the twisted Segre class $\mathfrak{s}(R, \det \mathcal{E})$. Indeed, by the blowup formula of [KT2, (4.5.6), p. 549, the twisted Segre class in $\mathrm{A}(V)$ is the difference,

$$\mathfrak{s}(R, \det \mathcal{E}) = s(\det \mathcal{E})[V] - \int_{\widetilde{V}/V} s(\det \widetilde{\mathcal{N}})[\widetilde{V}]. \tag{$*$}$$

Now, on $\widetilde{V}$ we have a surjection $F|\widetilde{V} \to \widetilde{\mathcal{N}}$, and the kernel, say $\mathcal{K}$, is locally free of rank 1. Hence $\mathcal{K} \otimes \det \widetilde{\mathcal{N}} = \det F|\widetilde{V} = \mathcal{O}_{\widetilde{V}}$, and it follows that $c(\widetilde{\mathcal{N}}) = s(\mathcal{K}^*) = s(\det \widetilde{\mathcal{N}})$. In particular,

$$\int_{/V} s(\det \widetilde{\mathcal{N}})[\widetilde{V}] = \int_{/V} c(\widetilde{\mathcal{N}})[\widetilde{V}] = \mathfrak{c}_{\mathrm{M}}(\mathcal{N}). \tag{$**$}$$

From (*) and (**), we obtain the equation,

$$\mathfrak{c}_{\mathrm{M}}(\mathcal{N}) = s(\det \mathcal{E})[V] - \mathfrak{s}(R, \det \mathcal{E}). \tag{8.4}$$

8.2 The Plücker Formula of Pohl. Let $f : X \to V$ be a proper birational map from a smooth variety X onto V. The cotangent complex d_f fits into the exact sequence,

$$\Omega^1_V|X \xrightarrow{d_f} \Omega^1_X \longrightarrow \Omega^1_{X/V} \longrightarrow 0,$$

and d_f is generically an isomorphism. Apply (8.1) with $V := X$ and $\alpha := d_f$, to obtain the following equation in the Chow group $\mathrm{A}(X)$,

$$\mathfrak{k}(f) = \mathfrak{c}(\Omega^1_X) - \mathfrak{c}_{\mathrm{M}}(\Omega^1_V|X), \tag{8.5}$$

where $\mathfrak{k}(f) := \mathfrak{k}(d_f)$. The ramification class $\mathfrak{k}(f)$ has support in the ramification locus R_f of f. Push it forward to obtain a class $\mathfrak{k}_V(f)$ in the Chow group of the image $f(R_f)$. From (8.5) we obtain the equation in the Chow group $\mathrm{A}(V)$,

$$\mathfrak{k}_V(f) = \int_{/V} \mathfrak{c}(\Omega^1_X) - \mathfrak{c}_{\mathrm{M}}(\Omega^1_V). \tag{8.6}$$

Assume that V is a subvariety of $\mathbb{P} = \mathbb{P}^N$. The relative invariant $\mathfrak{r} = \mathfrak{r}(V, \mathbb{P})$ is related to the Mather Chern class $\mathfrak{c}_{\mathrm{M}}(\Omega_V^1)$ by the equation (5.2). Use the same equation to define a relative ramification class $\mathfrak{k}_{V,\mathbb{P}}(f) := c(\mathcal{O}_{\mathbb{P}}(1))\mathcal{O}_{\mathbb{P}}(1)\star\mathfrak{k}_V(f)$. From (8.6) we obtain the following generalized Pohl–Plücker formula of (1.8):

$$\mathfrak{r} = c(\mathcal{O}_V(1))\,\mathcal{O}_V(1) \star \int_{/V} \mathfrak{c}(\Omega_X^1) - \mathfrak{k}_{V/\mathbb{P}}(f). \tag{8.7}$$

By (8.3), the codimension-1 part of (8.7) is the equation,

$$\mathfrak{r}_{n-1} = (n+1)h[V] + \int_{/V} \mathfrak{c}^1(\Omega_X^1) - \int_{/V} [R_f]_{n-1}.$$

from which (1.5) follows by taking the degree.

8.3 Piene's Plücker formula. Alternatively, we could have obtained (8.7) directly from the homomorphism $\tilde{d}_f$ in the exact sequence,

$$\mathcal{P}_V^1(1)|X \xrightarrow{\tilde{d}_f} \mathcal{P}_X^1(1) \longrightarrow \Omega_{X/V}^1(1) \longrightarrow 0.$$

Assume in particular that V is a hypersurface in $\mathbb{P}^N$. Then (8.4) applies to $\tilde{d}_f$, and it yields the following formula,

$$\mathfrak{r} = \int_{/V} s(\det \mathcal{P}^1(\mathcal{O}_X(1)))[X] - \int_{/V} \mathfrak{s}(R, \det \mathcal{P}^1(\mathcal{O}_X(1))). \tag{8.8}$$

Clearly,

$$\det \mathcal{P}^1(\mathcal{O}_X(1)) = \mathcal{O}_X(1) \otimes \mathcal{O}_X(1)^{\otimes n} \otimes \det \Omega_X^1.$$

Hence, by the formulas for twisted Segre classes, cf. (2.1), the formula (8.8) is equivalent to the following, with $\mathcal{K} := \mathcal{O}_V(n+1)$:

$$\mathfrak{r} = s(\mathcal{K})^{n+1}\mathcal{K}^* \star \int_{/V} \Big(s(\det \Omega_X^1)[X] - \mathfrak{s}(R, \det \Omega_X^1)\Big),$$

which is the formula of Piene [Pi1, Theorem (2.5), p. 259]:

$$\mathfrak{r} = \sum_i \Big(\frac{1}{1-(n+1)h}\Big)^{n+1-i} \int_{/V} \Big(c^1(\Omega_X^1)^{n-i}[X] - \mathfrak{s}_i(R, \det \Omega_X^1)\Big).$$

9 Example: Surfaces in 3-Space

Assume that $V : F = 0$ is a reduced surface of degree d in $\mathbb{P} = \mathbb{P}^3$, with ordinary singularities, that is, the singular (Jacobian) locus Z is a (non reduced) curve of general double points (at which Z is smooth) with a finite set T of triple points, and a finite set N of pinch points, see [F, Example 9.3.7, p. 168]. Let τ and ν be the numbers of triple and pinch points, and let e and $\check{e}$ be the degree and the class (the zeroth rank) of Z. Thus $\check{e} = 0$ if the dual of Z is not a surface, and $\check{e}$ is the degree of the dual otherwise. The reduced part $Y := Z_{\text{red}}$ is smooth except at the triple points.

The conormal module of V is $\mathcal{N} = \mathcal{O}_V(-d)$; hence, with $\mathcal{L} := \mathcal{O}_{\mathbb{P}}(d-1)$, the twisted normal module is $\mathcal{N}(1)^* = \mathcal{L}_V$. Thus the singular subscheme Z is the zero scheme of the homomorphism $E_V^* \to \mathcal{L}_V$ induced by the partial derivatives of F. The singular class $\mathfrak{e} = \mathfrak{e}(V, \mathbb{P})$ in $\mathrm{A}(Z)$ is the twisted Segre class $\mathfrak{s}(Z, \mathcal{L})$. It is related to the class $\mathfrak{v}(Z, \mathcal{L})$ obtained by the Stückrad–Vogel algorithm by the equation,

$$\mathfrak{e} = s(\mathcal{L})\mathfrak{v}(Z, \mathcal{L}).$$

The conormal scheme $C = CV$ is the blowup of V with center Z. The polar subschemes R_2, R_1, and R_0 (with R_i of dimension i) may be obtained from the Stückrad–Vogel algorithm:

$$R_2 = V, \qquad \partial V \cdot R_2 = \mathfrak{v}_1 + R_1, \qquad \partial V \cdot R_1 = \mathfrak{v}_0 + R_0,$$

where ∂V indicates the intersection with the surface defined by a general linear combination of the partials of F. Note that R_1 is an "outline" curve of V, represented by the closure of the set of smooth points x of V such that the tangent plane $T_V(x)$ passes through a given general point A_0; the degree, r_1, of R_1 is the 1st rank of V. Similarly, the finite set R_0 consists of the smooth points x of V such that the tangent plane $T_V(x)$ contains a given general line A_0A_1; the degree r_0 of R_0 is the 0th rank (classically called the class) of V. Finally, let R be the finite set of smooth points z on Z, such that one of the two tangent planes at z contains a given general point A_0; the degree $\rho := \deg[R]$ is called the *class of immersion* of Z in V, see [Pi2].

By the Plücker Formula, $\mathfrak{r} = \check{\mathfrak{s}} - \mathfrak{e}$, and $\check{\mathfrak{s}} = s(\mathcal{L})[V]$. Hence, with $\ell := c_1\mathcal{L}$, we have the equations,

$$\mathfrak{r}_2 = [V], \qquad \mathfrak{r}_1 = \ell\,[V] - \mathfrak{e}_1, \qquad \mathfrak{r}_0 = \ell^2\,[V] - \mathfrak{e}_0. \tag{9.1}$$

The 1-dimensional part of $\mathfrak{e}$ is easily determined: $\mathfrak{e}_1 = 2[Z]$. For the zero-dimensional part, we have three expressions,

$$\begin{aligned} \mathfrak{e}_0 &= 2\ell[Z] + [R] + 2[N], & (9.2)\\ &= 4\ell[Z] - V\cdot[Z] - 3[T] + 2[N], & (9.3)\\ &= 2\mathfrak{r}_0(Z) + 3[N] + 3[T] + V\cdot[Z], & (9.4) \end{aligned}$$

where $\mathfrak{r}_0(Z)$ is the zero-dimensional part of $\mathfrak{r}(Z)$; its degree is, by definition, equal to $\check{e}$. By taking the degrees, we obtain formulas,

$$r_2 = d, \quad r_1 = (d-1)d - 2e, \quad r_0 = (d-1)^2 d - \deg \mathfrak{e}_0,$$

where

$$\begin{aligned}\deg \mathfrak{e}_0 &= 2(d-1)e+\rho+2\nu,\\ &= (3d-4)e-3\tau+2\nu,\\ &= 2\check{e}+3\nu+3\tau+de.\end{aligned}$$

To prove (9.2), consider the preimage $c^{-1}z$ of a (closed) point z of Z. Clearly, if z is a smooth point of Z, then $c^{-1}z$ consists of two points; in particular, the preimage $c^{-1}Z$ has one component $\widetilde{Y}$ which is a $2 \mapsto 1$ cover of Z. Over a triple point z, the fiber $c^{-1}z$ consists of three points. Finally, if z is a pinch point, then $c^{-1}z$ consists (set theoretically) of the (line of) planes through the tangent line $T_Y(z)$. Hence, set theoretically, $c^{-1}N$ is a union, denoted $\widetilde{N}$, of lines in $\check{\mathbb{P}}$, one over each pinch point. Each pinch point is a double point on Z. Hence, for the fundamental class of the exceptional fiber $D := c^{-1}Z$, we obtain the equation,

$$[D] = [\widetilde{Y}] + 2[\widetilde{N}].$$

As a consequence,

$$\mathfrak{e}_1 = \mathfrak{v}_1 = c_*[D] = 2[Z],$$
$$\mathfrak{v}_0 = c_*\ell'[D] = [R] + 2[N], \quad \mathfrak{e}_0 = 2\ell\,[Z] + [R] + 2[N].$$

Thus (9.2) holds.

The formula (9.3) follows from the computation of the Segre class in Fulton [F, Example (9.3.7–8), p.169],

$$\mathfrak{s}(Z) = 2[Z] + \big(-V\cdot Z - 3[T] + 2[N]\big).$$

It follows, by using the relation (2.1) between ordinary and twisted Segre classes, that

$$\begin{aligned}\mathfrak{e} = \mathfrak{s}(Z,\mathcal{L}) &= 2(1-\ell)^{-3}(1-\ell)[Z] + (-V\cdot Z - 3[T] + 2[N])\\ &= 2[Z] + 4\ell[Z] - V\cdot Z - 3[T] + 2[N].\end{aligned}$$

Thus (9.3) holds.

Finally, to prove (9.4), note that Z, by the Euler formula (we assume that d is not divisible by the characteristic), is in fact the zero scheme of the homomorphism $\partial F : E^*_{\mathbb{P}} \to \mathcal{L}$ on $\mathbb{P}^3$. To obtain an expression for the twisted Segre class $s(Z,\mathcal{L})[\mathbb{P}]$, consider the blowup $\hat{b} : \widehat{B} \to \mathbb{P}$ with center Z. Then $\widehat{B}$ is a closed subscheme of $\check{\mathbb{P}} \times \mathbb{P}$; if $\mathcal{L}' := \mathcal{O}_{\check{\mathbb{P}}}(1)$, then, on $\widehat{B}$, the exceptional divisor $\widehat{D}$ is zero scheme of the natural injection $\mathcal{L}' \to \mathcal{L}$.

To describe the components of the exceptional divisor $\widehat{D}$, consider a (closed) point $z \in Z$. If z is a smooth point of Z, the fiber consists of the (line of) planes through $T_Z(z)$; in particular, one component of $\widehat{D}$ is the conormal variety CZ of Z. If z is singular on Z, i.e., a pinch point or a triple point, the (reduced) fiber consists of (the star of) planes through z; in particular, the reduced part of $\widehat{D}$ is a union, $\widehat{N}+\widehat{T}$, of lines, one over each singular point of Z. Hence we obtain a decomposition of the fundamental cycle,

$$[\widehat{D}] = [CZ] + a[\widehat{N}] + b[\widehat{T}]. \tag{$*$}$$

(It is not hard to see that $[CZ]$ appears with multiplicity 1; the multiplicities a and b are in fact equal to 2.) Consider similarly the fundamental cycle of $\hat{b}^{-1}V$. Then, cf. Baker [B, p. 176],

$$[\hat{b}^{-1}V] = [B] + 2[CV] + 3[\widehat{N}] + 3[\widehat{T}], \tag{$**$}$$

where B, the proper transform of V, is equal to the blowup of V with center Z. Of the components, $[B]$ is the only one not over Z. From (*) it follows that $s_1(Z,\mathcal{L})[\mathbb{P}] = [Z]$. Hence, by (**) and general properties of twisted Segre classes, cf. (7.3),

$$\begin{aligned} \mathfrak{e} = \mathfrak{s}(Z,\mathcal{L}) &= \int_{/Z} s(\mathcal{L}')[\hat{b}^{-1}V]^Z + V\cdot s(Z,\mathcal{L})[\mathbb{P}] \\ &= 2\mathfrak{r}(Z) + 3[N] + 3[T] + V\cdot[Z]. \end{aligned}$$

Thus (9.4) holds.

9.1 The classical formulas of Salmon. The formulas (see Cayley [C, p. 329]), as given by Baker [B, p. 159], are the following:

$$\text{(I)},\ d(d-1) = r_1 + 2e; \quad \text{(II)},\ r_1(d-2) = \kappa + \rho;$$
$$\text{(III)},\ e(d-2) = \rho + 3\tau; \quad \text{(IV)},\ 2\rho - 2\check{e} = \nu; \quad \text{(V)},\ r_0 + 2\nu = r_1 + \kappa.$$

In the classical notation for the invariants, $d = \mu_0$, $r_1 = \mu_1$, $r_0 = \mu_2$, $e = \varepsilon_0$, and $\check{e} = \varepsilon_1$. The number κ is the number of inflexional tangent lines through a given general point.

If Formula (II) as taken as a definition of κ, then the remaining formulas correspond exactly to the formulas for r_1 and r_0 and the three formulas for $\deg \mathfrak{e}_0$.

Recall, along the lines of Baker, the proof of (II) in the following form:

$$(d-2)(d-1)d = 2(d-2)e + \kappa + \rho. \tag{ii}$$

First, for a given general point p of $\mathbb{P}(E)$, let ∂V be the first polar (with the equation $\sum_i p_i\partial_i F = 0$), and let $\partial^2 V$ be the second polar (the Hessian, with the equation $\sum_{i,j} p_ip_j\partial_i\partial_j F = 0$). By Bezout, the intersection,

$$\partial^2 V\cdot\partial V\cdot V,$$

has degree $(d-2)(d-1)d$. On the other side, $\partial V\cdot V = 2[Z]+[R]$. Intersecting with $\partial^2 V$, we obtain $2(d-2)e$ from the component $2[Z]$. Consider the intersection $\partial^2 V \cdot [R]$. At a smooth point v of V, the zero locus of the Hessian in the tangent plane $T_V(v)$ consists of two lines, the *inflexional lines*. Hence a smooth point v belongs to $\partial^2 V \cap R$ if and only if one of the inflexional lines at v passes through p; the number of these points is κ. The pinch points belong to R, but, since p is general, they do not belong to $\partial^2 V$. Finally, at a smooth point v of Z, the Hessian vanishes in the two tangent planes at v. So, the part of $\partial^2 V \cdot R$ in Z consists of the smooth points v of Z such one of the tangent planes at v passes through p; the number of these points is ρ. Hence (ii) holds.

9.2 As noted above, for the coefficients a and b in the expression for $[\widehat{D}]$ in (**), we have in fact $a = b = 2$. Hence, for the twisted Segre class of Z in $\mathbb{P}$,

$$s(Z, \mathcal{L})[\mathbb{P}] = [Z] + \ell\,[Z] + \mathfrak{r}_0(Z) + 2[T] + 2[N],$$

in accordance with the expression in [A] for the ordinary Segre class of Z in $\mathbb{P}^3$,

$$\mathfrak{s}(Z) = [Y] - \left(de + \frac{3\nu}{2} - \tau\right)[H^3].$$

References

[AM] R. Achilles and M. Manaresi, Multiplicity for ideals of maximal analytic spread and intersection theory, *J. Math. Kyoto Univ.* **33–34** (1993), 1029–1046

[A] P. Aluffi, MacPherson's and Fulton's Chern classes of hypersurfaces, *Internat. Math. Res. Notices* (1994), 455–465

[B] H. F. Baker, *Principles of geometry VI: Algebraic surfaces*, Cambridge University Press, 1933

[BR] D. A. Buchsbaum and D. S. Rim, A generalized Koszul complex. II. Depth and multiplicity, *Trans. Amer. Math. Soc.* **111** (1964), 197–224

[C] A. Cayley, *Collected mathematical papers*, Cambridge University Press, 1893

[F] W. Fulton, *Intersection Theory*, Ergebnisse der Mathematik und ihrer Grenzgebiete, 3. Folge · Band 2, Springer–Verlag, Berlin, 1984

[FJ] W. Fulton and K. Johnson, Canonical classes on singular varieties, *Manuscripta Math.* **32** (1980), 381–389

[K1] S. L. Kleiman, The enumerative theory of singularities, *Real and complex singularities*, Proceedings, Oslo 1976, P. Holm (ed.), Sijthoff & Noordhoff, 1977, pp. 297–396

[K2] S. L. Kleiman, Intersection theory and enumerative geometry: A decade in review, *Algebraic Geometry*, Bowdoin 1985, Proc. Sympos. Pure Math., vol. 46, part 2, AMS, Providence, RI, 1987, pp. 321–370

[K3] S. L. Kleiman, A generalized Teissier–Plücker formula, *Classification of algebraic varieties*, Proceedings of the Conference in L'Aquila (Italy) 1992, C. Ciliberto, L. Livorne, and A. Sommese (eds.), Contemporary Mathematics **162**, AMS, Providence, RI, 1994, pp. 249–260

[KT1] S. Kleiman and A. Thorup, A geometric theory of the Buchsbaum–Rim multiplicity, *J. Algebra* **167** (1994), 168–231

[KT2] S. Kleiman and A. Thorup, Mixed Buchsbaum–Rim multiplicities, *Amer. J. Math.* **118** (1996), 529–569

[PP] A. Parusiński and P. Pragacz, *Characteristic classes of hypersurfaces and characteristic cycles*, Prépublication no. **51**, l'Université d'Angers, 1998

[Pi1] R. Piene, Polar classes of singular varieties, *Ann. Sci. École Norm. Sup.* (4) **11** (1978), 247–276

[Pi2] R. Piene, Some formulas for a surface in $\mathbb{P}^3$, *Algebraic Geometry*, proceedings, Tromsø, Norway 1977, Springer, Berlin, 1978, pp. 196–235

[Po] W. F. Pohl, Extrinsic complex projective geometry, *Proceedings of the Conference on Complex Analysis*, Minneapolis 1964, A. Aeppli, E. Calabi, H. Röhrl (eds.), Springer, Heidelberg, 1965, pp. 18–29

[S] C. Sabbah, Quelques remarques sur la géométrie des espaces conormaux, Systèmes différentiels et singularités, C.I.R.M. (France) 1983, A. Galligo, M. Granger, Ph. Maisonobe (eds.), *Astérisque* **130** 1985, pp. 161–192

[SGA6] P. Berthelot, A. Grothendieck, L. Illusie, *Théorie des intersections et théorème de Riemann–Roch (SGA6)*, Séminaire de géométrie algébrique du Bois Marie 1966/67, LNM 225, Springer-Verlag, Berlin, 1971

[T] B. Teissier, *Sur diverse conditions numériques d'équisingularité des familles de courbes (et un principe de specialisation de la dépendance intégrale)*, preprint, Centre de Math., École Polytechnique, 1975

[Th1] A. Thorup, Rational equivalence on general noetherian schemes, *Enumerative Geometry*, Sitges 1987, Proceedings, ed. S. Xambo-Descamps, Springer-Verlag, New York, 1990, pp. 256–297

[Th2] A. Thorup, *Plücker formulas for noetherian schemes*, preprint, Matematisk Afdeling, Københavns Universitet, 1998

Matematisk Afdeling
København Universitet
Universitetsparken 5
2100 København
Denmark
`thorup@math.ku.dk`